C. A. Barth, E. Schlimme (Eds.)

Milk Proteins

Nutritional, Clinical, Functional and Technological Aspects

Steinkopff Verlag Darmstadt
Springer-Verlag New York

The Editors:

Prof. Dr. C. A. Barth
Institut für Physiologie und
Biochemie der Ernährung
Bundesanstalt für Milchforschung
Hermann Weigmann-Str. 1
D-2300 Kiel 1

Prof. Dr. E. Schlimme
Institut für Chemie und Physik
Bundesanstalt für Milchforschung
Hermann Weigmann-Str. 1
D-2300 Kiel 1

CIP-Titelaufnahme der Deutschen Bibliothek

Milk proteins : nutritional, clinical, functional and
technological aspects / C. A. Barth ; E. Schlimme (eds.). –
Darmstadt : Steinkopff ; New York : Springer, 1988

ISBN-13:978-3-642-85375-3   e-ISBN-13:978-3-642-85373-9
DOI: 10.1007/978-3-642-85373-9

NE: Barth, Christian A. [Hrsg.]

Preface

This book reviews the state of knowledge and progress of research on food proteins, and in particular, milk proteins. Its basis is the Symposium on Milk Proteins that was held at the Federal Dairy Research Centre in Kiel, FRG, in June, 1988. Scientists from around the world attended and addressed pure, as well as applied fields of protein research and technology.

This book is divided into five sections, each adapted from the symposium's invited lectures, short communications, and poster presentations. New criteria for the "biological value" of dietary proteins and their relationships are considered according to:
- Milk Proteins and Nitrogen Equilibrium
- Milk Proteins and Ligands
- Milk Proteins: Structural and Genetic Aspects
- Milk Proteins: Technological and Functional Aspects
- Milk Proteins and Clinical Nutrition

Generally, different dietary proteins are classified according to their "biological value," i.e., their capacity to cause different retention of nitrogen in the body. But we think there are other intriguing leads worth studying that may help to identify which dietary proteins are best recommended for specific dietary situations or clinical conditions.

In addition, we have taken into consideration new fields such as attempts to determine the three-dimensional structure of proteins using two-dimensional NMR spectroscopy, and the application of genetic engineering to the lactating cell. In other words, we are on the way to the transgenic cow with customized milk constituents and composition.

We are indebted to the authors for the quality of their contributions and to the publishers, Steinkopff Verlag, Darmstadt, for their excellent cooperation.

We remember with deep gratitude the late Professor Werner Kaufmann, who organized the first symposium on milk proteins held in 1983.

Finally, the editors greatly appreciate the generous support by the European Community, the German Federal Ministry of Agriculture, and the German Dairy Industry, without which the publication of this book would not have been possible.

Kiel, FRG; November, 1988

C. A. Barth
E. Schlimme

Contents

Milk Proteins and Ligands

Milk Proteins: Structural and Genetic Aspects

Milk Proteins: Technological and Functional Aspects

X

Research Funding by the EEC

J. Gay, Principal Administrator

Commission of the European Communities, Directorate General VI (Agriculture),
Directorate D (Organization of markets in livestock products),
Division Milk products, Brussels, Belgium

Community framework program

On 28 September, 1987, the Council adopted the first framework program for
Community activities in the field of research and technological development for the
period 1987–1991. Taking into account research programs already decided on or
under way, a total of 6 480 MECU (i.e., about 13 Mrd. DM) has been made available for these five years. The total amount is subdivided into eight main headings as
follows:

Table 1. European Community framework program of research and technological development
(1987–1991)

Research field	Funds	
	Mio ECU	%
Quality of life	479	7.39
The large market and the information-communication society	2 465	38.05
Modernization of industrial sectors	989	15.26
Biological resources	310	4.78
Energy	1 752	27.04
Third World development	80	1.23
Marine resources	80	1.23
Improved European scientific and technical cooperation	325	5.02
Total	6 480	100.00

In elaborating and realizing the preparation and implementation of its policy in
regard to the stimulation of the Community's scientific and technical potential the
Commission will be supported by the Committee for the European Development of
Science and Technology which had already been installed on 6 December, 1982.

Community's support measures for dairy products

Due to steadily increasing surpluses in the dairy sector the Community introduced
on 17 May, 1977, a co-responsibility levy and measures to enlarge markets for milk
and milk products in order to achieve a better balance between production and
demand. The measures have been applied since 16 September, 1977, without inter-
ruption. The objective was not reached and a quota system had to be implemented

in 1984. For budgetary reasons the co-responsibility levy has been maintained. Thanks to this maintainance we were able to finance special support measures in the dairy sector, which are not included in the already mentioned framework program. One of those measures is the financial support of research measures whereby, as a general rule, 75% of total costs are borne by the Community.

Such work shall include:

(a) research into new or improved products;
(b) in cases of special need, research aimed at improving the marketing of dairy products in the Community;
(c) the search for new markets, or the possibility of extending existing markets outside the Community;
(d) scientific examination of the nutritional aspects of the consumption of milk and its constituents;

Within the scope of the six measures already undertaken, or underway, we have received 1 131 proposals for research projects of which 596 have been approved and some 37.15 MECU financial aid have been made available.

On 11 March, 1988, the Commission published the seventh action and made available 10 MECU for financial aid. Proposals were to have been made before 1 June, 1988, and the Commission will decide on these before 1 October, 1988. Research fields eligible for funding have not been changed.

Special promotion for milk protein

Now, as before, the Community valorizes a good fifth of total milk collection as skimmed milk inferior. In 1987 some 22 million tons of skim milk equivalent had to be subsidized and the weighted aid amounted to 60% of the net value of skim milk. In addition, in the 20 years of common market organization for milk the price ratio milk fat : milk protein has been changed from 71.4 : 28.6 to 48.2 : 51.8 in favor of protein. This change has increased the costs for protein utilization at the inferior valorization.

In this situation the Community's dairy industry, as well as the Commission, gave more attention to the valorization of milk protein. Since co-responsibility levy funds are available we have completed 41 research projects in this area.

Table 2. Research funding by co-responsibility levy funds in the milk sector in respect of milk protein since 1978

9	Projects research into the nutritional aspects of the consumption of milk protein
10	Projects research into product innovation in respect of milk protein
11	Projects research into new possibilities for using milk protein
5	Projects into the improvement of technology in respect of milk protein
6	Projects concerning analytical research in respect of milk protein
5	Congresses or symposia concerning milk protein
2	Actions concerning the dissemination of literature and promotion of milk protein
48	Total

Table 3. Research into new possibilities for using milk protein

1.	Sélection et utilisation des fractions protéiques laitières dans les boissons sans alcool (723/78 – B 13.1)
2.	Undersøgelses- og forsøgsaktiviteter for udviling og tilpasning af valleproteinkoncentrater til specifikke anvendelser som ingrediens i sammensatte levnedsmidler (2935/79 – 07)
3.	Untersuchungen über Möglichkeiten zur Verbesserung und Erweiterung des Absatzes von Milcheiweiß in der Gemeinschaft (2935/79 – 10, 271,82 – 15.1)
4.	Untersuchungen zur Erweiterung der Verwendungsmöglichkeiten von Milcheiweiß (2935/79 – 11, 271/82 – 15.2)
5.	Development and adjustment of whey protein concentrates for use in compound foods for human consumption (271/82 – 07)
6.	Onderzoek naar toepassingen van melkproteine in de voedingsnijverheid (271/82 – 4.2)
7.	Development of new functional milk proteins from caseins and whey proteins using genetically engineered enzymes (282/84 – 1.01)
8.	Milcheiweißprodukte für andere Lebensmittel (282/84 – 31.1)
9.	Milcheiweißstabilisatoren auf Buttermilchbasis (282/84 – 31.2)
10.	Concerned with the conversion of milk proteins for use in food processing (1150/86 – 59.3)
11.	Produit à base de protéins laitières à vocation diététique ou non (1150/86 – 84.2)

of which
– nine projects dealt with the scientific examination of the nutritional aspects of the consumption of milk protein;
– 10 projects were orientated to product innovations with milk protein;
– 11 projects dealt with new possibilities of using milk protein (Table 3);
– five projects for the improvement of technology in manufacturing milk protein;
– six projects were devoted to the improvement of analytical methods to determine milk protein.

In order to broaden experiences and knowledge on milk protein five congresses or symposia have been organized and mainly financed by co-responsibility levy funds.

Finally, we have supported two actions where new literature on milk protein has been widely disseminated. Indeed, this symposium belongs to the series of measures which have been financially supported by the Community.

New research initiative by the Community

With the adoption of the Single European Act which went into effect July, 1987, after deposition of the ratification documents, the EEC Treaty was amended by a Title VI "Research and Technological Development". Herewith the Community's aim shall be to strengthen the scientific and technological basis of European industry and to encourage it to become more competitive at an international level. In the new article 130 g) of the Treaty the realization of these objectives have been described as following:

"In pursuing these objectives the Community shall carry out the following activities, complementing the activities carried out in the Member States:

(a) implementation of research, technological development and demonstration pro-
grammes, by promoting co-operation with undertakings, research centres and
universities;
(b) promotion of co-operation in the field of Community research, technological
development, and demonstration with third countries and international organizations;
(c) dissemination and optimization of the results of activities in Community research, technological development, and demonstration;
(d) stimulation of the training and mobility of researchers in the Community."

The Industry's Interest in Protein Research

A. Nienhaus

German Milk Industry Association, Bonn, FRG

Despite the cut in EC milk supplies from 104 million tons in 1983 to 92 million tons in the dairy year 1988/89, and despite the great adjustment problems this dramatic turn in the EC milk policy entails for many processors, milk protein, with a share of some 20% in the European Community, continues to rely on production and marketing grants. As a result of the current increase of the skim-milk price, liquid skim-milk or skim-milk powder can only be used in the feed sector if a grant of some 40% of the milk value guarantees the competitiveness vis à vis the competing vegetable proteins – especially soy protein. This is why the German and European milk industry follows with extreme interest the efforts of scientists to research milk proteins and to expand the market; four research targets are of special importance:
1) Standards of valuation and investigation methods;
2) Valence under the aspect of the physiology of nutrition;
3) Technical possibilities of extraction;
4) Technological fields of application.

The Federal Dairy Research Centre with its various agencies is bent on simultaneous research and further development in all four sectors.

The standards of valuation for milk protein are safe; thanks are due in this connection to the EC Commission for the support it has given to two related research projects which have set new standards to value milk proteins vis à vis vegetable proteins.

The valence in regards to the physiology of nutrition is understood today because it has been possible, fortunately, to disprove the suspicion that milk protein, as distinct from vegetable proteins, might raise the cholesterol content.

The technique of protein extraction and the technological uses are in a fast and promising stage of development in which science and practice work hand-in-hand in mutual stimulation.

The broad range of uses of the various milk proteins were explained in detail as early as 1984 at the worldwide Protein Congress in Luxembourg. In 1986, the MIV Seminar "Milk Protein in Food" was a platform for the various research institutes to illustrate the use of milk protein in the fields of bread and confectionery, of meat and sausages, and in particular, in the promising diet sector. At the same time the seminar was a platform for the research and development departments of our milk industry plants to present the protein products already on the market and to convey an idea of possible future uses.

Since the Milk Protein Congress in Luxembourg and the Munich Seminar were oriented towards practical uses, major developments, some of them propitious, have taken place. It may well happen that rising prices for the raw material skim-milk and progressive findings in the field of the recombination of skim-milk powder turn the danger of artificially mixed skim-milk powder for the food and feed sector into reality again. It should also be borne in mind that customs tariff protection for imported casein is very slight and that, for this reason, caseins and caseinates produced in the EC are competitive only if supported by grants, so that this reasonably priced protein component will always be an incentive to get around the high-price level in the EC for milk and milk protein.

In the food sector the use of cheap vegetable proteins could lead to major competitive distortions, unless safe scientific methods are available to disclose such forgery.

In the production sector the development of a method to extrude the caseinates, opening the way to new products, results in a larger range of uses.

A more purposeful ultra-filtration of whey proteins will enable their use, not only in the fresh cheese sector but also in semi-hard and hard cheese. This promises completely new products.

The use of ultra-filtration in this sector opens new and unexpected aspects for the future of the liquid milk market. Research done at the Kiel Federal Institute has shown that the digestion of milk protein is changeable, almost at will, by changing the ratio of casein and whey protein. This enables the production of special liquid milk varieties, e.g., for athletes or convalescents, for young people and for the aged.

Our market research institutes are studying the potential market. The differentiation of products, as distinct from the relatively uniform offer of liquid milk in the past, will enable new advertising activities and will enhance the image of the overall offer.

In summation, the competition for the markets and the fact that the raw material milk protein is becoming scarcer, will require innovations to an extent so far unknown:

1) In the field of research, and with a view of deepening insight into the physiology of nutrition, new ways should be created to differentiate the products – and to guarantee safe test methods to supply evidence of various proteins;

2) In the sector of technique and technology, the prerequisites should be created for a broader application of technique and technology by a simplified and cost-saving production of the most diversified milk proteins, in pure or blended form, by safe cleaning of complicated plants, and by manufacturing products on a high level of hygiene and quality;

3) In the sector of product development future market areas must be occupied now by offering adjusted food for the various age groups, for students, for old people, for convalescents and athletes, for sedentary workers, and for laborers. Imitation not only courts the danger of losing markets but also of wasting the chance to establish milk products more firmly in the promising diet sector;

4) By the further development of investigation methods and control mechanisms, the consumer should feel sure that he receives what is declared on the label; the State and the EC should be protected from fraud; the economy as a whole should be kept safe from manipulations that distort competition.

I wish to convey my thanks and my appreciation to the Federal Dairy Research Centre for cooperating closely and successfully with the bodies engaged in the practical-side of the industry in all scopes mentioned. Our Symposium showed that new findings require international discussion and stimulation. In this regard the German Milk Industry conveys its best wishes to the Symposium – may its high expectations be met.

How to Evaluate Dietary Protein

V. R. Young and P. L. Pellett

Clinical Research Center and Laboratory of Human Nutrition, School of Science,
Massachusetts Institute of Technology, Cambridge, Massachusetts, USA
and Department of Food Science and Nutrition, University of Massachusetts,
Amherst, Massachusetts, USA

Introduction

The presentation of a paper on "How to evaluate dietary protein" might be approached in a number of ways, depending upon the depth and breadth of the coverage given to specific, relevant areas that could reasonably be included under this title. The primary nutritional function of dietary protein is to furnish the indispensable (essential) amino acids and total nitrogen required for synthesis of tissue and organ proteins and many other nitrogen-containing compounds necessary for normal growth and function of the organism. Hence, in the first instance, it is usual to consider the different food proteins and protein sources in relation to their capacity to meet the amino acid and nitrogen requirements of the host. On the other hand, a more comprehensive evaluation of dietary protein on the overall nutritional health of the individual and of populations, requires an assessment of the possible effects of various food protein sources on the utilization of, and requirements for, energy yielding substrates and other individual essential nutrients (e.g., [1]). Because later papers in this symposium will be devoted to milk proteins with reference to the utilization of, and requirements for, minerals and other micronutrients, as well as considering use of milk proteins in relation to various aspects of clinical nutrition, we have chosen the first instance above as our principal focus with respect to the "evaluation of dietary protein". In doing so, we will consider some recent research that, while still somewhat controversial (e.g., [2]), provides a new, and we believe, a more rational basis for judging the significance of milk proteins in human protein and amino acid nutrition.

The multiplicity of non-clinical (Table 1) and clinical methods and procedures (Table 2) that are used to evaluate the nutritional value, or quality, of dietary protein sources has been the subject of reviews by us [4, 5], and others [6, 7], and a number have given particular attention to milk proteins [8–11]. However, we will not discuss in any detail this particular aspect of dietary protein evaluation because there have been no fundamental advances in this area during the past few years. Furthermore, the suitability of a particular method will depend upon the use that will be made of the findings, as pointed out earlier by Bender [3].

Thus, in general, the systematic evaluation of dietary sources of protein involves initial chemical testing followed by biological assays. For the purposes of evaluating, for example, new or modified plant proteins made available by breeding new varieties of grains or legumes, the first step is usually a chemical analysis for protein content and amino acid profile. Sometimes the amino acid analysis may be for a single critical amino acid, or for a restricted number, such as tryptophan, sulphur amino acids, and lysine. For exploring new and novel sources of protein, the initial approach is to examine protein concentration and amino acid pattern from which an amino acid score can be derived followed by in vitro tests for digestibility. If these criteria suggest a potentially useful new protein source, then a bioassay to test for availability of amino acids and as a further test of quality, is undertaken. These biological tests may reveal a less favorable picture than that indicated by amino acid scoring alone. In this case, the product should be examined for non-available amino acids, as for example by use of tests for available lysine, or should be evaluated for possible toxic materials present in the foodstuff. The biological testing should include a measure of digestibility, because this can be a significant cause of discrepancy between chemical and biological evaluations of quality.

For a third use, namely the monitoring of variables introduced by food processing, it is common to begin with availability of lysine and of methionine, as these are usually the amino acids most sensitive to processing. For routine regulatory purposes, the examination of a protein sample should begin with chemical analysis for nitrogen, amino acids, and toxins, including microbial toxins. The regulatory requirements may also indicate bioassays of protein quality that are commonly specified in detail by the regulatory agency but that may not necessarily represent the choice of the investigator.

Finally, there have been a number of potential improvements concerning specific aspects of approaches for the assessment of protein quality including, for example, animal bioassays [12–15] and improved protocols for human studies [16, 17]. Nevertheless, we will focus our attention on recent developments relevant to a direct evaluation of the capacity of dietary protein(s) for meeting human protein and amino acids needs. These include an improved understanding of the nature of protein and amino acid metabolism in human subjects and how the body responds to altered protein and amino acid intakes. With this new information we can consider the question of the amino acid requirements of the human subjects and then assess them in relation to the amino acids supplied by milk and other food protein sources.

We will give emphasis in this paper to the level and biological availability of amino acids in dietary food proteins because we [18, 19] have concluded that use of dietary amino acid composition data (corrected for digestibility) provides a reasonably sound basis for evaluating the quality of dietary protein. Finally, against this background, an evaluation will be made of the contribution made by milk, as well as other food proteins, toward meeting the nutritional needs and wants of different population groups.

Some metabolic aspects of the protein and amino acid requirement

As noted above, the important physiological function of dietary protein, from a quantitative standpoint, is to provide substrate necessary for the maintenance of body-protein synthesis in the adult and for supporting an acceptable rate of net protein gain in the growing infant and child, for the formation of new tissue in the fetus and supporting tissues of the pregnant female and for milk production in the lactating woman. Thus, attention should be given first to some aspects of whole-body-protein synthesis and turnover, and how these change with normal growth and development and during aging in the human subject.

8

Briefly, protein synthesis rates are high in the newborn and, per unit of body weight, these rates decline with progressive growth and development (Table 1). Two points should be emphasized from the data shown here; first, a higher rate of protein synthesis in the young, as compared with the adult, is present even when a deduction is made for the net protein deposition associated with growth. Thus, not only is protein *synthesis* in the premature infant about twice as high as in the preschool child, and approximately three or four times as high as in the adult, but also rates of organ and tissue protein *breakdown* are considerably higher in the infant than in the adult. Second, at all ages the rates of whole-body protein synthesis and breakdown are considerably greater than the safe level of intake of dietary protein estimated to meet the need for maintenance of N balance or for the support of growth. It follows that there is an extensive reutilization within the body of the amino acids entering tissue pools during the course of protein breakdown. This recycling of amino acids, and the rates of synthesis and breakdown of body proteins, change in response to various stimuli, including alterations in the level and adequacy of protein and amino-acid intakes (e.g., [21–23]).

The importance of protein synthesis as a determinant of the dietary requirement for protein and for indispensable amino acids has been discussed previously [20, 21]. However, one point to be made here is that it appears necessary to achieve a concentration of free amino acids in tissue pools that is somewhat above that seen with low or grossly inadequate protein and amino acid diets (e.g., [24]), in order to maintain the integrity of the protein synthetic machinery of cells and organs. Because the oxidation of amino acids changes with alterations in their tissue level [25, 26] we have argued, therefore, that factors regulating the rate of amino acid oxidation are the primary determinants of the requirement for the indispensable amino acids [27]. The importance of the catabolic fate of amino acids in relation to requirements is, perhaps, well illustrated by the data shown in Table 2. Here, it may be seen that only a small percentage of the total requirement for leucine and lysine even in young, growing infants can be attributed directly to the amount of the amino acid associated with the net deposition of protein.

An additional point to be made about the turnover of body proteins and the extensive re-utilization of amino acids for the resynthesis of proteins is that this

Table 1. Rates of whole body and muscle protein synthesis compared to dietary protein allowances at various ages in man[a]

Group	Whole body protein synthesis (A)[b]	Muscle protein synthesis[b]	Protein Allowance (B)[b]	Ratio A/B
Infant (premature)	11, 14	~1–1.4	~3	4.5
Child (15 months)	6.3	~0.9	1.3	5
Child (2–8 years)	3.9		~1.1	4
Adolescent (~13 years)	~5		~1.0	5
Young Adult (~20 years)	~3.5	~1.1	~0.75	6
Elderly Men (~70 years)	~3.5	~0.7	~0.75	6

[a] Based on summaries by Young [8] and Young et al. [20]
[b] Values are g protein kg^{-1} day^{-1}

Table 2. The "requirement" for leucine and lysine in premature and older infants and children compared with the amount of these amino acids deposited in protein gain[a]

Parameter	Infant		Child (10 years)
	Premature	6 months	
"Requirement" (A) for			
Leucine	279	161	45
Lysine	198	103	60
Amount B (deposited)			
Leucine	137	21	3
Lysine	119	18	3
Ratio B/A (%)			
Leucine	49	13	7
Lysine	60	18	5

[a] All values are mg kg^{-1} day^{-1}. (See also [20])

process is not completely efficient. Hence, inevitably, amino acids are lost by oxidative catabolism. This loss includes the metabolism of the carbon skeleton of the indispensable amino acids and the nitrogen or both indispensable and dispensable amino acids. When the dietary protein level is reduced to a sub-maintenance intake, protein synthesis and breakdown are subsequently reduced, but a more immediate metabolic adjustment is a rise in the efficiency of re-utilization of the amino acids liberated during protein breakdown and a fall in the rate at which indispensable amino acids are irreversibly oxidized. This immediate response might be regarded as an *adaptive* mechanism [28, 29], with the somewhat later decline in rates of protein synthesis and breakdown rates representing an *accommodation* to a continued inadequate dietary intake. Consequently, the fall in the irreversible catabolism of indispensable amino acids leads to a lower output of nitrogenous compounds in urine and feces. At essentially protein-free intakes the output of nitrogen represents the so-called "obligatory" nitrogen losses.

As summarized by FAO/WHO/UNU [30] (Table 3), obligatory N losses in apparently healthy subjects in different countries (possibly with different habitual protein intakes) appear to be remarkably uniform. From these findings, it has been concluded [30] that the mechanisms for reducing the minimum requirement for N balance (or for the initial conservation of N) may be largely restricted, at least during the early phases of adaptation, to processes responsible for increasing the efficiency of amino-acid utilization for protein anabolism, including a reduction in the oxidation of specific indispensable amino acids. The practical implication of these observations is that the requirement of total N (protein), at least in consideration of the obligatory losses that must be balanced by an adequate level of dietary intake, is likely to be similar in all populations.

Finally, and to slightly extend the suggestions made by FAO/WHO/UNU [30], the requirements for protein (nitrogen) and for the specific, indispensable amino acids in an individual can be defined as the lowest level of intake that will balance the losses of nitrogen and of amino acids (via oxidative catabolism) from the body

Table 3. Obligatory urinary nitrogen losses in adult males[a]

Group (location)	Mean age (yr)	Mean Wt (kg)	Urinary N (mg N kg^{-1} day^{-1})
USA	20	71	38
USA	21	74	37
China[b]	23	55	33
India	27	46	38
Nigeria	26	54	34
Japan	–	63	33
Chile	27	63	36

[a] A partial summary from FAO/WHO/UNO [30], where original literature citations are given for each study
[b] Taiwan Province

without major changes in protein turnover and for a state of energy balance at a modest level of physical activity. In infants, children, pregnant and lactating women, the requirement also includes that amount of protein associated with the net deposition of new protein in tissues and the secretion of proteins in milk.

The starting point, therefore, for estimating total protein needs is the measurement of the amount of dietary nitrogen (N) needed for zero body N balance in adults during short-term metabolic studies. For determination of the needs for specific indispensable amino acids the approach might be the same or else one that is based on a direct estimate of measurement of rates of irreversible oxidation of these amino acids. In the following sections we will further examine the requirements for amino acids, taking into consideration the features of human protein and amino acid metabolism that were described briefly above.

Current estimates of needs for protein and amino acids

Most estimates of protein and amino acid requirements have been obtained directly, or indirectly, from measurements of N balance, and the use and application of the N balance technique for this purpose has been discussed in detail in the 1985 FAO/WHO/UNU [30] report. However, it should be noted that the nitrogen balance technique has serious limitations [31, 32] and it does not provide a totally secure basis for establishing the protein and amino acid needs of human subjects. Indeed, the functional significance of larger or smaller total body N pools or of faster or slower protein turnover rates that might be achieved, for the same degree of body nitrogen balance, when protein intake ranges somewhat above the so-called minimum physiological level, is uncertain.

Estimates of the requirements for protein and indispensable amino acids have been made by various groups, including those recently presented by the United Nations [30]. For example, the safe intake level for good-quality protein in each of the age groups has been proposed by FAO/WHO/UNU [30], and Table 4 presents a recent statement about the requirement for each of the indispensable amino acids.

Table 4. Some estimates of amino-acid requirements (mg kg^{-1} day^{-1}) in preschool and school-age children, and in adults

Amino acid	Preschool (2–5 years old)[a]	School children[b] (10–12 years old)	Adults (18 years +)
Histidine	?	?	8–10
Isoleucine	31	28	10
Leucine	73	44	14
Lysine	64	44	12
Methionine and cystine	27	22	13
Phenylalanine and tyrosine	69	22	14
Threonine	37	28	7
Tryptophan	12.5	3.3	3.5
Valine	38	25	10
Total ($-$His)	352	216	84
Total (per unit protein)[c]	32%	22%	11%

[a] From Tables 3 B and 4 in FAO/WHO/UNO [30]
[b] Based on NRC [33]
[c] Total indispensable amino acid requirement expressed as percent of safe protein allowance

A few points should be made about the figures presented in this summary table (Table 4). First, the recommendations for safe protein intakes include a factor for variation in protein requirements among apparently similar individuals. It has been estimated [30] that the biological variability in protein requirements amounted to a coefficient of variation of 12.5% and, thus, a value of 25% (2 SD) above the *mean* minimum physiological requirement of 0.6 g/kg/day for an adult would meet the needs of all but 2.5% of individuals within the adult population [30]. Hence, the mean minimum requirement should be increased to 0.75 g/kg/day to give a safe protein intake for healthy adults. Apparently, most individuals would require less than this to maintain an adequate state of protein nutrition; it follows that some subjects might require as little as 0.45 g high-quality protein per kilogram per day.

Second, a striking feature of current estimates of human amino-acid requirements is the marked decrease in indispensable amino acid needs expressed per unit of body weight, that occurs by the time adulthood is reached (Table 4). The lower requirement for indispensable amino acids, per unit of the total need for protein in the adult (see Table 4), could reflect biologically important differences in metabolism and, particularly, in the relative efficiency with which amino acids and nitrogen are used by the body in the younger and older age groups to maintain an adequate protein nutritional status. However, an equally plausible reason for these observations (Table 4) is that they are due to problems associated with the methods and approaches used to assess amino acid requirements, especially in adults. Although Millward and Rivers [2] consider that current estimations of the amino acid requirements are reasonable for adults, we do not. Hence, we will now examine this problem in further detail, particularly because it is relevant to the evaluation of dietary protein.

12

Reassessment of amino acid requirements

The above estimates of the requirements for indispensable amino acids have been derived largely from results of nitrogen (N) balance studies in healthy individuals, also as noted above. However, we have pointed out in previous reviews [27, 34, 35] several compelling reasons why these earlier metabolic balance studies should be viewed with considerable circumspection, particularly with respect to their use in providing determinations of the minimum physiological requirements for the indispensable amino acids. Furthermore, from a series of recent experiments on the kinetics of indispensable amino acid metabolism in which rates of amino acid oxidation were measured at various amino acid intake levels, we have concluded that the current amino acid requirement estimations (Table 4) are far too low for leucine [36], valine [37], lysine [38] and threonine [39], and perhaps for all of the indispensable amino acids. Our conclusions have important implications for practical aspects of human protein nutrition and assessment of dietary protein. Hence, in the next section we will develop an additional rationale, exploiting a current understanding of the physiology and regulation of amino acid and protein metabolism, as summarized earlier, to support our hypothesis that current requirements are too low and to seek further experimental evidence in support of this hypothesis.

Obligatory rates of amino acid oxidation

Earlier we discussed obligatory N losses which are achieved by a fall in the rates of amino acid catabolism and perhaps accomplished, in part, by adaptive changes in the levels and activities of enzymes of amino acid and urea cycle metabolism [40], as well as possibly via changes in rates of turnover of tissue and organ proteins [22]. However, because of continued protein turnover and because the activities of these enzymes are not reduced to zero, together with the free amino acids that remain at measurable but possibly diminished concentrations in the tissue amino acid pools [24], there is a continued, or obligatory, oxidative loss of amino acids. This occurs despite their more efficient reutilization for purposes of protein synthesis.

The rates of oxidative loss of the individual indispensable amino acids have not been measured directly under the precise conditions used to measure obligatory urinary nitrogen output, except perhaps for leucine (see below). However, an estimation of the obligatory oxidative losses of indispensable amino acids can be made on theoretical grounds, if it can be assumed that the oxidation rates of individual amino acids occur in proportion to the pattern of amino acids in mixed body proteins (see also [2]). Thus, as summarized in Table 5, the obligatory oxidation rates of leucine, lysine, and total sulfur amino acids (methionine + cystine) are estimated to be equivalent to approximately 27, 30, and 14 mg $\cdot$ kg^{-1} $\cdot$ day^{-1}, respectively. As discussed elsewhere [41], our earlier published and unpublished data offer further support for the predictions of amino acid losses that we have given in Table 5.

Another, and somewhat related approach might be taken to estimate the minimum rates of loss of indispensable amino acids. Thus, from a series of extensive studies concerned with determination of the rate of whole body protein turnover in healthy adults (e.g., [22] for review) it is reasonable to approximate this rate as being

Table 5. Calculation of "obligatory" amino acid losses (oxidation) and estimation of "minimum" amino acid losses from protein turnover

Amino acid	Amino acid losses predicted by:		
	"Obligatory N method"[a]	"Protein turnover method"	
		Flux	Oxidation
Isoleucine	16[b]	168[c]	17[d]
Leucine	27	283	28
Lysine	30	311	31
Methionine & Cystine	14	140	14
Phenylalanine & Tyrosine	27	280	28
Threonine	15	161	16
Tryptophan	4	42	4
Valine	17	175	18

[a] Total amino acid losses assumed to be equivalent to 50 mg N/kg/day and in proportion to the amino acid composition of beef [30]

[b] All figures expressed as mg amino acid kg^{-1} day^{-1}

[c] Based on assumption of a protein turnover of 3.5 g protein kg^{-1} day^{-1} with an amino acid composition equivalent to that of beef proteins (30)

[d] It is assumed that endogenous amino acids are recycled with 90% efficiency at low amino acid/protein intakes

between 3 and 4 g protein kg^{-1} day^{-1}. Furthermore, at submaintenance levels of protein or amino acid, intake of the recycling of amino acids following their release during tissue protein breakdown is higher than when dietary intakes meet or exceed requirements. From our studies [42] it appears that in the adult the amino acids are recycling with about a 90% efficiency when intakes via the diet are distinctly below requirement levels. This is further supported by the fact that at sub-maintenance protein intakes whole body nitrogen flux is approximately 600 mg N $kg^{-1} \cdot day^{-1}$ [43]. Hence, if obligatory nitrogen losses are taken to be 54 mg N $kg^{-1} \cdot day^{-1}$ [30], it is evident that the recycling of nitrogen is also about 90%. Therefore, assuming a protein turnover of 3.5 g $\cdot kg^{-1} \cdot day^{-1}$ for a well-nourished adult, the minimum oxidation losses of indispensable amino acids when recycling is 90% can be calculated. These calculations are also summarized in Table 5 and it can be seen that the values are close to those computed from obligatory nitrogen losses (Table 5, column 1). Perhaps, the most important conclusion is that the minimum rate of losses of indispensable amino acids are equal to or, in most cases, considerably higher than the estimates of the *upper range of amino acid requirements* as proposed in 1985 by FAO/WHO/UNU [30].

We have assumed, above, that a protein turnover of 3.5 g kg^{-1} day^{-1} occurs in well-nourished individuals when they consume protein and indispensable amino acids at a level just sufficient to maintain an adequate protein nutritional status. If this assumption is reasonable, then the predicted fluxes and oxidation rates of the indispensable amino acids (Table 5) should be consistent with fluxes and oxidation rates that have been measured at intakes of amino acids that we consider to be just sufficient to meet the mean minimum requirements. As the comparison in Table 6

Table 6. Comparison of predicted amino acid fluxes and oxidation rates of individual amino acids with published estimates

Amino acid	Predicted from a turnover of 3.5 g protein/kg/day [a]		Published estimates at "requirement" intakes of amino acid	
	Flux	Oxidation	Flux	Oxidation
		μmol kg^{-1} h^{-1}		
Leucine	89	9	88 [b]	11 [b]
Lysine	89	9	84 [c]	11 [c]
Threonine	56	6	40–59 [d]	4–5 [d]
Valine	62	6	50 [e]	8 [e]
Methionine (-cystine)	19	2	~18 [f]	2.7 [f]

[a] Based on estimates in Table 5, for a 90% efficiency of endogenous amino acid recycling
[b] From Young et al. [28]. A leucine requirement of 30 mg kg^{-1} day^{-1} is assumed
[c] From Meredith et al. [38]. A lysine requirement of 30 mg kg^{-1} day^{-1} is assumed
[d] From Zhao et al. [39]. A threonine requirement of 15 mg kg^{-1} day^{-1} is assumed. Figures are for 10 and 20 mg/kg/day intakes, respectively
[e] From Meguid et al. [37]. A valine requirement of 20 mg kg^{-1} day^{-1} is assumed
[f] From Young VR, Wagner DA, Burini R, Storch K., (unpublished results). A total methionine requirement (without dietary cystine) of 13 mg kg^{-1} day^{-1} is assumed. These flux and oxidation data are for the post-absorptive state

reveals, the predicted and measured (via ^{13}C-tracer studies) fluxes of the amino acids and their oxidation rates are in very good agreement.

Intakes of amino acids to balance obligatory amino acid oxidation

To maintain body amino acid balance, the obligatory amino acid oxidation rates must be compensated for by an appropriate dietary supply, just as in the case of balancing obligatory nitrogen losses via an adequate nitrogen intake [30]. The minimum intakes of the individual amino acids required to balance these losses can be calculated readily if the efficiency with which exogenous amino acids are utilized for this purpose is known. Again, there is a lack of direct, experimental data on this aspect of amino acid metabolism and nutrition. However, as an approximation of the efficiency of dietary indispensable amino acid retention, we can assume that it is similar to the efficiency with which high quality protein nitrogen is used to replace obligatory nitrogen losses. From nitrogen balance studies in healthy young adults by various investigators [44, 45], it is reasonable to estimate that the efficiency of nitrogen utilization from good quality dietary protein for replacing obligatory nitrogen losses is about 70%, although a number of studies suggest that the efficiency might be lower [30]. However, for the present, we will accept an estimate of 70% for our purposes, but in the future this value may require modification.

Taking, then, the obligatory amino acid oxidation rates and correcting these for an assumed 70% efficiency of retention of exogenous amino acids, an approximation of the minimum requirement for each of the indispensable amino acids is obtained.

The values for the specific amino acids are given in Table 7. For leucine, lysine and the sulfur-containing amino acids, these approximations amount to 39, 42 and 16 mg $\cdot$ kg^{-1} $\cdot$ day^{-1}, respectively. The argument could be made that the efficiency of utilization of dietary lysine for balancing the obligatory rate of lysine oxidation is not necessarily the same as that for other indispensable amino acids. Although this assumption of a common efficiency factor is also made in the assessment of dietary protein quality by the amino acid scoring procedure [2, 19], it is not certain whether it is valid. Some evidence [46, 47] suggests that lysine may be more effectively conserved or perhaps utilized, when consumed at low levels, than is so for threonine, for example. Furthermore, the addition of excess levels of non-limiting amino acids causes further changes in body weight gain and compositional parameters when diets limiting in different amino acids, such as threonine or lysine, are given to growing rats [48]. However, these effects appear to be mediated by changes in food intake [49, 50] rather than on changes in the efficiency of use of the limiting amino acid. Clearly, more data are needed on the in vivo kinetic characteristics of enzymes of amino acid metabolism in different tissues and on levels of amino acids in these tissues at different amino acid intakes, and how these various indices change in relation to the minimum physiological requirements of the human organism. With this knowledge we would be in a better position to state whether, and understand how, the various indispensable amino acids might differ in their efficiency of utilization for meeting minimum physiological needs. Nevertheless, it seems legitimate, at this stage, to apply a single figure for the efficiency with which the various indispensable amino acids coming from the diet are used to replace these obligatory oxidative losses. We must emphasize that this is a hypothesis and although it is thought to be a reasonable one it requires further experimental support.

Table 7. Prediction of minimum intakes of indispensable amino acids to balance losses via irreversible oxidation

Amino acid	Intake to balance losses predicted by:	
	"Obligatory N loss method"	"Protein turnover method"
Isoleucine	23[a]	24[a]
Leucine	39	39
Lysine	42	43
Methionine & Cystine	16	17
Phenylalanine & Tyrosine	39	39
Threonine	21	22
Tryptophan	6	5.6
Valine	24	25

[a] Intakes (mg kg^{-1} day^{-1}) required to balance predicted losses (see Table 5) assuming a 70% efficiency for amino acid retention. For methionine and cystine the amount of methionine converted to cystine at methionine requirement (i.e., 10 mg kg^{-1} day^{-1}) was taken to have a 100% efficiency and the remaining 4 mg (for cystine) was assumed to be used with a 70% efficiency

Comparison of calculated minimum requirements with those obtained from ^{13}C-tracer studies

The approximations of amino acid requirements (Table 7), obtained from the predictions of obligatory amino acid oxidation rates (Factorial or Obligatory N Loss Method) and from the estimates based on considerations of protein turnover (Protein Turnover Method), can be compared with those that can be derived from our recent studies involving use of ^{13}C-amino acid tracers. Table 8 gives this comparison for the five amino acids that have been studied in our laboratories to date with the aid of the ^{13}C-tracer approach. In view of the difficulties encountered in the design, conduct, and interpretation of amino acid kinetic studies, and the reasonable but not necessarily precisely accurate assumptions concerning amino acid recycling efficiency and retention, there is a remarkably good agreement between the estimated requirement levels as judged using these three new approaches. Furthermore, given the close agreement between the requirement estimations based on the obligatory loss, protein turnover, and ^{13}C-kinetic methods, this further highlights the much lower requirement figures previously accepted by FAO/WHO/UNU [30]. The discrepancy shown here (Table 8) must be viewed with considerable concern. Indeed, from this present analysis, the case for the inadequacy of current national and international amino acid requirement figures appears, to us, to be highly convincing and there is not any published and acceptable argument to the contrary in our opinion. In contrast, Millward and Rivers [2] have pointed out the limited strength of our knowledge about human amino acid requirements but they seem willing to accept current estimations, despite the problems raised above.

Table 8. Comparison of requirements for amino acids, in the adult, as estimated by various methods

Amino acid	Method of estimation			
	N balance[a] (FAO/WHO/UNU)	Factorial[b] obligatory N	Protein turnover[c]	^{13}C-Tracer studies
	mg kg^{-1} day^{-1}			
Leucine	14	39	39	30–40[d]
Lysine	12	42	43	30[e]
Threonine	7	21	22	15[f]
Valine	10	24	25	20[g]
Methionine (& Cystine)	13	16	17	13[h]

[a] From Table 4. These are *upper range* requirement estimates
[b] From Table 7
[c] From Table 7
[d] From [28 and 36]
[e] From [38]
[f] From [39]
[g] From [37]
[h] Young VR, et al. (unpublished data)

Summary of new predictions for amino acid requirements and relation to protein requirements

In summary, the foregoing estimates of the minimum physiological requirements for indispensable amino acids, as predicted from the factorial and protein turnover methods and as judged from studies involving ^{13}C-tracers, appear to be rational. The agreement between these various approches adds strength to our conclusions about the amino acid requirements in adult human subjects; the new estimates proposed above are more rational than those presented previously by international (e.g., [30]) groups. A summary of our revised estimates, expressed per kg body weight, are shown in Table 9.

To establish the relationship between the proposed, revised estimates of adult amino acid requirements and the requirement for protein we can arrive at a new adult amino acid requirement or amino acid scoring pattern (see [30]). Thus, the relationship between the approximate mean amino acid requirements, as estimated by the different approaches above, and the mean requirement of 0.6 g high quality, highly digestible protein $kg^{-1} \cdot day^{-1}$ (see [30]) can be determined. This has also been done, as shown in Table 9. It should be noted that the requirement values are somewhat lower when the requirement figures from ^{13}C-tracer studies are used.

Implications for evaluation of dietary protein

The proposed new estimates of amino acid requirements for the adult (Table 9) increases the levels, on average, by a factor of 2.5, relative to those proposed in 1985 by FAO/WHO/UNU [30] (see also Table 8). At first sight this would appear to have

Table 9. Proposed, revised estimates of the mean requirements for amino acids, in adults, and their relationship to the requirement for protein

Amino acid	Mean requirement (mg) per	
	kg body wt[a]	per g protein[b]
Isoleucine	23	38
Leucine	39 (40)[c]	65 (66)
Lysine	42 (30)	70 (50)
Total SAA	16 (13)	27 (22)
Total AAA	39	65
Threonine	21 (15)	35 (25)
Tryptophan	6	10
Valine	24 (20)	40 (33)

[a] Derived from factorial (Obligatory N) method (Table 7)

[b] A mean requirement for high quality, highly digestible protein of 0.6 g kg^{-1} day^{-1} is assumed [30]

[c] Value in parentheses based on mean requirement estimates from ^{13}C-tracer studies (see Table 8)

18

dramatic implications for all dietaries and possibly to recreate the "protein problem." The practical implications, however, while possibly significant for dietaries and food and agricultural policies in some developing countries, are less for diets of some other developing countries and the developed regions where diets are generally rich in foods of animal origin. In the following paragraphs we will develop and then explore the application of a new amino acid requirement, or *scoring* pattern for purposes of assessment of milk proteins in diets in both the developed and developing regions of the world.

Previous amino acid scoring systems

Before developing a new amino acid scoring pattern, it is worth noting that various amino acid scoring systems have been recommended previously by various national and international groups [30, 51–54]. They have been developed for purposes for predicting the capacity of food protein sources to meet human physiologic needs for nitrogen and indispensable amino acids and a number of the earlier amino acid scoring patterns are shown in Table 10. The significant changes that have occurred in the recommendations over the years may be seen clearly from this summary. Although the 1985 FAO/WHO/UNU [30] group was the first to explicitly recommend lower values for the adult, the earlier 1973 FAO/WHO group [51] had recognized that the adult requirement values (expressed per unit of protein) were indeed much lower but, nevertheless, based its scoring pattern on data for infants and young children.

Thirty years have now passed since the first FAO committee [52] considered protein, or amino acid, scoring. Provided that an adjustment is made for the digestibility of ingested proteins [55], the statement made by this international group in

Table 10. Some national and international recommended amino acid scoring patterns[a]

Amino acid	1957 FAO [52]	1973 FAO/ WHO [51]	1974 NAS- NRC [54]	1985 FAO/WHO/UNU [30]			
				Infant (< 1yr)	Pre-school child (2–5 yrs)	School-age child (6–12 yrs)	Adult 13 yrs +
Histidine	–	14	17	26	14	19	–
Isoleucine	42[a]	40	42	46	28	28	13
Leucine	48	70	70	93	66	44	19
Lysine	42	55	51	66	58	44	16
Total SAA	42	35	26	42	25	22	17
Total Aromatic	56	60	73	72	63	22	19
Threonine	28	40	35	43	34	28	9
Tryptophan	14	10	11	17	11	9	5
Valine	41	50	48	55	35	25	13
Total (w/o histidine)	314	360	356	434	320	222	111

[a] Figures expressed as mg amino acid per g protein (N × 6.25)

1957, concerning the role of scoring in the relation to an amino acid requirement patterns remains valid, and we quote:

"The concept of a desirable pattern of essential amino acid has one great advantage. By comparison with such a pattern, data on the amino acid content of food combinations can be appraised in a wide range of situations in terms of possible defects of the diet and of methods of improving it. Comparisons, no doubt rough and approximate, can be made directly by the use of tables showing the contents of foods in essential amino acids, regardless of the proportions in which individual foods are included in the diet. On the other hand, even when the biological value of each dietary component is known, a deduction cannot be made about the biological value of the diet as a whole."

We wish to point out here that as the use of amino acid scoring systems has progressed over the years the general expectation has developed that not only should an amino acid score be able to predict the potential nutritional value of a food or diet for humans but that such a score should (with or without digestibility considerations) also correlate directly with the results of animal assays, such as net protein utilization (NPU). Whether such correlations with animal assays are valid or even a desirable attribute of amino acid scoring systems intended for application in human nutrition remains questionable. With hindsight, our view is that scoring systems developed to predict the nutritional value of protein sources for humans should not necessarily be expected to agree with values obtained with growing rats. We now contend that the appropriate standard for dietary assessment is the human amino acid requirement pattern and that for animal bioassays to be useful they should be designed to give predictions in line with those based on human amino acid requirements rather than the reverse.

A proposed new amino acid scoring pattern

Thus, accepting the validity and usefulness of the amino acid scoring approach the newly proposed estimates of the amino acid requirements in adults (Table 9) provide us with a basis for making a further refinement in establishing a more satisfactory amino acid scoring pattern. It also follows that we have an opportunity to re-explore the nutritional value of milk proteins and protein quality of diets in various regions of the world.

To arrive at a new adult amino acid pattern the relationship between the mean amino acid requirements, as estimated by the different approaches discussed above and the mean requirement of 0.6 g protein kg^{-1} day^{-1} for high quality proteins must be established. This has been done in Table 9. For lysine and threonine values in the amino acid pattern based on requirement figures derived from ^{13}C-tracer studies will be used for the present purposes.

With respect to sulfur amino acids, the factorial prediction (Table 9) of the requirements for these amino acids in the adult leads to a somewhat higher level of methionine in the adult pattern than in that for the two-year old group as proposed by FAO/WHO/UNU [30] (see Table 10). However, nitrogen retention is not improved by methionine supplementation in healthy adults consuming, at intakes to 0.8 g protein kg^{-1}, isolated soy proteins as the sole source of dietary nitrogen and

20

supplying 25 mg sulfur amino acids per g protein [56]. Our preliminary ^{13}C-methionine tracer studies suggest that the mean requirement for methionine and cystine is close to 13 mg kg$^{-1} \cdot$ day^{-1} (Table 8). This would give a value of 22 mg per g protein. Thus, it seems resonable in the absence of more direct data, to suggest that an appropriate figure for the sulfur amino acid content in the adult amino acid requirement pattern would be no greater than the level tentatively established for the two-year-old group, i.e., 25 mg per g protein [30].

Finally, in terms of arriving at a new recommended amino acid pattern for the adult, the policy of the arlier expert groups was to round the figures for each amino acid to the nearest value that is divisible by 5. If we also follow this policy, together with the above refinements, a recommended adult amino acid pattern can be proposed. This is given in Table 11. It can be seen that this new amino acid scoring pattern is essentially the same as that for the 1985 FAO/WHO/UNU two- to five-year old pattern (Table 10), with the single exception that threonine is lower in the proposed adult pattern and lysine is somwhat higher in the 1985 FAO/WHO/UNU [30] pre-school child pattern. On this basis, and in relation to the practical problems of the evaluation of dietary protein and for purposes of adjusting safe practical allowances for protein intakes for usual diets [30], it appears to us that a single tentative amino acid scoring pattern could be proposed for the entire age range, covering pre-school children through adults. Therefore, we will use the new recommended pattern shown in Table 11 for this purpose.

In Table 12 we compare the amino acid pattern of cow's milk and the 1985 FAO/WHO/UNU [30] for infants, which is based on the amino acid composition of human breast milk, with the new amino acid scoring pattern we have proposed. From this comparison the amino acid concentration of cow's milk proteins exceeds the individual amino acid requirements (per unit of protein) for individuals of pre-school age and older and, thus, it can be regarded as a "complete" protein source. For infants, this comparison in Table 12 suggests that the sulfur amino acid content of cow's milk proteins is somewhat lower than an optimal. Therefore, an adjustment would need to be made to the proposed safe practical intake for this age

Table 11. Recommended amino acid scoring pattern for individuals from pre-school age through adulthood[a]

Amino acid	Amount (mg per g protein)
Isoleucine	35
Leucine	65
Lysine	50
Total SAA	25
Total Aromatic	65
Threonine	25
Tryptophan	10
Valine	35

[a] For derivation of pattern see revised amino acid requirement estimates discussed in the text

Table 12. Comparison of the amino acid pattern of cow's milk proteins with suggested patterns of amino acid requirements (mg per g protein)

Amino acid	Infant[a] (FAO/WHO/UNU 1985)	Pre-school[b] to adult	Cow's milk[a]
Histidine	26	–	27
Isoleucine	46	35	47
Leucine	93	65	95
Lysine	66	50	78
SAA	42	25	33
AAA	72	65	102
Threonine	43	25	44
Tryptophan	17	10	14
Valine	55	35	64
Amino acid score		Infant 79 (SAA) Adult > 100	

[a] From FAO/WHO/UNO [30]
[b] From Table 11

group (1.48 g · kg^{-1} day^{-1} for infants age 0.75–1 year) [30], if cow's milk, as compared to breast milk, was the sole source of dietary protein. However, the relative value of milk protein in diets as consumed by children and adults is of greater interest here and in subsequent sections we explore the application of the new pre-school-to-adult amino acid requirement pattern for assessment of milk proteins and of their significance in regional and national diets. Indeed, use of dietary amino acid composition data, in comparison with an appropriate reference pattern, is the only practical approach that we can take for evaluating the protein value of diets.

World Dietary Protein Patterns

To evaluate the protein of diets consumed in broad geographical areas of the world, the amino acid composition can be obtained by either direct analysis of sample diets or, more conveniently, from various levels of information concerning the overall dietary composition of the particular population in question. We are of the view that this information can be obtained relatively easily.

When mean dietary patterns in various countries and regions of the world are examined it has been noted that as wealth increases not only is more food consumed but also that there are major changes in the *pattern* of foods selected. These changes were tabulated in 1969, in a classic report [57] in which it was demonstrated that as the proportion of energy intake from fats rose so did income, while it declined as the proportion from total carbohydrates rose. Simultaneously, however, there was a positive correlation between the proportion of the total protein supply from animal sources and income.

Among the broad relationships illustrated by data from 130 countries (Table 13) it can be seen that as wealth (indicated by per capita GNP) increases both the mortality rate of children under fiye years and the prevalence of births of low birth

Table 13. Health, wealth and dietary data for 130 countries grouped on the basis of their under-five mortality rates

Group	No. of countries	Popu-lation (million)	GNP per caput (US $)	Under-five mortality rate	% of children below 5 yrs.	Low birth weight %	Total protein (TP) g/day	Animal protein (%) (AP/TP)	Milk protein (%) (MP/TP)	Fat g/day	Retinol µg/day	Food energy (% of reg)
1	32	462	295	227	18	16	53	21	6	38	140	91
2	32	1 498	1 623	130	17	13	62	25	7	51	190	101
3	30	1 692	2 207	58	14	11	70	37	10	63	260	111
4	36	1 165	7 817	15	8	6	94	57	19	126	620	129

Source of Data: UNICEF [58] and FAO [59–61]

weight infants decline. The dietary changes that accompany these measures of health are increases in total protein per day, animal and milk protein per day, the percentage of animal protein in relation to total protein and in total fat and retinol intakes.

A further elaboration of these relationships is shown in Table 14, derived from linear regression analysis of data from the same 130 countries. All correlation coefficients tabulated were highly significant ($p < 0.001$). Further details involving social and health variables have been described elsewhere [62, 63]. As per capita GNP increases, the population percentage of children under five years and the infant mortality rate (IMR) both decline. At the same time total protein, animal protein, milk protein, dietary lysine concentration and utilizable protein availability all increase. In addition, as protein increases the greater is the amount and proportion (not shown) of animal protein. Lysine concentration also increases as more protein and (more significantly) more animal protein is available and, presumably, consumed. It must be emphasized that the significant relationships presented here between IMR and the dietary factors examined here *must not be considered causal,* since all are directly or indirectly linked with wealth.

Table 14. Correlation matrix between wealth, demographic and dietary data for 130 countries[a]

	IMR	Under 5 yrs %	Protein g/d	Animal protein g/d	Milk protein g/d	Fat (g/d)	Lysine[b] (m/g protein)	Utilizable protein[c] g/d
GNP US $ per caput	−0.61	−0.59	0.68	0.78	0.72	0.78	0.73	0.69
Infant mortality rate		0.78	−0.70	−0.75	−0.65	−0.70	−0.74	−0.74
Percentage of population under 5 yrs.			−0.67	−0.74	−0.66	−0.74	−0.69	−0.71
Protein g/day				0.88	0.81	0.85	0.73	(0.99)[d]
Animal protein g/day					0.90	0.91	0.93	0.93
Milk protein g/day						0.88	0.84	0.84
Fat g/day							0.81	0.87
Lysine mg/g protein								0.81

[a] Number of data pairs vary between 108–130, depending on the variable: All R values $P < 0.001$. Data obtained from [58–61]

[b] Lysine (mg/g protein) calculated from multiple regression equation Lys = 0.57 AP% + 0.33 PP% + 31.3, derived from data for 20 countries ($R^2 = 0.86$): AP% and PP% are percentages of total protein from animal protein and pulse protein (includes soybean), respectively

[c] Utilizable protein = protein availability (g/day) × protein quality score × digestibility. Score and digestibility expressed as fractions. Score derived in relation to new scoring pattern (table 11) (Lysine is limiting amino acid in all cases). Digestibility is taken to be 0.85 for diets high in coarse grain cereals and 0.90 for other diets

[d] Total protein is major determinant of utilizable protein

The further relationships between IMR, GNP, and dietary data with percentages of children below five years also demonstrate that population growth is highly correlated with overall nutritional status. The inverse nature of these relationships may also be of interest, particularly concerning milk availability. On a world basis, the higher the proportion of the population that is represented by children below five years of age, the lower are the average amounts of animal and milk protein that are available (Tables 13, 14). This is, perhaps, ironic but presumably reflects the fact that the poorest countries show high population growth and are those least able to afford animal protein foods. In sharp contrast, data for the United States (Table 15) demonstrate that average milk protein consumption is about 20% of total protein (compared to 6% in the poor countries), with 30% for the one–five year age group and as high as 54% for the formula-fed infant below one year of age.

Also summarized in Table 15 are the patterns of amino acid supply in the various age and sex groups of the US population, shown here as an example of dietary data for the technologically advanced nation. The constancy of the dietary amino acid pattern as a function of age and sex is a striking feature of this analysis. There is, notwithstanding this relative constancy, a considerable variation in the consumption of different protein foods by the various age and sex groupings. This can be seen by comparison of the coefficients of variation (CV) for these same data. The CVs for the eight indispensable amino acids (expressed as mg/g protein) across the same 21 age and sex groupings averaged only about 2.0%, while the dietary use of milk and meat products had CVs in excess of 30%. This is because the major changes in consump-

Table 15. Amino acid and protein supply in the U.S. diet for various age and sex groups[a]

Nutrient	Age in years						Mean ± SD[b] (all ages and both sexes)
	<1 M/F	1–5 M/F	19–22		65–74		
			M	F	M	F	
	mg/g protein						
Isoleucine	54	54	53	52	51	51	52.3 ± 1.02 (2.0)[c]
Leucine	86	81	78	77	76	75	77.0 ± 1.96 (2.5)
Lysine	69	70	72	72	70	69	70.6 ± 1.16 (1.6)
Total SAA	31	35	35	35	34	34	34.7 ± 0.48 (1.4)
Total AAA	84	81	79	78	77	76	78.5 ± 1.75 (2.2)
Threonine	40	40	40	40	39	39	39.6 ± 0.59 (1.5)
Tryptophan	12	12	12	12	12	12	12.0 ± 0.35 (2.8)
Valine	59	55	55	55	54	53	54.9 ± 1.20 (2.2)
Total Kcal/day	950	1363	2240	1560	1895	1455	1757 ± 407 (23.2)
Total protein (g/day)	41.7	53.6	94.5	64.4	80.0	62.1	73.9 ± 14.2 (19.2)
% meat/poultry/fish	17	33	51	48	49	47	41.0 ± 13.8 (33.8)
% milk/milk products	54	30	17	18	16	17	19.3 ± 6.3 (32.8)
% grain products	9	22	18	18	17	18	19.0 ± 2.4 (12.7)

[a] Calculated from 1983 USDA National Food Consumption survey [64], using amino acid data from FAO [65] and USDA [66]
[b] Mean ± SD for all 21 age and sex groups (excludes <1 yr)
[c] Value in parenthesis is coefficient of variation

tion pattern, as affected by age and sex, involve different patterns of supply of the animal protein foods which do not differ greatly in their amino acid composition.

The diets of economically poor countries reveal a marked contrast to the situation in developed countries because animal foods may account for only 10% of the protein and cereals may contribute more than 60% of the dietary protein supply. The very low levels of animal protein and animal foods consumed in the developing regions are shown in Table 16. Milk supplies, in relation to DES (dietary energy supply) for example, are three to four times greater in the developed market economies than in the poor countries of Africa and the Far East. Milk and dairy production data [67] show an even more extreme picture, with per capita milk production values showing a 10-fold difference between developed and developing countries.

Amino acid compositional data are now available for many foods [65, 66] and, hence, the amino acid pattern of regional and country diets (Table 17) can be easily calculated from Food Balance Sheet data [60]. It can be seen that the nutritionally important differences between the amino acid pattern of the diets in developing and developed regions lies in the lysine content. This is due, primarily, to the consistently much lower levels of lysine in cereal products (about 29 mg/g protein) as compared to animal foods (about 85 mg/g protein). The important role of both animal foods, and of legumes (about 65 mg/g protein) in supplying lysine to the dietary, we believe, remains relevant in dietary planning and assessment as well as for agricultural policy [68], despite implications to the contrary that have been drawn from the estimates of adult amino acid needs in the FAO/WHO/UNU [30] report on energy and protein requirements. Also, it should be noted that, from a practical perspective, it is lysine rather than the sulfur amino acid content that seems to be important. Evaluation of milk proteins (Table 12) should be made in this context.

Table 16. Per capita dietary composition in selected world regions[a]

	Developed market economies	Africa	Far East	Near East	Latin America	Least developed countries
Total food energy (Kcal)	3390	2260	2160	2840	2620	2070
Total protein (g)	99	56	51	78	67	53
Animal protein (g)	56	11	8	19	28	10
Share of total DES[b] (%)						
Vegetable products	68.3	93.5	94.2	89.3	82.9	93.3
Animal products	31.7	6.5	5.8	10.7	17.1	6.7
Food groups						
Cereals	26.4	46.7	67.4	58.2	40.2	65.0
Pulse nuts seeds	2.4	7.0	5.9	3.8	4.9	6.5
Roots, tubers	3.7	19.8	3.0	1.8	5.6	8.3
Meat	15.3	2.8	1.2	4.0	8.1	2.7
Milk	8.5	2.0	2.5	3.7	5.5	2.5
Eggs	1.6	0.2	0.3	0.4	0.8	0.2
Fish	1.6	0.8	0.7	0.3	0.7	0.5

[a] Adapted from FAO [61]

[b] DES = Dietary energy supply. Values expressed as % of total energy supply. Only those food groups providing significant levels of dietary protein are shown. Totals are, thus, less than 100%

26

Table 17. Amino acid pattern (mg/g protein) for diets in selected regions: comparison with infant and adult amino acid requirement patterns[a]

| | Region | | | | | Requirement pattern | | |
| | Developed countries | Africa | Far East | Near East | Latin America | Infant[b] | Adult[c] | |
							A	B
Isoleucine	48	42	47	44	47	46	13	35
Leucine	77	75	81	73	82	93	19	65
Lysine	65	47	46	41	56	66	16	50
Total SAA	36	34	40	36	34	42	17	25
Total AAA	81	72	86	79	79	72	19	65
Threonine	40	35	39	34	39	43	9	25
Tryptophan	12	12	12	12	11	17	5	10
Valine	53	48	57	49	52	55	13	35

[a] Generated from data summarized in Table 16 using amino acid composition data of FAO [65] and USDA [66]
[b] Human milk pattern for infants below 1 year; see table 12
[c] A. FAO/WHO/UNU (30) see Table 10; B. from Table 11

Special needs for infants

The pattern of the amino acid requirement for infancy is based on human breast milk, as noted above. The values are much higher than those for the older age groups. Mean dietary patterns for the various regions are all inadequate for one or more indispensable amino acids, in relation to the requirement pattern for infancy (Table 17). Special consideration must be given to the protein value of diets for meeting the nutritional needs of infants.

Milk has been widely used as a supplementary protein in vegetable mixtures for use in developing countries. Many mixtures have been proposed that will meet protein and energy needs of infants at low cost [69–71]. Some recent studies [72] confirmed that the protein value of 14 selected weaning food mixtures based on wheat and rice with milk, bean, or meat as the supplementary protein sources when assayed by standard rat biological assay procedures met the goals for protein quality recommended in the formulations. Agreement was also found [73] to be high when the amino acid composition was determined and the results compared with the FAO/WHO/UNU [30] reference patterns for infants and young children. In addition, comparison was made between estimated intakes and requirement values for essential amino acids. It was concluded that all mixtures would meet protein and amino acid needs if they were supplied at levels sufficient to meet food energy needs.

The importance of quite small quantities of animal foods, such as milk and meat, in the diet of children is illustrated in Table 18. In addition to its importance in protein complementation, milk is also a significant source of vitamin A and riboflavin. Its potential role in diets of developing countries has been recently reviewed [9].

Quite small quantities of animal protein foods can, thus, dramatically improve the overall nutritional value of diets that are based heavily on cereals. This is particularly

Table 18. Estimated child requirements of protein and other nutrients in relation to intakes of milk and meat

| Child age group | Daily requirement | | | | | Approximate percentage of daily requirement supplied by: | | | | | | | | | |
| | Kcal | Protein g | Lysine mg | Calcium mg | Vitamin B_{12} µg | 200 ml whole milk | | | | | 25 g cooked lamb | | | | |
						Kcal	Protein	Lysine	Calcium	Vitamin B_{12}	Kcal	Protein	Lysine	Calcium	Vitamin B_{12}
6–11.9 m	950	14	900	540	0.5	14	46	59	44	>100	5	50	63	<1	80
1–2 yr	1 150	14	800	800	0.7	10	46	66	30	100	4	50	71	<1	57
2–4 yr	1 350	16	900	800	1.0	9	41	58	30	70	3	44	63	<1	40
4–6 yr	1 550	18	1 000	800	1.0	8	37	53	30	70	3	39	56	<1	40

Source: Protein, food energy and lysine requirement values FAO/WHO/UNU [30]. Calcium requirements (NAS-NRC; 74); vitamin B_{12} requirement [75]. Milk and lamb composition Pellett and Shadevian [76] and Pennington and Church [77]. From Pellett and Young [62]

important when considering the nutritional needs of infants and also needs for pregnant and lactating women. Recommendations encouraging breast feeding of young infants should not, however, be neglected. This is based mainly on consideration of the presence of protective substances acting against diarrheal disease on the potential contraceptive effects of breast feeding and on the psychological benefits to the mother/child dyad [78, 79]. Nevertheless, in good environments, human infants grow and develop when they receive a wide variety of modified and synthetic formula diets based on cow's milk [80, 81].

Amino Acid Supply Patterns

When comparison of dietary amino acid patterns is made with the current internationally derived amino acid requirement values for all ages above infancy (Table 10) it will be noted (Table 17) that for all regions, including Africa and the Near East, the amino acid pattern of regional diets exceeds the FAO/WHO/UNU [30] amino acid requirement pattern for the adult by a wide margin. This is not true when comparison is made with our newly proposed pattern (Table 11). While the pattern of requirements is exceeded for most amino acids in all regional diets the *mean level of dietary lysine* remains low for Africa, Far East, and Near East regions (Table 17). This implies that many individuals may be at risk of an inadequate protein nutritional status because of dietary lysine limitation. As can also be seen (Table 16) these regions are those associated with the lowest levels of animal protein availability.

The amino acid composition of average diets in 20 countries has also been calculated by us [82]. Using food balance sheet [60] and amino acid composition data [65, 66] linear and multiple regression analyses were performed between the amino acid composition of the diets, the percentages of the protein from animal protein (AP%), pulse protein (PP%) and cereal protein (CP%). For this purpose, pulse protein (PP%) was defined as the FAO [60] category of pulses (mainly beans and peas) together with soybean from the nuts and oilseeds category. For the indispensable amino acids, only lysine showed high correlations ($p < 0.01$) with the other variables, confirming the observations made above concerning the amino acid pattern in regional diets.

From multiple regression analysis of these kind of data, a prediction equation:

$$\text{Lysine (mg/g protein)} = 0.57\,AP\% + 0.33\,PP\% + 31.3$$

was derived [82]. About 76% of the correlation was due to AP% ($Lys = 0.47\,AP\% + 37.5$), increasing to 86% by including PP%. Inclusion of CP% into the equation did not increase the prediction ability ($R^2 = 0.861$) despite the fact that there was a high negative correlation ($R = -0.77$, $p < 0.001$) between lysine concentration and CP% when simple regression analysis was used.

Animal protein was then desegregated into its individual components of meat, milk, fish, and egg protein and these values were then used for multiple regression analysis, together with pulse and cereal proteins. The results of this analysis showed that very little was gained by this procedure, since R^2 only increased from 0.86 to 0.89 accompanied, however, by a considerable fall in the overall F value (52.5 to

Table 19. Consumption of various sources of protein and lysine value for selected countries

Country	GNP U.S. $ per caput	Protein g/day	Animal protein g/day	Meat protein g/day	Milk protein g/day	Fish protein g/day	Legume[a] protein g/day	AP/TP %	Lysine[c] mg/g protein
Bangladesh	130	39.7	5.3	1.6	1.2	2.2	1.3	13.4	44
Zaire	140	33.2	6.3	4.2	0.3	1.8	2.7	19.0	46
Nigeria	730	54.6	10.6	4.5	1.1	4.5	5.2	19.4	42
Thailand	860	46.9	12.2	5.1	0.8	5.7	1.4	26.0	51
Guatemala	1 160	58.2	14.8	7.6	5.5	0.3	7.0	25.4	46
Syria	1 620	84.5	21.6	8.1	10.8	0.5	5.8	25.6	46
United Kingdom	8 570	90.1	53.9	27.4	18.5	3.9	2.1	59.8	65
Japan	10 630	88.6	46.4	10.5	5.5	25.2	9.9	52.4	63
United States	15 390	105.6	71.5	41.1	22.0	3.5	1.4	67.7	70

Source: FAO [60] and UNICEF [58]
[a] Includes soybean from nuts and oilseeds category
[b] Animal protein as percentage total protein
[c] Calculated from food balance sheet using amino acid composition data

23.7). As can be seen from the selected country data shown in Table 19, despite much variation in the distribution of the animal protein, as wealth (per capita GNP) increased so did a) the amount and proportion of animal protein, and b) the dietary lysine value.

As with the evaluation of the US diet discussed earlier, a strong relationship between animal protein and lysine value can be explained by the rather similar content (mg/g protein) of lysine in all of the major animal protein foods.

The lysine value of world dietaries

By use of the prediction equation involving only AP% and PP% ($R^2 = 0.86$) shown above, the lysine contents of the diets for all 146 countries in the FAO Food Balance Sheets [60] were estimated. Only 43 countries had lysine value below the proposed requirement (scoring) pattern value of 50 mg/g protein (Table 11). Many of these countries were, however, among the poorer, more populous countries with their combined population representing more than half of the world's total population.

Because amino acid score is not the only variable affecting the protein value of a diet, an approach involving the concept of utilizable protein (quality × quantity) was used to evaluate these world dietaries. The quantity of protein was represented by daily protein availability (g/day) while quality induced digestibility and amino acid score. The digestibility was taken as 85% when the diets involved consumption of high levels of coarse grain cereals and 90% for all other diets [30]. The amino acid *score* was judged in relation to the new requirement pattern (Table 11) where, in all cases, the limiting amino acid was lysine (50 mg/g protein). The dietary supply of utilizable protein was then compared with protein requirements.

30

For this type of comparison the average requirement should be used rather than safe practical allowances. Average protein needs for selected reference male (65 kg) and female (55 kg) adults would be 36 g (highly digestible, high quality) protein daily [30]. Although 43 countries had been found to have lysine values below 50 mg/g/ protein only 15 of these countries had mean utilizable protein supplies at or below the average adult requirements of 36 g/daily while a further 11 countries had values below 45 g/day (the safe pratical allowance) and could be considered marginal by some criteria. The diets of the remaining countries all provided a *mean* utilizable protein in excess of the safe practical allowance, by virtue of the higher levels of total protein.

To place these values into perspective, the range (Table 20) for protein supply, animal protein and estimated protein quality for the 146 countries in the FAO [60] food balance sheets should be considered. Total protein is the most important determinant and the range of values between the rich and poor countries is large. Nevertheless, average protein supply exceeds average requirements for reference adults [30] for all countries except Mozambique and Zaire. If protein quality considerations are included, diets with little animal protein and legumes which also contain coarse grain cereals could have an overall (including digestibility) protein value of 0.64 (0.75 × 0.85) in relation to the adult pattern. Thus, average protein supplies below 56 g protein daily (36/0.64) could be nutritionally inadequate. For the 146 countries whose diets are tabulated in FAO [60], 39 countries have average supplies below this value. In addition, many of the very poor in countries where the average supply is somewhat higher would also be at risk since their dietary supplies of protein in both quality and quantity are likely to be below the mean. In contrast it must be recognized that when protein supplies are high then protein quality considerations (including digestibility) become much less significant.

In summary, therefore, the countries, and the individuals within those countries, most at risk are those with low intakes of diets based heavily on coarse grain cereals

Table 20. Mean and range for selected dietary protein indicators from food balance sheet data (146 countries)

	Mean ± SD	Range
Total protein g/day	70 ± 21	32–129
Animal protein[a] (%)	35 ± 18	6– 80
Legume protein[b] (%)	6 ± 6.6	1– 45
Digestibility[c] (%)	ca. 90	85– 95
Protein quality score[d]		
Human milk pattern	92 ± 14	55–109
Adult pattern	109 ± 19	75–144

Source: FAO [60] for protein, animal protein and legume protein
[a] Expressed as percentage of total protein
[b] Expressed as percentage of total protein, includes soybean
[c] Estimated from digestibility data of FAO/WHO/UNU [30]
[d] Lysine value calculated from multiple regression equation (see text) and compared to breast milk for the infant and to new amino acid requirement pattern for the adult (Table 11)

where both the quantity and quality (including digestibility) of the protein available for consumption are likely to be low. The overriding importance of the total quantity of food, as represented by both dietary energy supply and total protein are the major causes for concern, but the protein quality of the diet, even for the adult, remained important and can be readily evaluated in terms of the contents of animal and legume protein. In many cases evaluation in terms of animal protein percentage alone may be of sufficient accuracy.

As a practical guide, we conclude, that providing total protein and food energy needs are met, there would be no major problems of protein quality (amino acid supply for pre-school age to adults) if about one-third of the dietary protein is provided from animal sources, including milk proteins. For children below one year this must be increased to a little above one-half in order to meet the higher lysine need of 66 mg/g protein. When animal protein percentages are below these values the well known concepts of protein complementation should be applied. In general, a combined percentage of about 40–45% of the protein from the high lysine sources of animal proteins and legume proteins implies that daily lysine needs are met adequately for adults. For young children it is perferable to use the various published guides for food mixtures [69–71], since as much as 65% for the combined percentages of protein from animal foods and from legumes may be necessary to meet the relatively high nutritional requirements of the infant. Nevertheless, for both adults and young children the greater the degree that total protein intake exceeds requirements, the less important are protein quality considerations.

Conclusion

The direct evaluation of the capacity of dietary protein(s) to meet the needs for indispensable amino acids in human subjects is the principal focus of this paper. It is our thesis that amino acid composition data, considered in relation to human amino acid requirement estimates, provide an acceptable basis for evaluating dietary protein with reference to practical aspects of human protein and amino acid nutrition. Hence, we briefly review some aspects of protein and amino acid metabolism in humans and recent observations concerned with a re-evaluation of the amino acid requirements in adult subjects. The relevance of this discussion is in terms establishing appropriate amino acid requirement (scoring) patterns for comparison with amino acid composition data of protein foods. Indeed, the 1985 FAO/WHO/UNU [30] report on energy and protein requirements proposed use of age-specific amino acid scoring patterns. However, the very low adult amino acid requirement values tabulated have major policy and nutrition implications, particularly since it would appear from FAO/WHO/UNU [30] that almost any diet could provide indispensable amino acids well in excess (two- to threefold) of apparent adult amino acid requirements. In this paper new data on amino acid requirements are reviewed. With this background a new tentative amino acid scoring pattern is proposed for application for the age range from pre-school through adults. This pattern is then used to evaluate the protein component of regional diets, incuding an assessment of the role of milk proteins. From this analysis the important practical conclusion is that diets based principally on cereals, especially in various areas of Africa, the Near and Far

East, are likely to be marginal with respect to meeting lysine requirements, while diets typical of developed regions are more than adequate. Total dietary protein availability is a major determinant of the nutritional quality of the diets but the amino acid pattern also is of importance especially in those areas where animal protein foods contribute less than about one-third of total dietary protein. Furthermore, these considerations underscore the high nutritional value of milk proteins and the fact that relatively small amounts of them would fill the predicted "lysine gap". Because of the practical value of the approach used here to evaluate dietary protein we recommend that a new advisory group under UN auspices should be convened to discuss the amino acid requirement issues raised in this paper and make appropriate, authoritative recommendations. As an interim measure we do not believe that the FAO/WHO/UNU [30] adult pattern should be used to evaluate dietary protein. The newly proposed pattern described above would seem to provide a more rational and safe basis for this purpose.

References

1. Young VR (1981) Food protein sources: Implications of nutrient requirements. In: Selvey N, White PL (eds) Nutrition in the 1980s: Constraints on our knowledge. Alan R. Liss, Inc., New York, NY, pp 189–206
2. Millward DJ, Rivers JPW (1988) The nutritional role of indispensable amino acids. Eur J Clin Nutr 42 (in press)
3. Bender AE (1983) Protein evaluation. In: Kaufmann W (ed) Role of milk proteins in human nutrition. Kieler Milchwirtschaftliche Forschungsberichte 35:267–273
4. Young VR, Pellett PL (1985) Wheat proteins in relation to protein requirements and availability of amino acids. Am J Clin Nutr 42:1077–1090
5. Pellett PL, Young VR (eds) (1980) Nutritional value of protein foods. United National University, World Hunger Program; Food and Nutrition Bulletin, Supplement 4. United Nations University, Tokyo, pp 154
6. Porter JWG, Rolls BA (eds) (1973) Proteins in human nutrition. New York, Academic Press, pp 560
7. Bodwell CE, Adkins JS, Hopkins DT (eds) (1981) Protein quality in humans: Assessment and in vitro estimation. AVI Publishing Co., Inc., Westport, Connecticut, pp 435
8. Young VR (1983) Human studies for evaluation of protein quality with particular reference to milk proteins. In: Kaufmann W (ed) Role of milk proteins in human nutrition. Kieler Milchwirtschaftliche Forschungsberichte 35:247–265
9. Young VR, Pellett PL (1985) Milk proteins with reference to human needs, and world food supply and nutritional situation. In: Galesloot TE, Tinbergen BJ (eds) Milk proteins 1984. Pudoc, Wageningen, Netherlands, pp 18–46
10. Räihä NCR (1980) Protein in the nutrition of the pre-term infant: Biochemical and nutritional considerations. Adv Nutr Res 3:173–206
11. Hambraeus L (1985) Importance of milk proteins in human nutrition: physiological aspects. In: Gallsloot TE, Tinbergen BJ (eds) Milk proteins '84. Pudoc, Wageningen, Netherlands, pp 63–79
12. Finke MD, DeFoliart GR, Benevenga NJ (1987) Use of simultaneous curve fitting and a four-parameter logistic model to evaluate the nutritional quality of protein sources at growth rates of rats from maintenance to maximum gain. J Nutr 117:1681–1688
13. Phillips RD (1981) Linear and non-linear models for measuring protein nutritional quality. J Nutr 111:1058–1066
14. Phillips RD (1982) Modification of the saturation kinetics model to produce a more versatile protein quality assay. J Nutr 112:468–473
15. Mercer LP (1982) The quantitative nutrient-response relationship. J Nutr 112:560–566

16. Young VR, Rand WM, Scrimshaw NS (1977) Measuring protein quality in humans: a review and proposed method. Cereal Chem 54:929–948
17. Fomon SJ, Ziegler E, Nelson SE, Edwards BB (1986) Requirement of sulfur-containing amino acids in infancy. J Nutr 116:1405–1422
18. Pellett PL, Young CR (1984) Background paper 4: Evaluation of the use of amino acid composition data in assessing the protein quality of meat and poultry products. Am J Clin Nutr 40:718–736
19. Young VR, Pellett PL (1984) Background paper 5: Amino acid composition in relation to protein nutritional quality of meat and poultry products. Am J Clin Nutr 40:737–742
20. Young VR, Fukagawa NK, Storch KJ, Hoerr R, Jaksic T (1988) Stable isotope probes: Potential for application in studies of amino acid utilization in the neonate. In: Lindblad BS (ed) Perinatal nutrition. Academic Press, Inc., San Diego, California, pp 221–241
21. Young VR, Moldawer LL, Hoerr R, Bier DM (1985) Mechanisms of adaptation to protein malnutrition. In: Blaxter KL, Waterlow JC (eds) Nutritional adaptation in mMan. John Libbey, London, pp 189–215
22. Waterlow JC, Garlick PJ, Millward DJ (1978) Protein turnover in mammalian tissues and in the whole body. North Holland, Amsterdam, pp 804
23. Young VR, Fukagawa NK, Bier DM, Matthews D (1986) Some aspects of in vivo human protein and amino acid metabolism, with particular reference to nutritional modulation. Verhandlungen der Deutschen Gesellschaft für Innere Medizin, 92. Band. Bergmann, München, pp 639–665
24. Munro HN (1970) Free amino acid pools and their role in regulation. In: Munro HN (ed) Mammalian Protein Metabolism, vol. IV, pp 229–386
25. Kang-Lee TA, Harper AE (1977) Effect of histidine intake and hepatic histidase activity on the metabolism of histidine in vivo. J Nutr 107:1427–1443
26. Harper AE, Benjamin A (1984) Relationship between intake and rate of oxidation of leucine and α-ketoisocaproate in vivo in the rat. J. Nutr 114:57–62
27. Young VR, Meredith C, Hoerr R, Bier DM, Matthews DE (1985) Amino and kinetics in relation to protein and amino acid requirements: the primary importance of amino acid oxidation. In: Garrow JS, Halliday D (eds) Substrate and energy metabolism in man. John Libbey, London, pp 119–133
28. Young VR, Gucalp C, Rand WM, Matthews DE, Bier DM (1987) Leucine kinetics during three weeks at submaintenance-to-maintenance intakes of leucine in men: Adaptation and accommodation. Hum Nutr Clin Nutr 41 C:1–18
29. Waterlow JC (1985) What do we mean by adaptation? In: Blaxter KL, Waterlow JC (eds) Nutritional adaption in man. John Libbey, London, pp 1–10
30. FAO/WHO/UNU (1985) Energy and protein requirements. Tech Rept Ser No 724, World Health Organization, Geneva
31. Hegsted DM (1976) Balance studies. J Nutr 106:307–311
32. Young VR (1986) Nutritional balance studies: indicators of human requirements or of adaptive mechanisms? J Nutr 116:700–703
33. NRC (1974) Improvement of protein nutriture. Food and Nutrition Board, National Academy of Sciences, Washington DC
34. Young VR, Meguid M, Meredith C, Matthews DE, Bier DM (1981) Recent developments in knowledge of human amino acid requirements. In: Waterlow JC, Stephen JML (eds) Nitrogen metabolism in man. Applied Science Publishers, London, pp 133–153
35. Meredith C, Bier DM, Meguid MM, Matthews DE, Wen Z, Young VR (1982) Whole body amino acid turnover with ^{13}C-tracers: a new approach for estimation of human acid requirements. In: Wesdorp RIC, Soeters PB (eds) Clinical nutrition '81. Churchill Livingston, London, pp 42–59
36. Meguid MM, Matthews DE, Bier DM, Meredith CN, Soeldner JS, Young VR (1986) Leucine kinetics at graded leucine intakes in young men. Am J Clin Nutr 43:370–380
37. Meguid MM, Matthews DE, Bier DM, Meredith CN, Young VR (1986) Valine kinetics at graded valine intakes in young men. Am J Clin Nutr 43:781–786
38. Meredith CM, Wen Z-W, Bier DM, Matthews DE, Young VR (1986) Lysine kinetics at graded lysine intakes in young men. Am J Clin Nutr 43:787–794
39. Zhao X-L, Wen Z-W, Meredith CM, Matthews DE, Bier DM, Young VR (1986) Threonine kinetics at graded threonine intakes in young men. Am J Clin Nutr 43:795–802

34

40. Waterlow JC (1986) Metabolic adaptation to low intakes of energy and protein. Ann Rev Nutr 6:495–526
41. Young VR, Bier DM, Pellett PL (1988) A theoretical basis for increasing current estimates of the amino acid requirements in adult man, with experimental support. Am J Clin Nutr (submitted)
42. Motil KJ, Matthews DE, Bier DM, Burke JF, Munro HN, Young VR (1981) Whole-body leucine and lysine metabolism: response to dietary protein intake in young men. Am J Physiol 240:E712–E721
43. Steffee WP, Goldsmith RS, Pencharz PB, Scrimshaw NS, Young VR (1976) Dietary protein intake and dynamic aspects of whole body nitrogen metabolism in adult humans. Metabolism 25:281–297
44. Calloway DH, Margen S (1971) Variation in endogenous nitrogen excretion and dietary nitrogen utilization as determinants of human protein requirement. J Nutr 101:205–216
45. Young VR, Taylor YSM, Rand WM, Scrimshaw NS (1973) Protein requirements of man: Efficiency of egg protein utilization at maintenance and submaintenance levels in young men. J Nutr 103:1164–1174
46. Yoshida A, Ashida K (1969) Pattern of essential amino acid requirement for growing rats fed on a low amino acid diet. Agric Biol Chem 33:43–49
47. Said AK, Hegsted DM (1970) Response of rats to diets of equal chemical essential amino acids. J Nutr 100:1363–1376
48. Cieslak DG, Benevenga NJ (1886) Response of rats to diets of equal chemical score: Effect of lysine and threonine as the limiting amino acid and of an amino acid excess. J Nutr 166:969–977
49. Cieslak DG, Benevenga NJ (1984) The effect of amino acid excess on utilization by the rat of limiting amino acid–threonine. J Nutr 114:1871–1877
50. Cieslak DG, Benevenga NJ (1984) The effect of amino acid excess on utilization by the rat of limiting amino acid–lysine and threonine at equalized food intakes. J Nutr 114:1878–1883
51. FAO/WHO (1973) Energy and protein requirements. Report of a Joint FAO/WHO ad hoc Expert Committee. WHO Tech Rep Ser No. 522. World Health Organization, Geneva
52. FAO (1957) Protein requirements. FAO Nutr Studies No. 16. Food and Agriculture Organization, Rome
53. FAO/WHO (1965) Protein requirements. FAO Nutr Meetings Rep Ser No. 37. Food and Agriculture Organization, Rome
54. Williams HH, Harper AE, Hegsted DM, Arroyave G, Holt LE Jr (1974) Nitrogen and amino acid requirements. In: Improvement of protein nutriture. US Food and Nutrition Board, National Academy of Science, Washington DC, pp 23–63
55. FAO/WHO (1975) Energy and protein requirements: recommendations by a joint FAO/WHO informal gathering of experts. Food and Nutrition 1(2):11–19
56. Young VR, Puiz M, Querioz E, Scrimshaw NS, Rand WM (1984) Evaluation of the protein quality of an isolated soy protein in young men: relative nitrogen requirements and effect of methionine supplementation. Am J Clin Nutr 39:16–24
57. Perissé J, Sizaret F, François P (1969) The effect of income on the structure of the diet. Nutrition Newsletter 7:1–10. FAO, Rome
58. UNICEF (1987) The state of the world's children 1987. UNICEF, New York, NY 1987
59. FAO (1980) Food Balance Sheets 1975–1977. Average and Per Caput Food Supplies 1961–1965 and 1967–1977. Food and Agricultural Organization, Rome
60. FAO (1984) Food Balance Sheets 1979–1981. Food and Agriculture Organization, Rome
61. FAO (1985) The Fifth World Food Survey. Food and Agriculture Organization, Rome, 129 pages
62. Pellett PL, Young VR (1988) The contribution of livestock products to human dietary needs with special reference to West Asia and North Africa. In: Workshop on Small Ruminants in Mediterranean Areas. International Center for Agricultural Research in Dry Areas–ICARDA, Aleppo, Syria (in press)
63. Pellett PL (1987) Determinants of nutritional status. Food and Nutrition 13:2–14. FAO, Rome
64. USDA (1983) Food intakes: Individuals in 48 states, year 1977–1978. Consumer Nutrition Division Human Nutrition Information Service, US Dept. of Agriculture. Hyattsville, MD 20782. Nationwide Food Consumption Survey 1977–1978. Report No. I-1
65. FAO (1970) Amino acid content of foods and biological data on proteins. FAO Nutritional Studies No 24. Food and Agriculture Organization, Rome

66. Agricultural Handbook No. 8-1 (1976); 8-2 (1977); 8-5 (1979); 8-6 (1980); 8-7 (1980); 8-8 (1982); 8-9 (1982); 8-10 (1983); 8-11 (1984), 8-12 (1986); 8-13 (1986) and 8-14 (1986). Agriculture Research Service, United States Department of Agriculture, Washington DC
67. FAO (1986) Country tables: basic data on the agriculture sector. Economic and Social Policy Department. Food and Agriculture Organization, Rome
68. Pellett PL, Young VR (1988) Protein and amino needs for adults. Ecol Food Nutr (in press)
69. Cameron M, Hofvander Y (1983) Manual of Feeding Infants and Young Children, 3rd Edition. UN/ACC Subcommittee on Nutrition – Oxford University Press, Oxford, p 214
70. Pellett PL, Mamarbachi D (1979) Recommended proportions of foods in home-made feeding mixtures. Ecol Food Nutr 7:219–228
71. Pellett PL, Pellett AY (1983) Food mixtures for combatting childhood malnutrition. In: Rechcigl M (ed) Handbook of nutritional supplements, vol 1. Human Use. CRC Press, Boca Raton, Florida, pp 411–480
72. Mathew SP, Pellett PL (1986) Protein Quality of homemade weaning food mixtures: 1. Biological evaluation and quality. Ecol Food Nutr 19:31–40
73. Pellett PL, Mathew SP (1986) Protein quality of homemade weaning food mixtures: 2. Amino acid composition versus biological methods. Ecol Food Nutr 19:41–49
74. NAS/NRC (National Academy of Sciences – National Research Council) (1980) Recommended dietary allowances. 9th edition. National Academy of Sciences, Washington DC, pp 185
75. Herbert V (1987) Recommended dietary intakes (RDI) of vitamin B_{12} in humans. Am J Clin Nutr 45:671–678
76. Pellett PL, Shadarevian S (1970) Food composition tables for use in the Middle East. 2nd Edition. American University of Beirut, Lebanon, pp 116
77. Pennington JAT, Church HN (1985) Bowes and Church's food values of portions commonly used. 14th Edition. JB Lippincott Co., Philadelphia, pp 257
78. Jelliffe DB, Jellife EFP (1979) The uniqueness of human milk. Am J Clin Nutr 24:968–1024
79. Hambraeus L, Sjölin S (1979) The mother-child dyad – nutritional aspects. Almquist and Wiksell International, Stockholm, Sweden
80. Fomon SJ (1974) Infant nutrition. 2nd Edition. WB Saunders Co., Philadelphia
81. Cockburn F (1983) Milk composition; the infant human diet. Pro Nutr Soc 42:361–372
82. Pellett PL, Young VR (1988) National and international implications of revised estimates of human adult needs for indispensable amino acids (submitted for publication)

Milk Proteins and Tissue Nitrogen Equilibrium

P. J. Reeds

USDA/ARS Children's Nutrition Research Center, Department of Pediatrics, Baylor College of Medicine and Texas Children's Hospital, Houston, USA

Introduction

The period of life immediately after birth is marked by a rapidity of growth that, in terms of weight specific rates, is exceeded only in late fetal life. During this time, the infant also is maturing physiologically and metabolically. Postnatal development involves important changes in the chemical composition of the body, the distribution of the mass of the body between cellular and extracellular components, and the contributions of the various tissues to body protein mass. The assessment of growth during this period can be complex and we should recognize that the assessment of the adequacy of a given diet should include the degree to which it supports development as well as macromolecular accretion.

Growth, protein deposition, and amino acid requirements

Protein requirements can be estimated by the factorial method in which the utilization of protein is partitioned between the processes that require an input of amino acids. The use of this approach is based on the assumption that if these processes can be quantified, an estimate of the minimal requirement can be made, and investigators can determine how best to provide these requirements.

Two factors receive primary consideration: 1) the relative requirements of essential (indispensable) and nonessential (dispensable) amino acids, and 2) the division of these needs between those that support the accretion of new tissue and those required for the replacement of the basal or endogenous losses [11].

There are a number of problems when this approach is applied in the newborn infant. First, the division of amino acids into essential and non-essential categories in immature individuals may not be as simple as that in more mature individuals. Preterm infants have a limited ability to synthesize glycine (Jackson et al., 1981) [28] and perhaps other amino acids [55, 1]. Second, during a period of maturation, processes that are not necessarily directly associated with protein accretion may make metabolic demands on specific amino acids, and thereby potentially limit their utilization for the accretion of protein (see [65, 44] for discussion). Some of these processes (e.g., heme, nucleic acid, and creatine synthesis) result in the retention of amino nitrogen, but are not necessarily associated with a substantial gain of water and, hence, weight. Whether the end products of these processes contain nitrogen or represent the productive utilization of amino acid-carbon is important in the assessment of the true efficiency with which the dietary protein is utilized, because protein accretion is usually measured as nitrogen accretion.

"

One important element in the factorial approach is the concept of an obligatory amino acid requirement for protein deposition. This is the product of the rate of protein accretion and the amino acid composition of the proteins that are deposited. For the human, a substantial body of data exists on rates of nitrogen retention in full-term infants fed human milk. At "normal" rates of human milk ingestion, rates of nitrogen retention range from 200 to 240 mg/kg/d at two weeks of age [53] through 170 to 190 mg/kg/d at one month to between 90 and 130 mg/kg/d at three months of age. On the basis of comparisons of nitrogen balance data [9, 13, 14] with interpolated estimates of the changes in body protein mass, based either on measurements of whole body water and potassium [12] or on estimates of body water from total body electrical conductivity measurements [8], the majority of this nitrogen is retained as protein. For reasons that are unclear, the relationship between nitrogen retention and calculated protein accretion appears to break down at higher protein intakes [10].

The amino acid composition of this gain in body protein is a separate question. Although extensive information on amino acid composition is available in young animals [40], no directly comparable information exists for human infants. Widdowson et al. [62] have reported direct analyses of the amino acid composition of fetuses of gestational age 79 to 280 days (Table 1). With the exception of two amino acids, there is a remarkable similarity between the patterns in animals and humans. The proportions of glycine ($+51\%$) and proline ($+59\%$), however, are much higher in the human fetus than in the weanling rat, thereby suggesting a substantially higher contribution of collagen to fetal body protein mass. Although this conclusion is supported by the high umbilical uptake of glycine by fetal sheep [12, 30], it does not necessarily reflect the limited information on collagen contents of human fetal tissues [60, 61]. Even the collagen concentration in the skeleton does not appear to be markedly high, and the soft tissues, especially the skin, show an accelerated deposition of collagen after, rather than before, birth [62].

The increment in nitrogen retention per unit increment in nitrogen intake in full-term infants fed human milk suggests a biological value between 0.71 and 0.75 (see, e.g., [13]). When nitrogen retention is expressed as a proportion of absorbed nitrogen, the efficiency of utilization (less than 50%) is even poorer. When the amino acid composition of the body and that of milks (Table 1) are compared, interesting anomalies are revealed. Milk proteins are rich in branched-chain amino acids, but have low levels of methionine and, especially, glycine. The basis for this imbalance remains unexplained, but it seems a common feature of the milks of a variety of species [44].

The relatively poor net utilization of milk proteins by the infant, however, emphasizes a further feature of human growth. When the manner in which humans gain weight and deposit protein is compared with that of other mammals, it is clear that the processes occur more slowly in humans. Consequently, dietary protein used for the maintenance of body protein mass to accommodate continual and obligatory losses is of particular importance. Table 2 presents a summary of estimates of basal *nitrogen* losses in pigs, rats, and humans. When these values are normalized to weight raised to the power 0.75 (i.e., per unit basal energy expenditure), only small differences are apparent among the three species; the difference between infants and adults is also small. Based on these results, at least 50% of the infant's *minimal* requirement

Table 1. Comparison of the amino acid composition of the young rat and human fetus with the amino acid composition of human and bovine milk

Amino acid	Body composition		Milk composition		GI protein[a]
	Rat	Man	Rat	Bos	
	(mg amino acid/g protein)				
Isoleucine	37	37	55	56	41
Leucine	76	83	101	102	31
Valine	47	48	61	66	58
Lysine	72	78	74	78	48
Methionine	20	21	15	24	9
Cystine	19	ND[b]	21	10	23
Threonine	46	45	50	46	65
Phenylalanine	39	45	41	47	34
Tryptophan	12	ND	24[c]	14[c]	ND
Tyrosine	36	31	39	41	28
Arginine	78	84	38	36	35
Histidine	30	28	30	28	20
Alanine	62	79	41	37	65
Aspartate	87	89	89	79	91
Glutamate	138	141	185	215	115
Serine	46	47	48	54	46
Glycine	85	128	24	21	140
Proline	58	92	92	95	151

[a] For reference an estimate of the amino acid composition of endogenous gastrointestinal protein loss is included

[b] ND = Not determined

[c] One determination

(The amino acid compositions of the human and rat were taken from Widdowson et al. (1979) and Pellett & Kaba (1972). Milk compositions are averages of Svanberg et al. (1977), Widdowson et al. (1979), Janas et al. (1985), and Harzer & Bindels (1985). The estimated composition of ileal amino acid loss is taken from data in pigs (Low, 1980).)

Table 2. Basal losses of nitrogen in mammals receiving low or protein free diets

Species	Basal nitrogen loss		
	(mg/day)	(mg/kg/day)	$(\text{mg/kg}^{0.75}/\text{day})$
Immature rat	17	242	125
Immature pig	2 784	70	175
Human: infants	458	62	110
children	1 117	71	140
adults	3 433	50	145

(Rat data: Lin & Huang (1986); Pig data: Fuller & Crofts (1977); Infant and children: Fomon et al. (1965): Adults: Young & Scrimshaw (1968), Huang et al. (1972), Nicol & Phillips (1975), Fujita et al. (1984).)

of nitrogen during the first six months of life appears to be expended on the replacement of basal losses.

Although endogenous nitrogen losses have been defined with some certainty, the amino acid composition of the losses of body protein is an important and unanswered question. A related question is whether the pattern of amino acids required to replace the losses is the same as that required for optimum growth. There is evidence to suggest that methionine and threonine may be of particular significance to the amino acid economy of mammals that receive protein-deficient diets [64]. It seems reasonable to conclude, therefore, that the removal of some specific amino acids into nonprotein pathways of metabolism may play a role in restricting the reutilization of amino acids released from protein breakdown.

In addition to oxidative losses of amino acids, however, the loss of nitrogen in the feces under basal conditions can account for up to 50% of the total [11, 15]. Whether the source of the nitrogen is of dietary or endogenous origin is of considerable nutritional importance. When no dietary protein is supplied, the ultimate source is body protein. It is important to recognize, however, that because much of fecal nitrogen consists of microbial protein, its amino acid composition bears no direct relationship to the proteins from which it has been derived [35]. There are two major sources for endogenous nitrogen, plasma urea and protein lost into the gut lumen.

Table 3. Calculated rates of amino acid accretion in the first three months of life compared with the amino acids supplied by human milk adjusted for the amino acid loss in feces

	Amino acid gain	Amino acid supply	Supply/Gain
	(mg per day)		
Isoleucine	196	391	1.99
Leucine	424	801	1.89
Valine	307	411	1.33
Lysine	398	540	1.36
Methionine	106	106	1.00
Cystine	106	125	1.17
Threonine	238	264	1.10
Phenylalanine	223	254	1.13
Tryptophan	80	ND	ND
Arginine	430	166	0.38
Histidine	154	207	1.34
Alanine	371	227	0.61
Aspartate	466	585	1.25
Glutamate	742	1 356	1.82
Serine	249	322	1.29
Glycine	562	−58	[a]
Proline	398	399	1.00

(The data were calculated assuming a daily rate of protein gain of 5.5 g and fecal protein loss of 2.1 g (Fomon & Owen, 1962; Southgate & Barrett, 1966) and that the fecal nitrogen loss was derived originally from endogenous protein having the composition shown in Table 1.)

[a] Note that minimum glycine requirement exceeds supply

40

The capture of urea nitrogen by the gut flora [63] represents an alternative route whereby nitrogen, derived from the catabolism of amino acids, is eliminated from the body. As such, the pattern of amino acids that contribute to urea loss in urine and to its capture by the gut flora is probably the same. The amino acid composition of the *net* losses of endogenous protein in the gut has not been established in man. Available data for pigs receiving protein-free diets [34] are shown in Table 1, and demonstrate that the secretions are relatively rich in cystine, threonine, proline, and glycine. Despite these observations, the question of the relative contributions of endogenous and exogenous (i.e., dietary) components has proved difficult to answer under practical feeding conditions, and may be particularly complex in infants. During the neonatal period, changes occur with time in the qualitative nature of the milk proteins [24] in their relative digestibilities, as well as in the bacterial population [48] and the fermentative processes in the large bowel [31]. Furthermore, specific milk proteins appear in the feces of infants who are fed human milk [23]. These considerations emphasize the complicated nature of the problem. The extent of the complication can be realized if we assume that the fecal losses are derived from proteins secreted into the gut (Table 3). We then discover that the human milk-fed infant receives a *negative* supply of glycine! In other words, all the glycine deposited in the body may have to be synthesized de novo. Although such a conclusion is supported by evidence that suggests virtually no catabolism of glycine to urea in the premature infant [28, 5], these results support the crucial importance of basal nitrogen metabolism to man. Understanding the phenomenon and the factors by which it is influenced remains a challenge to those concerned with protein metabolism and requirements.

Tissue protein synthesis and its relationship to protein deposition

Because of ethical constraints, most information on protein turnover in man relates to whole-body protein turnover. Furthermore, in reference to infants, the literature deals largely with the turnover of nitrogen in low birthweight infants. Caution must be exercised in extrapolating this information to full-term neonates and infants.

It is important to recognize that whole-body protein turnover represents the sum of tissue protein turnover and that each tissue has a characteristic relationship between the rates of protein synthesis and deposition (Table 4). Furthermore, protein synthesis continues when the individual is either in a state of nitrogen equilibrium [15, 27] or in a state of active protein depletion [17, 37].

In addition to these considerations, the immediate postnatal period in mammals appears to be marked by accelerated growth of the gastrointestinal tract and a specific increase in skin protein. Both these tissues have rapid rates of protein synthesis (Table 4). These observations also appear to apply to infants during the first three months of life [61, 52]. During the perinatal period, therefore, these two components may contribute disproportionately to the turnover of body protein (Table 5), and may explain, in part, why high rates of protein turnover per unit of protein intake are generally observed in human neonates. In a recent experiment, Duffy and Pencharz [7] have shown that intravenously-fed neonates have significantly lower rates of body protein turnover than those fed enterally. Total parenteral

Table 4. Fractional rates of protein synthesis and the relationship between protein synthesis and deposition in the tissues of young rats and pigs

	Rat		Pig	
	Protein synthesis (%/d)	Synthesis/ Deposition	Protein synthesis (%/d)	Synthesis/ Deposition
Gut[a]	100	10	79	26
Skin	64	6	ND	
Bone	90	18	65	22
Liver	86	9	48	16
Muscle	17	2	10	3

(Rats growing at 10% per d: McNurlan et al. (1979), Garlick et al. (1983), Preedy et al. (1983); Pigs growing at 3% per d: Seve et al. (1986).)

[a] Jejunal mucosa

Table 5: Contribution of different tissues to whole body protein synthesis in rats before and after weaning

	% Protein mass	% Protein synthesis
	Neonatal	
Tissue		
Muscle[a]	12	6
Liver	13	30
Intestine	5	10
Skin[b]	30	43
	Weaned	
Muscle[a]	31	14
Liver	6	11
Intestine	11	32
Skin[b]	20	42

(Data on the liver, intestine, and muscle protein synthesis taken from Goldspink and Kelly (1984) and Goldspink et al. (1984).)

[a] Total muscle mass calculated from the relationship between the protein content of the leg muscles and total muscle protein (Reeds et al., 1987)

[b] The contribution of skin calculated from the volume/surface area relationship of animals of different sizes and assuming a rate of protein synthesis of 60%/d

nutrition normally is associated with a reduced gut mass, and these results may reflect the high contribution of the visceral tissues to body protein turnover in the neonate.

A significant relationship can be found in young weaned animals between the rates of whole body protein synthesis and protein accretion [19, 45, 33]. When we consider the available data for human neonates, however, the relationship is much less clear; protein turnover is much more variable than protein (nitrogen) retention (Table 6). A critical point to consider is that the between-individual variations, both in body protein turnover and in the ratio of turnover to deposition, are higher in premature infants who receive human milk than in those who receive formulas. This difference may be a reflection of three factors: 1) the high contribution of tissues with rapid rates of protein synthesis, 2) contributions of nonprotein pathways of amino acid metabolism, and 3) the possibility that milk contains trophic factors that influence protein turnover in these and other tissues.

There is increasing interest in this last possibility, particularly as it relates to components of colostrum and immature milk. Considerable attention has been paid to the role of trophic factors in stimulating intestinal mucosal growth and development [32, 25, 62, 39]. Others have proposed (for example Klagsbrun, 1978; Brown and Brakely, 1983; Shing and Klagsburn, 1984) that some of these factors may play systemic roles.

When we analyze the significance of these developments a number of important points merit emphasis. First, an important distinction must be made between colostrum and milk produced at later stages of lactation, because the levels of mitogenic molecules fall rapidly during the early stages of lactation in man [17], cattle [51], and pigs [6]. Second, the trophic factors that have been identified show interspecific differences (Table 7). Epidermal growth factor (summarized by Berseth, [3]) appears to be present in a number of milks, including human (Carpenter, 1980), rat [3] and mouse [22], but not bovine [51]. Recent work by Nichols et al. [39] suggests that lactoferrin, a major component in human milk, may play other roles in addition to

Table 6. Protein deposition (Dep) and body protein synthesis (Synth), determined with ^{15}N-glycine, in premature infants receiving either human milk (H) or formula (F)

Study	Feed-ing	Synth (mg/kg/d)	CV[a]	Dep (mg/kg/d)	CV	Synth/ Dep	CV
Jackson et al. (1981)	H	11.3	27	1.55	15	7.27	37
Pencharz et al. (1983)	H[b]	18.4	26	1.78	13	10.32	ND
	F[b]	13.1	18	1.65	13	7.91	ND
Catzeflis et al. (1985)	H	12.1	33	1.66	19	7.28	25
	F	11.4	19	1.80	22	6.31	20
Duffy & Pencharz (1986)	F[b]	12.6	5	1.53	17	8.22	ND
Stack et al. (1988)	H	14.2	36	1.41	14	10.07	35
	F	12.8	12	1.76	12	7.27	25

[a] CV = Coefficient of variation
[b] These measurements were based on the labeling of urinary urea; all others were based on labeling of urinary ammonia

Table 7. Trophic factors identified in colostrum or milk

Source	Factor identified	Assay	Reference	
Human	unidentified	fibroblasts	Klagsbrun	(1978)
Human	EGF	fibroblasts	Carpenter	(1980)
Human	insulin EGF	immunoassay and L6 myoblasts	Read et al.	(1984)
Human	IGF-I	immunoassay	Baxter et al.	(1984)
Human	lactoferrin	crypt cell	Nichols et al.	(1987)
Bovine	unidentified	various cells	Steimer et al.	(1981)
Bovine	unidentified	fibroblasts	Shing & Klagsbrun	(1984)
Bovine	PDGF	radio receptor	Shing & Klagsbrun	(1987)
Bovine	IGF-I, IGF-II	immunoassay	Francis et al.	(1988)
Porcine	unidentified	GI growth	Widdowson	(1984)
Porcine	unidentified	various cells	Cera et al.	(1987)
Porcine	unidentified	GI growth	Nichols	(per commun)
Porcine	bombesin neurotensin prostaglandins	immunoassay	Westrom et al.	(1987)
Rat	unidentified	GI growth	Berseth et al.	(1983)
	EGF	GI growth	Berseth	(1987)
Mouse	EGF	immunoassay	Grueters et al.	(1985)
Goat	EGF	fibroblasts	Brown & Blakely	(1984)

EGF = Epidermal growth factor
IGF = Insulin-like growth factor
PDGF = Platelet-derived growth factor

its involvement in iron transport and bacteriostasis. These authors have shown that lactoferrin stimulates thymidine incorporation into rat crypt cells. Shing and Klagsbrun (1987) [51] have identified a platelet-derived growth factor-like molecule in bovine colostrum. Similarly, insulin-like growth factor-I has been isolated both from human milk (Baxter, 1984) and bovine colostrum (Francis et al., 1988). Observations such as these suggest that each species may have evolved species-specific milk-born growth factors.

A third and critical point is that the degree to which the gastrointestinal tract, and perhaps other tissues, respond to these trophic factors may be critically dependent on the stage of maturity of the newborn. The degree of expression of certain hydrolases and the degree to which intestinal growth accelerates after birth varies widely between species. Rats and mice are born with an extremely immature gastrointestinal tract and respond in a marked way to the ingestion of homologous milk [58] and EGF [2]. The full-term guinea pig, however, shows little or no response [57]. The human is born with a comparatively mature gut, and it may be extremely significant that the studies of protein turnover with which we introduced this section were performed entirely in premature infants. Whether similar considerations apply to the full-term neonates remains to be proven.

The final point relates to the systemic effects of these trophic factors. Presumably, the systemic effects will be critically dependent on the degree of gut closure, i.e., the speed with which infants lose the ability to take-up proteins by pinocytosis. The ability is retained in rats and mice up to 15 days after birth, while it is lost in the pig within 48 hours of receiving colostrum.

These new developments may have an important influence on our views of the role played by milk in the early growth and development of nursing infants. Diet is too easily viewed solely as a source of nutrients. In addition, it is a variable environmental factor that interacts initially with the surface of the gut. Elsewhere in this volume, the immunologic role of milk proteins is discussed and the evidence presented above suggests other trophic roles for milk proteins. The earlier sections of this paper emphasized the peculiarities of milk seen simply as a source of amino acids. These later developments suggest that milk may have evolved, in addition, as a source of critical factors that initiate certain aspects of postnatal development.

Acknowledgements:

This work is a publication of the USDA/ARS Children's Nutrition Research Center, Department of Pediatrics, Baylor College of Medicine and Texas Children's Hospital, Houston, Texas. This project has been funded in part with federal funds from the U.S. Department of Agriculture, Agricultural Research Service under Cooperative Agreement number 58-7MNI-6-100. The contents of this publication do not necessarily reflect the views or policies of the U.S. Department of Agriculture, nor does mention of trade names, commercial products, or organizations imply endorsement by the U.S. Government.

The author is grateful to Dr. B. L. Nichols, Jr., Dr. R. Shulman, and Dr. M. Fiorotto for stimulating discussions and permission to consider unpublished results. The editorial assistance of Mrs. E. R. Klein has been much appreciated.

References

1. Battaglia FC, Meschia G (1979) Principal substrates of fetal metabolism. Physiol Rev 58:499–527
2. Berseth CL (1987) Enhancement of intestinal growth in neonatal rats by epidermal growth factor in milk. Am J Physiol 253:G662–G665
3. Berseth CL, Lichtenberger LM, Morriss FM Jr (1984) Comparison of the gastrointestinal growth promoting effects of rat colostrum and mature milk in newborn rats in vivo. Am J Clin Nutr 37:52–60
4. Brown KD, Blakely DM (1984) Partial purification and characterization of a growth factor present in goat's colostrum. Biochem J 219:609–617
5. Catzeflis C, Schutz Y, Micheli J-L, Welsch R, Arnaud MJ, Jequier E (1985) Whole body protein synthesis and energy expenditure in very low birth weight infants. Pediatr Res 19:679–687
6. Cera K, Mahan DC, Simmen FA (1987) In vitro growth-promoting activity of porcine mammary secretions: initial characterization and relationship to known peptide growth factors. J Anim Sci 65:1149–1159
7. Duffy B, Pencharz PB (1986) The effect of feeding route (IV or oral) on the protein metabolism of the neonate. Am J Clin Nutr 43:108–111
8. Fiorotto ML, Cochran WJ, Klish WJ (1987) Fat-free mass and total body water in infants estimated from total body electrical conductivity measurements. Pediatr Res 22:417–421
9. Fomon SJ (1960) Comparative study of adequacy of protein from human milk and cow's milk in promoting nitrogen retention by normal full-term infants. Pediatrics 26:51–61
10. Fomon SJ (1961) Nitrogen balance studies with normal full-term infants receiving high intakes of protein. Pediatrics 28:347–361

11. Fomon SJ, DeMaeyer EM, Owen GM (1965) Urinary and fecal excretion of endogenous nitrogen by infants and children. J Nutr 85:235–246
12. Fomon SJ, Haschke F, Ziegler EE, Nelson SE (1982) Body composition of reference children from birth to age 10 years. Am J Clin Nutr 35:1169–1175
13. Fomon SJ, May CD (1958) Metabolic studies of normal full-term infants fed pasteurized human milk. Pediatrics 22:101–115
14. Fomon SJ, Owen GM (1962) Retention of nitrogen by normal full-term infants receiving an autoclaved formula. Pediatrics 29:1005–1011
15. Fujita Y, Okuda T, Rikumaru T, Ichikawa M, Miyatami S, Kajijiwara NM, Yamaguchi Y, Oi Y, Koishi H, Alpers MP, Heywood PF (1984) Endogenous nitrogen excretion in male highlanders of Papua New Guinea. J Nutr 114:1997–2002
16. Fuller MF, Crofts RMJ (1977) The protein sparing effect of carbohydrate. Br J Nutr 38:479–488
17. Garlick PJ, Clugston GA, Waterlow JC (1980) Influence of low-energy diets on whole-body protein turnover in obese subjects. Am J Physiol 238:E235–E244
18. Garlick PJ, Fern M, Preedy VR (1983) The effect of insulin infusion and food intake on muscle protein synthesis in postabsorptive rat. Biochem J 210:669–676
19. Golden M, Waterlow JC, Picou D (1977) The relationship between dietary intake weight change nitrogen excretion and protein turnover in man. Am J Clin Nutr 30:1345–1348
20. Goldspink DF, Kelly FJ (1984) Protein turnover and growth in the whole-body liver and kidney of the rat from fetus to senility. Biochem J 217:507–516
21. Goldspink DF, Lewis SEM, Kelly FJ (1984) Protein synthesis during developmental growth of the small and large intestine of the rat. Biochem J 217:527–535
22. Greuters A, Alm J, Lakshmanan J, Fisher DA (1985) Epidermal growth factors in mouse milk during early lactation: lack of dependency on submandibular glands. Pediatr Res 19:853–856
23. Hambraeus L, Fransson GB, Lonnerdal B (1984) Nutritional availability of breast-milk protein. Lancet II:167–168
24. Harzer G, Bindels JG (1985) Changes in human milk immunoglobulin A and lactoferrin during human lactation. In: Schaub J (ed) Composition and Physiological Properties of Human Milk. Elsevier, Amsterdam, pp 285–295
25. Heird WC, Schwartz SM, Hansen IH (1984) Colostrum-induced enteric mucosal growth in beagle puppie. Pediatr Res 18:512–515
26. Huang PC, Chong HE, Rand WM (1972) Obligatory urinary and fecal nitrogen losses in young Chinese men. J Nutr 102:1605–1614
27. Jackson AA, Byfield R, Jahoor F, Royes J, Soutter L (1983) Whole body protein turnover and nitrogen balance in young children at intakes of protein and energy in the region of maintenance. Hum Nutr Clin Nutr 37C:433–466
28. Jackson AA, Shaw JCL, Barber A, Golden MHN (1981) Nitrogen metabolism in pre-term infants fed human donor breast milk: the possible essentiality of glycine. Pediatr Res 15:1454–1461
29. Janas LM, Picciano MF, Hatch TF (1985) Indices of protein metabolism in term infants fed human milk whey predominant formula or cow's milk formula. Pediatrics 75:775–784
30. Lemons LJ, Adcock EW III, Jones D Jr, Naughton MA, Meschia G, Battaglia FC (1976) Umbilical uptake of amino acids in the unstressed fetal lamb. J Clin Invest 58:1428–1434
31. Lifschitz CH, Chen M, Nichols BL (1987) In vivo and in vitro response to lactose by the flora of breast- and bottle-fed infants. Pediatr Res 22:100
32. Lin CP, Huang P-C (1986) Comparison of control diets containing various protein levels for determining net protein utilization in rats. J Nutr 116:216–222
33. Lobley GE, Connell A, Buchan V (1987) Effect of food intake on protein and energy metabolism in finishing beef steers. Br J Nutr 57:457–465
34. Low AG (1980) Nutrient absorption in the pig. J Sci Food Agric 31:1087–1130
35. Mason VC, Palmer RM (1973) The influence of bacterial activity in the alimentary canal of rats on fecal nitrogen excretion. Acta Agric Scand 23:141–150
36. McNurlan MA, Tomkins AM, Garlick PJ (1979) The effect of starvation on the rate of protein synthesis in rat liver and small intestine. Biochem J 178:373–379
37. Motil KJ, Matthews DE, Bier DM, Burke JF, Munro HN, Young VR (1981) Whole body leucine and lysine metabolism: response to dietary protein intake in young men. Am J Physiol 240:E712–E721

38. Nichol BM, Phillips PG (1975) Endogenous nitrogen excretion and utilization of dietary protein. Br J Nutr 35:181–188
39. Nichols BL, McKee KC, Henry JF, Putman M (1987) Human lactoferrin stimulates thymidine incorporation into DNA of rat crypt cells. Pediatr Res 21:563–567
40. Pellett PL, Kaba H (1972) Carcass amino acids of the rat under conditions of determination of net protein utilization. J Nutr 102:62–68
41. Pencharz PB, Farr L, Papageorgiou A (1983) The effect of human milk and low-protein formula on the rates of total body protein turnover and urinary 3-methyl histidine excretion of preterm infants. Clin Sci 64:611–616
42. Preedy VR, McNurlan MA, Garlick PJ (1983) Protein synthesis in skin and bone of the young rat. Br J Nutr 49:517–523
43. Prentice A, Ewing G, Roberts SB, Lucas A, MacCarthy A, Larjou A, Whitehead RG (1987) The nutritional role of breast-milk IgA and lactoferrin. Acta Paediatr Scand 76:592–598
44. Reeds PJ (1988) Nitrogen metabolism and protein requirements. In: Blaxter KB, MacDonald IA (eds) Comparative Nutrition. John Libbey & Co, London, In press
45. Reeds PJ, Cadenhead A, Fuller MJ, Lobley GE, McDonald JD (1980) Protein turnover in growing pigs. Effects of age and food intake. Br J Nutr 43:445–455
46. Reeds PJ, Hay SM, Dorward PM, Palmer RM (1987) The effect of b-agonists and antagonists on muscle growth and body composition of young rats. Comp Biochem Physiol 89C:337–341
47. Ried LC, Upton FM, Francis GL, Wallace JC, Dahlenberg DW, Ballard FJ (1984) Changes in the growth promoting activity of human milk lactation. Pediatr Res 18:133–139
48. Roberts AK, Van Biervliet J-P, Harzer G (1985) Factors in human milk influencing the bacterial flora of infant feces. In: Schaub J (ed) Composition and Physiological Properties of Human Milk. Elsevier, Amsterdam, pp 259–269
49. Seve B, Reeds PJ, Fuller MF, Cadenhead A, Hay SM (1986) Protein synthesis and retention in some tissues of the young pig as influenced by dietary protein intake after early-weaning. Possible connection to the energy metabolism. Reprod Nutr Develop 26:849–861
50. Shing YW, Klagsbrun M (1984) Human and bovine milk contain different sets of growth factors. Endocrinology 115:273–282
51. Shing Y, Klagsbrun M (1987) Purification and characterization of a bovine colostrum-derived growth factor. Mol Endocrinol 1:335–338
52. Siebert JR (1980) Small-intestine length in infants and children. Am J Dis Child 134:593–595
53. Southgate DAT, Barrett IM (1966) The intake and excretion of caloric constituents of milk by babies. Br J Nutr 20:363–372
54. Stack T, Reeds PJ, Preston T, Hay S, Lloyd DJ, Aggett PJ (1988) ^{15}N-tracer studies of protein metabolism in low birth weight pre-term infants: a comparison of ^{15}N-glycine and ^{15}N-yeast protein hydrolysate and of human milk and formula fed babies. Hum Nutr Clin Nutr, Submitted
55. Sturman JA, Gaull G, Raika NCR (1970) Absence of cystathionine in human fetal liver: is cystine essential. Science 169:74–78
56. Svanberg V, Gebre M, Ljungqvist B, Olsson M (1977) Breast milk composition in Ethiopian and Swedish mothers. Am J Clin Nutr 30:499–507
57. Thompson JF, Singh M, Wang Y, Zucker C, Heird WC (1986) Developmental differences in the effect of natural feeding on early enteric mucosal growth in guinea pigs. J Pediatr Gastroenterol Nutr 5:643–647
58. Tonkiss J, Smart JL, Auestad NS, Edmond J (1985) Type of milk substitute influences growth of the gastrointestinal tract in artificially reared rat pups. J Pediatr Gastroenterol Nutr 4:818–825
59. Widdowson EM (1984) Development of the digestive system: comparative animal studies. Am J Clin Nutr 41:384–390
60. Widdowson EM, Dickerson JWT (1962) The effect of growth and function on the chemical composition of soft tissues. Biochem J 77:30–43
61. Widdowson EM, Dickerson JWT (1964) Chemical composition of the body. In: Comar CL, Bonner F (eds) Mineral Metabolism: An advanced Treatise. Academic Press, New York, Vol IIA, pp 2–247
62. Widdowson EM, Southgate DAT, Hey EN (1979) Body composition of the Fetus and Infant. In: Visser HKA (ed) Nutrition and Metabolism of the Fetus and Infant. Martinus Nijhoff, The Hague, pp 169–177

63. Wrong OM, Vinci AJ, Waterlow JC (1985) The contribution of endogenous urea to fecal ammonia in man determined with ^{15}N-labelling of plasma urea. Clin Sci 68:193–199
64. Yoshida A, Moritoki K (1974) Nitrogen sparing action of methionine and threonine in rats receiving a protein-free diet. Nutr Rep Int 9:159–168
65. Young VR (1987) Kinetics of human amino acid metabolism: nutritional implications and some lessons. Am J Clin Nutr 46:709–725
66. Young VR, Scrimshaw NS (1968) Endogenous nitrogen metabolism and plasma free amino acids in young adults given a "protein-free" diet. Br J Nutr 22:9–20

The Endocrine Response to Dietary Protein: the Anabolic Drive on Growth

D.J. Millward

Nutrition Research Unit (London School of Hygiene & Tropical Medicine), St. Pancras Hospital, London

Introduction

In this review, I examine the endocrine response to dietary protein in the context of the *anabolic drive* on growth, our term for the regulatory influence of dietary protein on all aspects of the maintenance of optimal organ function and growth [21].

Hormonal mechanisms are a central component of anabolic drive. The role of four major anabolic hormones which are sensitive to the nutritional state: growth hormone, insulin, IGF-1, and thyroid hormones, will be examined in terms of regulation of their levels and a brief review of the nature of their targets. The primary role of insulin, which we believe to be of pivotal importance to the anabolic drive, will be emphasized, demonstrating its role in mediating both the response to dietary protein and changes in peripheral hormone metabolism. The identification of what appears to be separate regulation of growth of muscle myoplasm and connective tissue gives rise to a speculative model for muscle growth regulation – the "bag" theory of muscle growth – which involves description of muscle growth in terms of bag filling and bag enlargement, and which we believe accounts for much of the existing phenomenology of muscle growth.

1. The concept of the anabolic drive

The introduction of the concept of the anabolic drive [21] arises out of our attempts to both rationalize the magnitude of the protein and indispensable amino acid (IAA) requirements and to provide a coherent framework within which the regulatory mechanisms by which dietary factors influencing growth can be evaluated [19, 22]. The main observations which connect the requirements for IAA and their regulatory function and those which led us to develop the concept of the anabolic drive are,

1) Protein and IAA requirements for growth and maintenance are more than can be accounted for factorially

Factorial analysis of the best estimates of protein and amino acid requirements [11] shows that at all stages average requirements are at least 40% more than that needed for growth and replacement of endogenous losses (Fig. 1). This led us to search for an explanation for the inefficiencies of protein utilization.

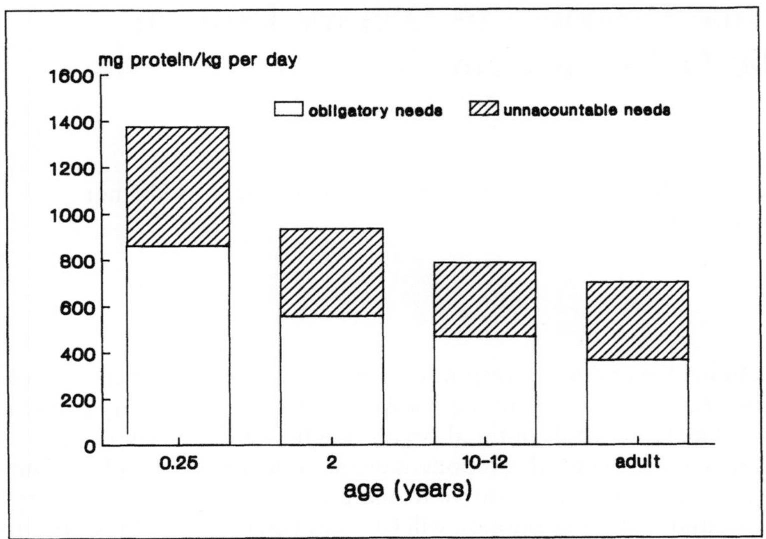

Fig. 1. Age-related change in human protein requirements, showing the component which can be accounted for on a factorial basis (growth and replacement of obligatory losses) and the magnitude of the remaining unaccountable component

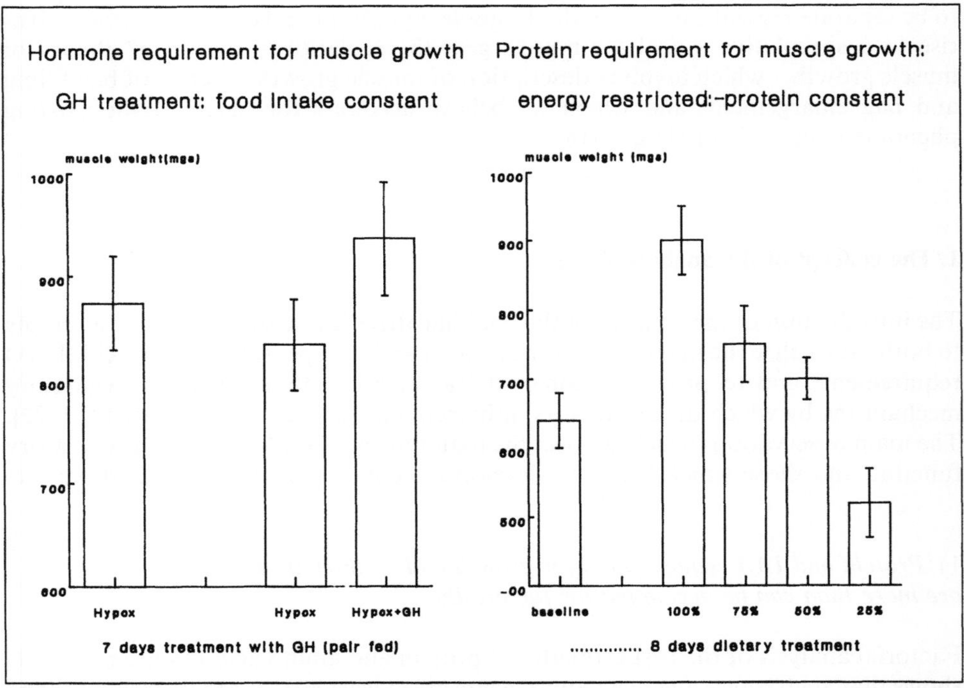

Fig. 2. Regulatory influences on muscle growth. In each case the regulatory stimulus (growth hormone or a high concentration of dietary protein) induces muscle growth even in the absence of sufficient dietary energy (Bates, Yahya and Millward, unpublished results)

50

2) *Dietary protein, especially IAAs, exert a regulatory influence on growth*

Although growth in terms of protein deposition cannot occur without dietary protein, it is also the case that a) protein deposition cannot occur without an intact system of hormonal regulatory responses, b) that these regulatory responses are so powerful that in the absence of sufficient dietary energy body fat will be mobilized to provide the energy, and c) that high concentrations of dietary protein will elicit muscle and bone growth even when the organism is in negative energy balance. Figure 2 illustrates these phenomena. Hypophysectomized rats will not grow without growth hormone treatment, but when treated with growth hormone and not allowed to increase food intake they will increase muscle mass. These increases in muscle mass occur without increases in body weight – implying decreases in body fat, demonstrating the power of the regulatory influence flowing from growth hormone. Normal rats fed restricted intakes of diets of increasing protein concentrations (i.e., maintaining constant protein intakes) continue to deposit muscle even with 50% food restriction (again resulting in loss of body fat and falling body weights). This demonstrates the power of the anabolic drive of protein.

The fact that IAAs play a central role in these phenomena would be predicted from the large literature on the influence of protein quality on growth, from specific investigations of the influence of selected IAA depletions on growth retardation [4], as well as studies showing effects of IAA as opposed to dispensable amino acids on nitrogen balance and IGF-1 levels during refeeding [8].

3) *Most IAAs cannot be tolerated by the organism and are rapidly oxidized if concentrations rise*

The problem posed by any regulatory anabolic mechanism requiring substantial intakes of IAAs is the organism's desire to maintain only very low tissue concentrations of most IAAs, exemplified by the fact that oxidative catabolism of IAAs is stimulated by feeding [29]. Any regulatory influence is therefore only likely to be transient since dietary IAAs are rapidly removed either by net protein synthesis or oxidation. This dual role of IAAs is shown in Fig. 3. Clearly, the extent to which the

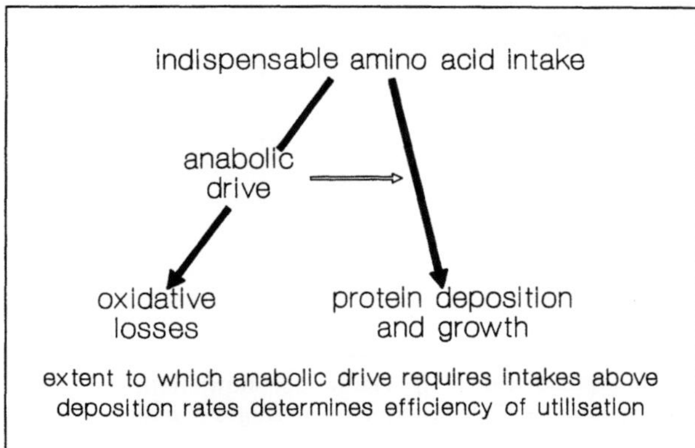

Fig. 3. Dual role of indispensable amino acids as substrates and as a regulatory influence as part of the anabolic drive

dual role of IAAs as substrates and as a regulatory influence can be served without promoting excessive oxidation will determine both the efficiency of protein utilization and the overall protein requirement. From the magnitude of the "unaccountable" component of protein requirements (Fig. 1), it would appear that the intake necessary to exert sufficient anabolic drive is 40% more than substrate requirements. The complete model is shown in Fig. 4 and is discussed in detail elsewhere [21].

2. Nature of the anabolic drive

In Fig. 5, the anabolic drive is represented as a combination of direct regulatory influences of dietary protein on tissues together with indirect hormonal influences. The hormones shown to mediate the anabolic drive are growth hormone, IGF-1, T_3, and insulin. This is not to ignore the large number of other hormones known to exert both positive and negative influences on growth, but rather to focus on those hormones known both to mediate anabolic effects and to be nutritionally sensitive. Distinction is made in Fig. 5 between nutritional influences exerted centrally, acting on growth hormone and (through TSH) T_4 production, or peripherally, influencing the GH-mediated synthesis of IGF-1, T_4–T_3 conversion, and insulin production.

Central mechanisms

The importance of growth hormone in growth regulation is beyond dispute. What is less well understood in the present context is the nature and extent of the nutri-

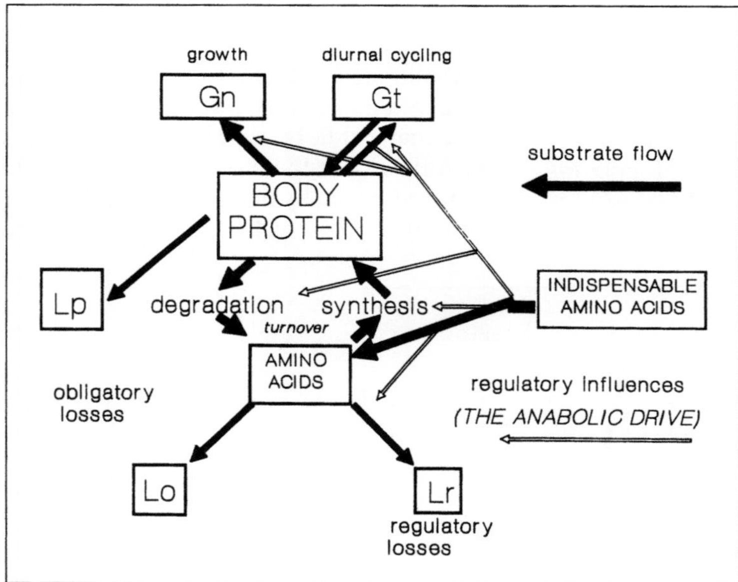

Fig. 4. Model for utilization of dietary amino acids (see [21]). The model shows both the fate of amino acids as substrates as well as their regulatory influence acting on all stages of amino acid and protein metabolism

tional regulation of growth hormone production. Growth hormone secretion is pulsatile [35], making interpretation of simple measurements of its plasma concentration difficult. Work in rodents suggests that the amplitude and frequency of its secretion determines its action [35]. In humans, growth hormone concentration increases during sleep, on fasting, and in PEM [23, 24], whereas in rodents it falls. However, even in rodents the fall in plasma concentrations with fasting or food restriction is much slower than with insulin [5] and not correlated with growth inhibition [4]. Our own attempts to study the nutritional regulation of growth hormone secretion in the rat have involved measuring the hypothalamic content and the K^+ stimulated release in vitro of growth hormone releasing hormone (GHRH) and somatostatin from 3d starved, protein deficient, and well fed rats. As shown in Table 1, we observed no dietary effects.

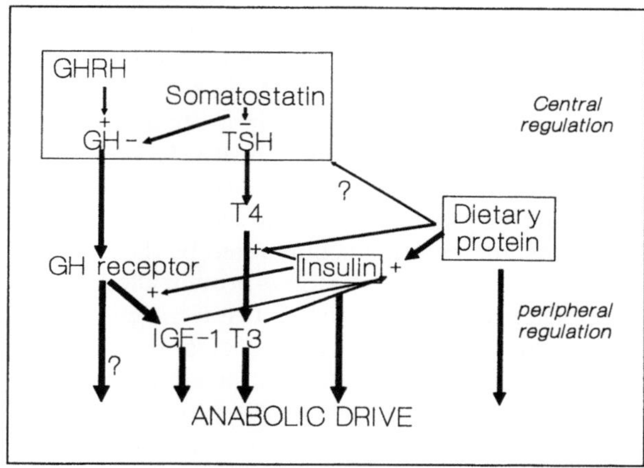

Fig. 5. Hormonal components of the anabolic drive

Table 1. Influence of diet on hypothalamic content and in vitro release of growth hormone releasing hormone and somatostatin

Experiment 1. Content

Diet	GHRH	Somatostatin
	ng/hypothalamus	
Control	3.52 ± 1.07	82 ± 26
3-day fasted	3.63 ± 1.26	95 ± 46
0.5% protein diet	3.78 ± 1.52	85 ± 50

Experiment 2. Release

Diet	GHRH	
	Basal	K^+ stimulated
Control	50 ± 10	229 ± 86
3-day fasted	48 ± 4	162 ± 40
0.5% protein diet	46 ± 5	151 ± 68

Data of Tsagarakis, Bates, Price, and Millward (unpublished)

In the case of thyroid hormones, while TSH and T_4 might reasonably be assumed to be sensitive to nutritional factors, there is in fact little information of the extent to which amino acids are involved in regulation.

Peripheral mechanisms

In Fig. 5 growth hormone is shown as contributing to the anabolic drive independently, as well as indirectly through IGF-1. This is consistent with the *dual effector theory* which arises out of Isaksson's work identifying chondrocyte populations which proliferate in response to either GH or IGF-1 [17] and which exhibit different responses to the two hormones when administered separately [18]. Clearly, within the context of an autocrine mechanism of IGF-1 action, the possibility that IGF-1 mediates GH action on any particular cell type in bone or muscle [7], is most difficult to exclude.

Notwithstanding the long-established importance of insulin in stimulating protein synthesis in muscle [34], it is not generally regarded as an anabolic hormone when defined in terms of mediating cellular proliferation. Nevertheless, we would place insulin in a central position in the regulatory scheme. First, more than any other anabolic hormone, insulin is able to respond acutely and sensitively to nutrient intake. Its secretion reflects a complex interaction between the content of glucose and amino acids in the diet and the prevailing state of the organism [3]. The importance of amino acids as a stimulus for insulin secretion is illustrated by the fact that, in our recent studies of the responses of rats to various low-protein diets fed ad libitum, plasma insulin levels reflected protein intake and not non-protein energy [15]. In addition to this, in these experiments, insulin levels reflected T_3 status and we have confirmed this statistical relationship in other studies of insulin secretion in response to refeeding fasted rats with low plasma T_3 levels [20]. Not only was the increase in plasma insulin in response to refeeding dependent on the plasma T_3 level, but pre-treatment with thyroid hormones of these fasted rats which had low T_3, increased the insulin response to refeeding (Fig. 6). This influence of T_3 on the insulin response to feeding is consistent with the studies of Okajima & Ui [26].

Recent studies show that in addition to thyroidal influences on insulin secretion, β cells of the pancreas are also sensitive to IGF-1 and this influences insulin secretion [33]. The importance of these thyroidal and IGF-1 influences on insulin secretion means that, to the extent that these hormones are a reflection of the more chronic nutritional state, insulin secretion is also sensitive to chronic nutritional state as well as to acute changes in food intake.

The second unique feature of insulin in our regulatory scheme is its involvement in the peripheral metabolism of the other anabolic hormones. Results of a recent analysis of the statistical relationships (by partial correlation analysis) between dietary protein energy, plasma insulin and free T_3, and muscle protein turnover are shown in Fig. 7. The apparent dependence of T_3 on insulin concentration in this scheme (see [15]) is explained by the fact that in many tissues the T_4 deiodenase is an insulin-dependent hormone [12, 30]. Thus, the adaptation in thyroid hormone metabolism in starvation which results in low T_3 levels [20] reflects, at least in part, the low insulin levels which occur in response to the absence of food intake. As would be expected from this, T_3 levels fall markedly in diabetes [6].

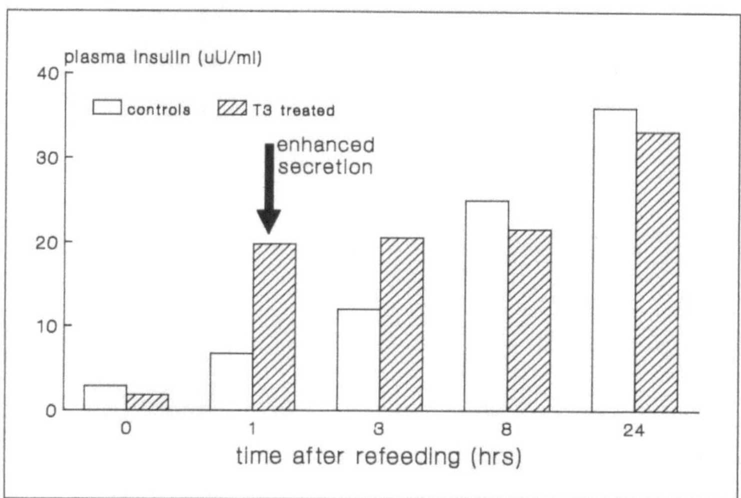

Fig. 6. Influence of thyroid hormone pre-treatment on the insulin release in response to refeeding in fasted rats (see [20])

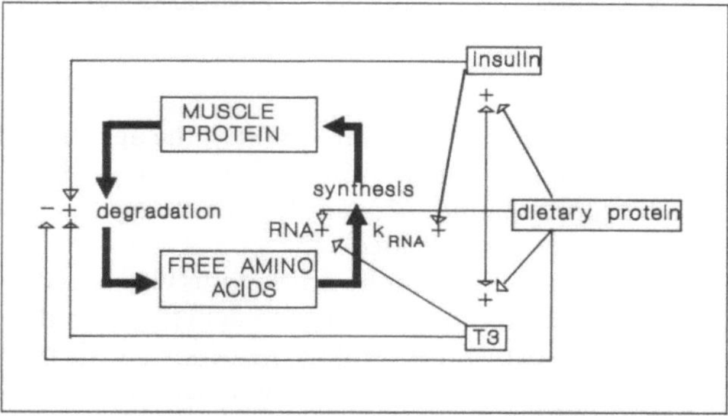

Fig. 7. Influence of dietary protein on protein metabolism in skeletal muscle as indicated by partial correlation analysis of the interactions of the variables measured in vivo. Results were obtained in rats fed diets of varying protein intakes ad libitum (see [15])

The influence of insulin on IGF-1 production is less well understood but indirect evidence suggests it to be very important. Thus diabetes is associated with low levels of IGF-1 [27] which persists with growth hormone treatment [32]. Although the influence of dietary protein on plamsa IGF-1 concentrations is often stressed [28], with evidence for specific influences of dietary IDDAs [8], such relationships could well reflect the influence of insulin on IGF-1 production. We have recently examined this question by feeding rats restricted amounts of diets containing increasing protein concentrations, so that protein intakes are constant. As shown in Fig. 8, the fall in plasma insulin with the energy restriction was accompanied with parallel falls in IGF-1, so that plasma IGF-1 and insulin are highly correlated. In other experiments we have measured insulin and IGF-1 in well-fed rats which exhibit increased levels

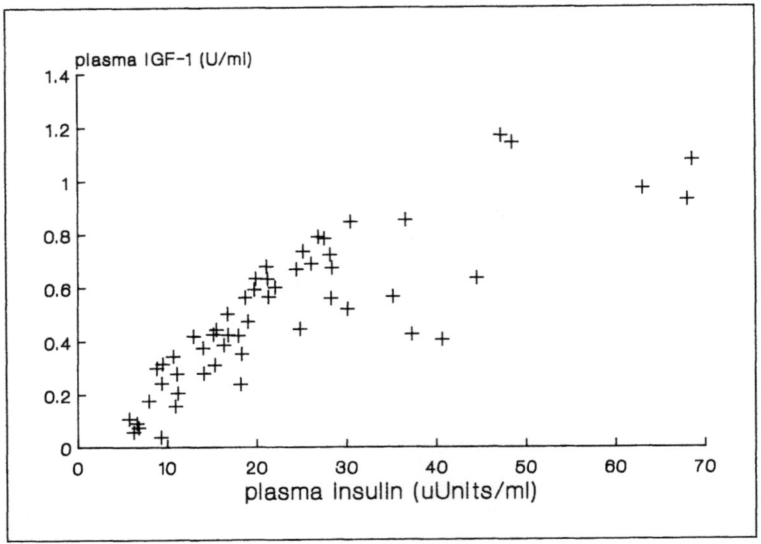

Fig. 8. Interrelationship between plasma IGF-1 levels and insulin in rats fed restricted intakes containing increased protein concentrations. In these experiments, dietary protein intake was constant and the fall in insulin was dependent on the fall in energy intake (Yahya and Millward, unpublished results)

of insulin with age [19, 37]. In rats with no change in dietary intakes, the age-related increase in insulin is accompanied by parallel increases in IGF-1. Clearly, causality is not proven but dietary amino acid supply cannot account for the changes, and a role of insulin in IGF-1 production is strongly indicated.

A major problem in attempting to evaluate the anabolic drive in this way concerns the relevance of change in plasma concentration of hormones to net effects on their various targets. Changes in receptor number and sensitivity associated with both stimulatory and inhibitory influences from other factors mean that measurements of plasma hormone concentrations provide at best a crude, qualitative guide to their actual physiological role in growth. For both T_3 and IGF-1, the problem can be particularly difficult. Given the intracellular location of both T_3 receptors and the T_4 de-iodenase, T_3 can be considered to exert its action in an autocrine manner and T_3 concentrations at the receptor could be quite different from plasma levels. In addition, as we have described [9], concentrations of free T_3 in the protein-deficient rat change in the opposite direction to total T_3 due to increases in binding proteins. Binding proteins are also involved in IGF-1 action, possibly increasing their half-life [38] and modifying their receptor interaction [31]. Given the possibility that plasma IGF-1 reflects liver-produced hormone and that extrahepatic target tissues such as muscle and bone can produce their own IGF in paracrine or autocrine fashion, the relevance of changes in plasma IGF-1 is increasingly under scrutiny. We recently examined the extent to which the rapid fall in plasma IGF-1 on feeding a protein-free diet was mirrored by changes in tissue levels and by extraction of the hormone from muscle [36]. In fact, while plasma IGF-1 falls within 24 h of the diet changes, muscle concentrations were not lowered for three days (Fig. 9) [36]. Thus, the acute changes

56

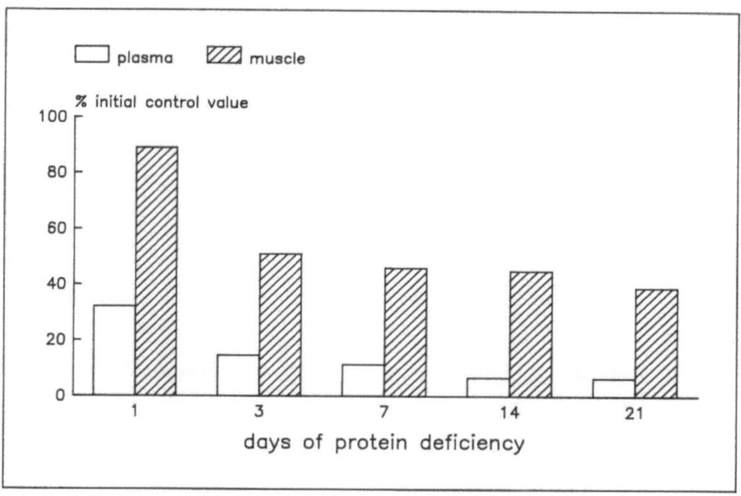

Fig. 9. Changes in plasma and muscle IGF-1 concentration in protein deficient rats [36]

in plasma IGF-1 in response to dietary variation may overestimate the actual nutritional sensitivity of IGF-1 at the target receptors.

3. Targets for the anabolic drive

While growth involves coordinated increase in size of all organs, increases in bone length (height) and muscle mass can be considered as primary targets since to a greater or lesser extent the growth of all other organs is secondary to their growth, being dependent (mainly) on functional demand [13].

The targets include mitosis, ribosome synthesis, the translational phase of protein synthesis and protein degradation in the muscle myofiber, muscle connective tissue, and bone. Analysis of changes in muscle growth and protein turnover, plasma insulin and free T_3 in rats fed low-protein diets results in statistical relationships shown in Fig. 7 and discussed elsewhere [15]; there is experimental confirmation for all these influences. In a more recent study of the changes in plasma insulin, muscle IGF-1, protein and proteoglycan synthesis in rats fed a protein-deficient diet, insulin and IGF-1 appear to be independently associated with changes in protein and proteoglycan synthesis, respectively [36]. Separate studies of the changes in insulin and IGF-1 in relation to protein and proteoglycan synthesis in the tibial epiphysis of diabetic rats with and without insulin indicate that in bone, protein and proteoglycan synthesis are maintained for several days after insulin withdrawal [2]. This is in contrast with the prompt fall in protein synthesis in muscle. Putting these studies together, the changes in Fig. 10 are suggested; this is not intended to represent a comprehensive scheme. As discussed elsewhere, given that the proliferation of satellite cells, the source of myonuclei, is more sensitive to IGFs rather than insulin [10], IGF-1 may well have an important role in mediating myofiber growth as well as connective tissue growth in muscle.

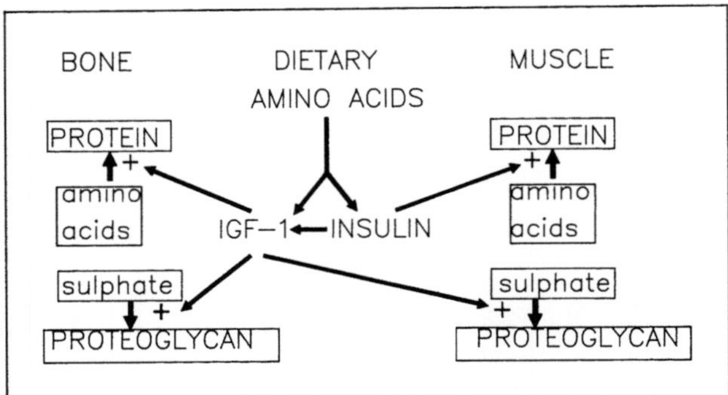

Fig. 10. Relationship between insulin and IGF-1 in mediation of protein deposition in muscle and bone. From the statistical relationships between changes in protein, plasma insulin, tissue IGF-1, and proteoglycan synthesis in muscle and bone in rats fed a low-protein diet, the above influences are indicated [36]

The important feature of Fig. 10 is that it emphasizes the complementary but quite different role of insulin and IGF-1, and that insulin responds acutely to each individual burst of food (especially protein) intake to mediate protein deposition in muscle. As introduced earlier, dietary amino acids (especially IAAs) need to be very quickly deposited as protein following a meal, to avoid oxidation. The ability of insulin to respond immediately to each meal and to promote protein deposition in muscle (augmenting the deposition of hepatic protein) allows this to be achieved. Since the rapid insulin-mediated increases in muscle protein synthesis are matched by equally rapid falls in the postabsorptive state [29], muscle protein is very easily lost if food intake is not maintained. However, for growth to occur in terms of irreversible structural changes in muscle connective tissue and bone, our scheme would imply dependence on a sustained nutrient input in which insulin levels are elevated for long enough to increase levels of IGF-1 and the other growth factors which together mediate such responses. These roles of insulin are shown in Fig. 11. Thus, to some extent growth hormone, T_3, and IGF-1 levels represent an integrated response to food intake with sufficient torque in their regulation that several meals need to be missed before malnutrition is sensed and growth is shut down. This identification of separate targets for growth regulation in muscle has another important implication for growth regulation.

4. Bag theory of muscle growth

While there is no doubt that dietary protein intake exerts the major role in the nutritional regulation of growth, it is also the case that this control is only apparent within certain limits. Rates of growth in height and rates of protein deposition in animals and man can be optimized by nutritional means but cannot be increased above particular limits which appear to be genetically determined. These limits change at different stages of maturity in man, increasing in puberty and ceasing altogether in early adult life. The three observations which have led to the "bag" theory of muscle growth are

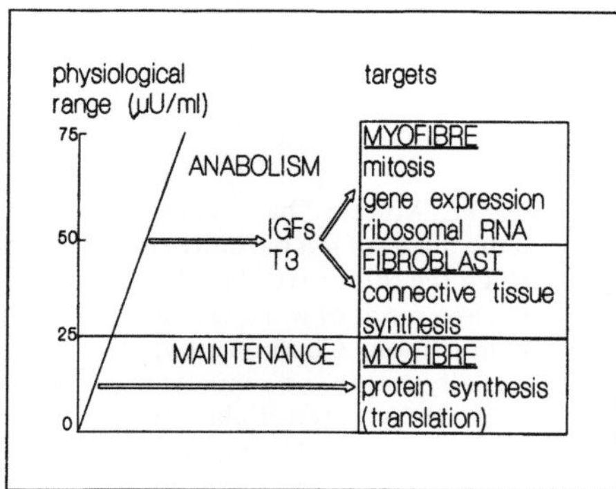

Fig. 11. Role of insulin in mediating the anabolic drive. This scheme shows that at low physiological concentrations insulin exerts a maintenance role, maintaining muscle protein synthesis, whereas at higher concentrations its role is extended to include regulation of the production of other anabolic hormones, namely T_3 from T_4 and IGF-1 from growth hormone, which mediate growth

1) the apparently unlimited extent to which increased food intake can promote protein deposition during catch-up growth in malnourished infants exhibiting a deficit in weight for height [1];
2) the cessation of muscle growth with epiphyseal fusion and the cessation of long-bone growth;
3) the ability of isometric tension or maximal work exercise, or mechanical stretching to promote muscle growth and the fact that this is often associated with damage to muscle, as indicated by marked local edema [16].

We explain these disparate observations in the following way: muscle comprises two elements – the connective tissue sheath (the endomysium, perimysium, and epimysium), in effect collagen bags, surrounding the second element of muscle – the myofibers [14]. The connective tissue sheath regulates and transmits the contractile force to the structure (usually bone) to which the muscle is attached by the tendons, and has minimum elasticity. Protein deposition in muscle can therefore be conceived as "bag filling", which with an inelastic bag might be limited by physico-chemical factors such as viscosity and ionic strength which would increase if protein was deposited in a fixed volume. Growth would require, in this context, bag filling and bag enlargement. If the regulation of bag enlargement involved factors which linked it to bone growth, such as growth hormone and IGF-1, this would provide a mechanism for the coordination of bone and muscle growth. Catch-up growth would be explained in terms of bag filling, i.e., a malnourished child with wasted muscle would have empty bags of a size which corresponded to that necessary to maintain appropriate muscle mass for that particular bone length, and the "bag full" signal would coincide with restoration of normal weight for height (see Fig. 12). Stretch-induced hypertrophy is explained in terms of bag tearing or splitting in response to mechanical factors, and subsequent enlargement. Collagen synthesis is an early event in stretch-induced hypertrophy [16]. The cessation of muscle growth with cessation of long-bone growth is more difficult if the link between muscle and bone growth is simply growth hormone and/or IGF-1. These hormones do not vanish in adults so that inhibition or marked down-regulation of their receptors in muscle connective tissue would need to be postulated and a reason or mechanism by which this might occur is not at all obvious. It may be that for growth hormone or IGF-1 to induce bag enlargement, a combination of hormonal signals and other signals deriving from

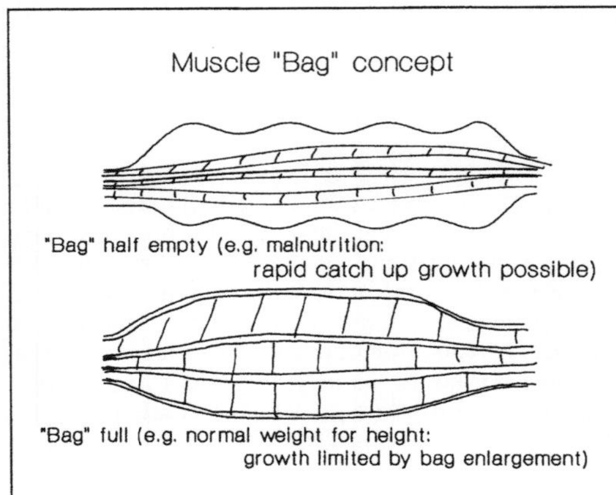

Fig. 12. Muscle "bag" concept. Shown in the diagram is a partially empty bag in which the three myofibers have lost protein as a result of wasting associated with malnutrition. Consequently, the bag is partially empty, i.e., muscle myofiber is replaced by extracellular water. Below this is shown a full bag in which the three myofibres have expanded to occupy all the volume within the bag. It is postulated that no further growth is possible without increasing the bag size

muscle stretching consequent on bone growth is needed. As suggested elsewhere [1] these other signals could involve prostaglandins or other prostanoids produced consequent to the membrane deformation by stretching. However, this is further complicated by the observation that growth hormone treatment can induce protein deposition in muscle in adults. This may involve maximal bag filling in individuals where some capacity still exists, but if this involves bag enlargement then this raises the question of why the level of growth hormone observed in normal adults is ineffective in inducing muscle growth. Notwithstanding these last pieces of the puzzle, we believe that the bag theory of muscle growth is a useful concept if for no other reason than its focus on muscle connective tissue which is generally overlooked in considerations of the regulation of muscle growth.

Conclusions

In this paper the discussion of the endocrine response to dietary protein has been limited to a brief review of selected aspects of the involvement of four anabolic hormones which are likely to be nutritionally sensitive, and their respective targets. One practical implication of this discussion in the context of the anabolic drive and the new model for amino acid requirements [21] is the possibility of identification of indicators of nutritional adequacy of dietary protein. Certainly, both the insulin secretion in response to a particular meal and the plasma levels of IGF-1 as suggested by Munro [25] would, according to the data presented here, be worthy of evaluation. However, given the age-related changes in insulin sensitivity and plasma concentrations which could be responsible at least in part for the age-related increases in IGF-1 concentrations, plasma concentrations deemed to indicate nutritional adequacy of dietary regimes would need to increase with age.

60

Acknowledgements

Results from the author's laboratory described in this paper include the work of Peter Bates, Margaret Jepson, Gillian Price, John Rivers, Stelios Tsagarakis, and Zainal Yahya. We gratefully acknowledge support from the Wellcome Trust and the Nestlé Nutrition Programme.

References

1. Ashworth A, Millward DJ (1986) Nut Revs 44(5):157–163
2. Bates PC, Donachie PA, Yahya ZAH, Millward DJ (1987) J Endocr 115 (Suppl):69
3. Berthoud HR (1984) Metabolism 33:18–25
4. Bolze MS, Reeves RD, Lindbeck FE, Elders JM (1985) J Nutr 115:782–787
5. Borer KT (1987) Growth 51:103–117
6. Brown JG, van Beuren J, Millward DJ (1983) Biochem J 214:637–640
7. Cameron CM, Kostyo JL, Adamafio NA, Brostedt P, Roos P, Skottner A, Forsman A, Fryklund L, Skoog B (1988) Endocrinology 122:471–474
8. Clemmons DR, Seek MR, Underwood LE (1985) Metabolism 34:391–395
9. Cox MD, Dalal SS, Heard CRC, Millward DJ (1984) J Nutr 114:1609–1616
10. Dodson MV, Allen RE, Hossner KL (1985) Endocrinology 117:2357
11. FAO/WHO/UNU (1985) Energy and protein requirements. Technical Report Series. No. 724. World Health Organization, Geneva.
12. Gavin LA, Moeller M (1983) Metabolism 32(6):543–551
13. Goss RJ (1978) Adaptive mechanisms of growth control. In: Falkner F, Tanner JM (eds) Human Growth VI. Bailliere Tindall, London, pp 17–44
14. Gould RP (1973) The microanatomy of muscle. In: Bourne, Geoffrey H (ed) The Structure and Function of Muscle. Academic Press, New York London, pp 186–243
15. Jepson MM, Bates PC, Millward DJ (1988) B J Nutr 59:397–415
16. Laurent GJ, Millward DJ (1980) Fed Proc 39:43–48
17. Lindahl A, Isgaard J, Carlsson L, Isaksson OGP (1987) Endocrinology 121:1061–1069
18. Lindahl A, Isgaard J, Isaksson OGP (1987) Endocrinology 121:1070–1075
19. Millward DJ (1988) Aquaculture (in the press)
20. Millward DJ, Brown JG, van Bueren J (1988) J Endocr 118:417–422
21. Millward DJ, Rivers JPW (1988) Eur J Clin Nut 42:367–393
22. Millward DJ, Rivers JPW (1988) Diabetes-Metabolism Reviews (in the press)
23. Milner RDG (1971) Arch Dis Child 46:301–305
24. Milner RDG (1971) Pediat Res 5:33–39
25. Munro HN (1985) Historical perspective on protein requirements: objectives for the future. In: Blaxter Sir K, Waterlow JC, John Libbey (eds) Nutritional adaptation in man. London, Paris, pp 155–168
26. Okajima F, Ui M (1978) Am J Physiol 234(2):E106–E111
27. Phillips LS, Fusco AC, Unterman TG (1985) Metabolism 34:765–770
28. Prewett TEA, D'Ercole AJ, Switzer BR, van Wyk JJ (1982) J Nutr 112:144–150
29. Rennie MJ, Edwards RHT, Emery PW, Halliday D, Matthews DE, Wolman SL, Millward DJ (1982) Clin Sci 63:519–523
30. Sato K, Robbins J (1981) J Clin Invest 68:475–483
31. Schoenle E, Zapf J, Hauri C, Steiner T, Froesch ER (1985) Acta Endocr 108:167–174
32. Schweiller E, Guler H-P, Merryweather J (1986) Nature 323:169–171
33. Sieradzki J, Fleck H, Chatterjee AK, Schatz H (1988) J Endocrin 117:59–62
34. Waterlow JC, Garlick PJ, Millward DJ (1978) Protein Turnover in Mammalian Tissues & the Whole Body. Elsevier/North-Holland Biomedical Press, Amsterdam
35. Weigle DS (1987) Diabetes 36:764–775
36. Yahya ZAH, Bates PC, Millward DJ (1988) Biochem Soc Trans 16:624–625
37. Yahya ZAH, Tirapegui JO, Bates PC, Millward DJ (1987) J Endocrin (Suppl) 115:71
38. Zapf J, Hauri M, Waldvogel M, Froesch ER (1986) Clin Invest 77:1768–1775

Endocrine Response to Animal and Vegetable Protein

C. A. Barth, K. E. Scholz-Ahrens, M. Pfeuffer, and M. de Vrese

Institut für Physiologie und Biochemie der Ernährung, Bundesanstalt für Milchforschung, Kiel, FRG

Introduction

The intake of dietary protein provokes an endocrine response. A rise of glucagon [16], adrenocorticotropic hormone and cortisol [21], corticosterone [19] and insulin [18] has been reported.

There is considerably less knowledge about whether different dietary proteins cause different patterns of endocrine response. It might be that such a pattern causes the specific metabolic reactions which finally result in what we call the "biological value."

Moreover, it is up to now not completely clear why animal proteins raise and vegetable proteins lower serum cholesterol levels in rodents as reported by numerous investigators [3, 23]. As hormones modulate serum lipids (Table 1) we studied whether a different endocrine response is caused by the intake of different dietary proteins.

Methods

Plasma hormone concentrations were analyzed with radio-immunoassay. Free thyroid hormones were determined with a low-affinity antibody method (Mallinckrodt Diagnostika, Dietzenbach, FRG, SPAC ET). The rationale for this method is discussed extensively in [10]. We compared the results of this method with equilibrium dialysis. A satisfying agreement of both methods (Table 2) shows that free thyroid hormone concentrations have been determined accurately.

Table 1. Hormonal effects on serum cholesterol levels

Hormone	Effect	Reference
Insulin	LDL decrease	[17]
	HDL increase	[20]
	Decrease	[4]
	No effect	[5]
Glucagon	Decrease	[2, 9]
Growth hormone	Decrease	[7]
	No effect	[15]
Cortisol	Increase	[14, 24]
Thyroid hormones	Decrease	[1, 11, 13]

Table 2. Free T_4: evaluation of method

Species	Sample	Equilibrium dialysis	Radio-Immunoassay
		pmol/l	
Pig	#356	0	1.300
Pig	#366	4.247	4.827
Rat	#124	11.58	10.62

Pig experiments

Experiment I was performed with six adult female miniature pigs of the Göttingen strain that for six weeks consumed a semisynthetic euenergetic diet containing 20 wt.% protein and 1 wt.% cholesterol, (for further details see Fig. 1).

Experiment II was done with young male Göttingen miniature pigs weighing from 10.3–16.0 kg; they consumed 0.55 MJ ME $\times$ kg$^{-0.75}$ $\times$ d^{-1}. Groups of eight animals consumed for six weeks a semisynthetic diet containing either 15 wt.% casein or soy isolate.

Rat experiments

Experiments were performed with eight to 10 young male rats per group. In experiment III diets contained 20 wt.% casein or soy isolate, 1 wt.% corn oil, and 69 wt.% sucrose and were given ad lib. for 10 days. In experiment IV the semisynthetic diets contained 10 wt.% protein, 5 wt.% soy oil and 74.8 wt.% sucrose, respectively. Diets were fed ad lib. for 17 days.

Results

Experiment I shows that soy protein isolate caused 15% higher plasma total thyroxine concentrations in the pig. This difference was significant at the $p < 0.01$ level as demonstrated by multivariate statistics performed with 21 measurements in six adult pigs over 24 h (Table 3 and Fig. 1).

We wondered whether free thyroxine levels paralleled total thyroxine levels and therefore performed experiments II and IV.

Experiment II resulted in higher total and free thyroxine concentrations following soy; however, the difference did not reach statistical significance because of the low number of observations (Table 3).

In experiment IV casein-fed rats had the same food consumption as soy-fed animals and the protein efficiency ratios conformed with earlier reports [22]. The corrected PER values amounted to 2.5 and 1.59 for casein and soy protein isolate, respectively. Significant differences between casein- and soy-fed animals were

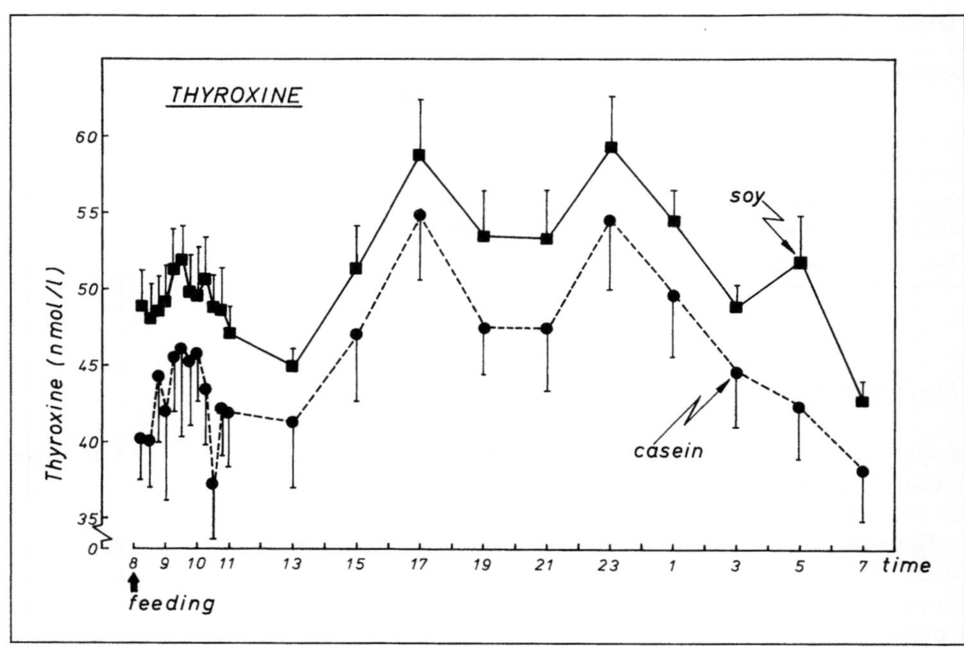

Fig. 1. Influence of casein or soy protein on plasma thyroxine in pigs obtained in a cross-over trial (experiment I). Plasma samples were taken at times of day indicated on the abscissa (hours), following a meal shown by the arrow (n=6, averages ± S.E.M.). Multivariate statistics show that casein and soy protein caused highly significant differences of thyroxine concentrations ($p < 0.01$)

Table 3. Synopsis of experiments: comparison soy (S)/casein (C)

Experiment	Species		Total T_4 nmol/l	Free T_4 pmol/l	Total T_3 nmol/l	Free T_3 pmol/l
I	Pig	S	52.00	N.D.	1.075	N.D.
		C	46.85	N.D.	1.187	N.D.
			$p < 0.01$	—	N.S.	—
II	Pig	S	39.77	8.933	2.146	2.673
		C	35.78	8.225	2.149	2.811
			N.S.	N.S.	N.S.	N.S.
III	Rat	S	43.12	N.D.	1.293	N.D.
		C	38.87	N.D.	1.352	N.D.
			N.S.	—	N.S.	—
IV	Rat	S	53.16	12.343	2.375	5.483
		C	32.05	9.036	1.705	4.639
			$p < 0.001$	$p < 0.01$	$p < 0.01$	$p < 0.01$

N.D. = Not determined
T_3 = Triiodothyronine (3,5,3′-triiodo-L-thyronine)
T_4 = Thyroxine (3,5,3′,5′-tetraiodo-L-thyronine)

observed regarding total and free thyroxine and triiodothyronine concentrations (Table 3).

In experiment III total thyroxine concentrations tended to be higher in soy-fed animals; however, the number of animals studied was too small to demonstrate statistical significance (Table 3).

Conclusion

In two out of four experiments in which two different species were studied we observed a significant rise of plasma total thyroxine by dietary soy protein as compared to casein. Statistical significance was not reached in the other two experiments, although values tended to be higher in soy-fed animals as well. This is in accordance with data of Forsythe, who reported higher thyroxine levels in the gerbil consuming soy protein [12] and with data of Cree and Schalch [8] who did observe higher thyroxine levels in rats consuming gluten as compared to casein. The latter finding suggests that vegetable proteins in general may cause higher plasma thyroxine concentrations as compared to animal proteins.

In two experiments free thyroxine was determined and shown to parallel the changes of total thyroxine. In one of these experiments statistical significance was not reached though the values tended to be different.

The implications of these observations are twofold. First, the differences of serum cholesterol levels following casein vs soy protein can easily be explained by this soy-induced hyperthyroxinemia. Hyperthyroidism is accompanied by lower serum cholesterol [1, 13] and all biochemical sequelae reported so far in soy-fed animals contributing to a lowering of plasma cholesterol are also metabolic consequences of hyperthyroidism (Fig. 2).

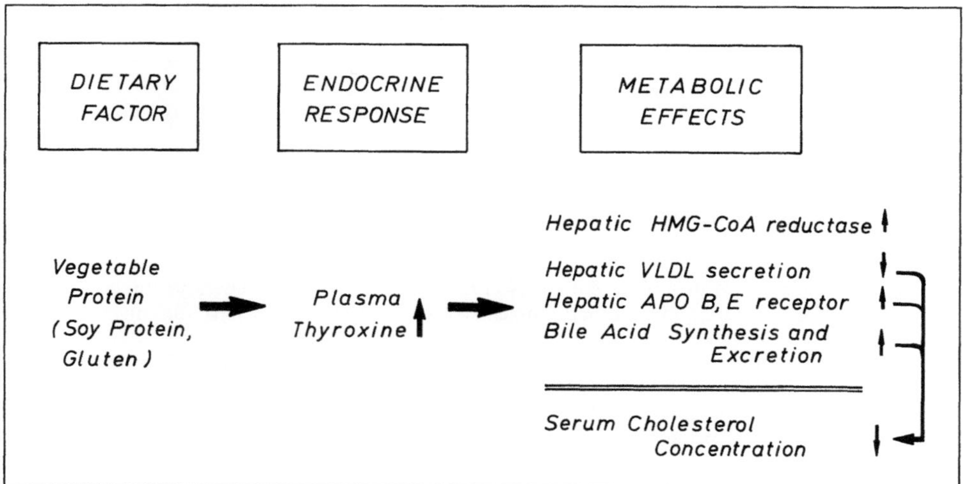

Fig. 2. Proposed unifying concept for mechanism of protein-induced serum cholesterol changes

Second, it is known that thyroid hormones cause a rise of the rate of proteolysis [6]. It remains to be established whether dietary proteins causing higher thyroid function also produce higher rates of proteolysis.

Acknowledgements

We thank Dr. Agergaard (National Danish Institute of Animal Physiology, Foulum) for performing the thyroxine determinations of experiment I, Mrs. F. Repenning for technical assistance, and Mrs. M. Reese for secretarial help.

References

1. Agdeppa D, Macaron C, Mallik T, Schnuda ND (1979) Plasma high density lipoprotein cholesterol in thyroid disease. J Clin Endocrin Metab 49:726–729
2. Aubry F, Marcel YL, Davignon J (1974) Effects of glucagon on plasma lipids in different types of primary hyperlipoproteinemia. Metabolism 23:225–238
3. Barth CA, Pfeuffer M (1988) Dietary protein and atherogenesis. Klin Wochenschr 66:135–143
4. Bennion LJ, Grundy SM (1977) Effects of diabetes mellitus on cholesterol metabolism in man. N Engl J Med 296:1365–1371
5. Briones ER, Mao SJT, Palumbo PJ, O'Fallon WM et al. (1984) Analysis of plasma lipids and apolipoproteins in insulin-dependent and noninsulin-dependent diabetics. Metabolism 33:42–49
6. Burini R, Santidrian S, Moreyra M, Brown P, Munro HN, Young V (1981) Interaction of thyroid status and diet on muscle protein breakdown in the rat, as measured by N^7-methylhistidine excretion. Metabolism 30:679–687
7. Byers SO, Friedman M, Rosenman RH (1970) Prevention of hypercholesterolemia in thyroidectomized rats by growth hormone. Nature 228:464–465
8. Cree TC, Schalch DS (1985) Protein utilization in growth: effect of lysine deficiency on serum growth hormone, somatomedins, insulin, total thyroxine (T_4) and triiodothyronine, free T_4 index, and total corticosterone. Endocrinology 117:667–673
9. Dugan RE, Ness GC, Lakshmanan MR et al. (1974) Regulation of β-hydroxy-β-methylglutaryl-coenzyme A-reductase by the interplay of hormones. Arch Biochem Biophys 161:499–504
10. Ekins RP (1985) Principles of measuring free thyroid hormone concentrations in serum. Nuc Compact 16:305–313
11. Friedman M, Byers SO, Rosenman RH (1952) Changes in excretion of intestinal cholesterol and sterol digitonides in hyper- and hypothyroidism. Circulation 5:657–660
12. Forsythe III WA (1986) Comparison of dietary casein or soy protein effects on plasma lipids and hormone concentrations in the gerbil (Meriones unguiculatus). J Nutr 116:1165–1171
13. Hansson P, Valdemarsson S, Nilsson-Ehle P (1983) Experimental hyperthyroidism in man: effects on plasma lipoproteins, lipoprotein lipase and hepatic lipase. Horm Metab Res 15:449–452
14. Henze K, Seidl O, Wolfram G, Zöllner N (1981) Der Einfluß einer Therapie mit Kortikosteroiden auf die Serumlipide. Verhandlungen der Deutschen Gesellschaft für innere Medizin 87:982–984
15. Heubi JE, Burstein S, Sperling MA et al. (1983) The role of human growth hormone in the regulation of cholesterol and bile acid metabolism. J Clin Endocrinol Metab 57:885–891
16. Jarrousse C, Lardeux B, Bourdel G, Girard-Globa A, Rosselin G (1980) Portal insulin and glucagon in rats fed proteins as a meal. J Nutr 110:1764–1773
17. Lopes-Virella ME, Wohltmann HJ, Loadholt CB, Buse MG (1981) Plasma lipids and lipoproteins in young insulin-dependent diabetic patients: Relationship with control. Diabetologia 21:216–223

66

18. Nuttall FQ, Gannon MC, Wald JL, Ahmed M (1985) Plasma glucose and insulin profiles in normal subjects ingesting diets of varying carbohydrate, fat, and protein content. J Am Coll Nutr 4:437–450
19. Rabolli D, Martin RJ (1977) Effects of diet composition on serum levels of insulin, thyroxine, triiodothyronine, growth hormone and corticosterone in rats. J Nutr 107:1068–1078
20. Sadur CN, Eckel RH (1983) Insulin-mediated increases in the HDL cholesterol/cholesterol ratio in humans. Arteriosclerosis 3:339–343
21. Slag MF, Ahmed M, Gannon MC, Nuttall FQ (1981) Meal stimulation of cortisol secretion: A protein induced effect. Metabolism 30:1104–1108
22. Steinke FH, Prescher EE, Hopkins DT (1980) Nutritional evaluation (PER) of isolated soybean protein and combinations of food proteins. J Food Sci 45:323–327
23. Terpstra AHM, Hermus RJJ, West CE (1983) The role of dietary protein in cholesterol metabolism. World Rev Nutr Diet 42:1–55
24. Troxler R, Sprague EA, Albanese RA, Fuchs R, Thompson AJ (1977) The association of elevated plasma cortisol and early atherosclerosis as demonstrated by coronary angiography. Atherosclerosis 26:151–162

Reliability and Limitations of the Homoarginine Method for Evaluation of Protein Digestibility in the Pig

H. Hagemeister[1], M. Schmitz[1], and H. Erbersdobler[2]

[1] Institut für Physiologie und Biochemie der Ernährung, Bundesanstalt für Milchforschung, Kiel, FRG
[2] Institut für Humanernährung der Universität Kiel, Kiel, FRG

Introduction

Traditional methods to follow protein digestibility do not take into account that two variables interfere with the results: 1) the microbial activity in the gut, and 2) the variation of endogenous protein secretion.

The first influence can be eliminated by measuring the prececal digestibility, i.e., the disappearance of nitrogen up to the ileum because its influence in the small intestine is only of minor importance. But the second variable is more difficult to control. Even the use of ^{15}N as marker for exogenous protein gives no correct results. Label from dietary protein appears in the secreta within 25 min after intake, proving that amino acids of food protein are rapidly incorporated into endogenous secreta [1].

Within the homoarginine method [3] it becomes possible to follow the metabolic fate of dietary protein and to distinguish between exogenous and endogenous protein in the chyme. By guanidination of the dietary protein lysine is chemically transformed into homoarginine [2] according to the following reaction:

$$R-NH_2 + CH_3O-C(NH)NH_2 \rightleftarrows R-NH-C(NH)NH_2 + CH_3OH.$$

More than 99% of lysine in a protein can become guanidinated if proper conditions are chosen [4].

Monitoring of the homoarginine content of the digesta at the ileum gives a precise estimate of the non-absorbed protein if we assume that the absorption of homoarginine corresponds to that of the other amino acids of the protein. That presupposes 1) the distribution of homoarginine in the protein, 2) the degradation of the guanidinated protein, and 3) the absorption of homoarginine must be representative for the whole protein.

Methods

Five adult Göttingen minipigs fasted overnight and carrying T-shaped permanent jejunal (proximal or distal) or ileal cannulae, were given a balanced semi-synthetic diet containing 15% protein. Half of the protein supplied was guanidinated protein in which more than 95% of lysine was transformed to homoarginine. The proteins

68

studied were in trial 1 acid casein (15.1% N of dry matter), and in trial 2 either toasted (105 °C, 2 bar, 20 min) or untoasted soy beans (6.8% N of dry matter). Digesta were continuously sampled for 34 h after the morning meal and analyzed for nitrogen and homoarginine. Homoarginine was quantified by conventional amino acid analysis following acid hydrolysis.

Homoarginine as marker for exogenous protein

Absorption and secretion of free homoarginine in the intestine: when 9 mmol homoarginine were infused into the proximal part of the jejunum, there was a 99.9% disappearance of the free amino acid homoarginine up to the ileum. In addition, following the infusion of 9 mmol homoarginine into the vena jugularis less than 0.2% of the infused homoarginine appeared in the intestine.

We concluded from these data that with the homoarginine method endogenous protein in the chyme of the intestine can be distinguished from the food protein as homoarginine is not incorporated into endogenous secreta and free homoarginine is completely absorbed.

Differentiation between dietary and endogenous protein

Figure 1 shows the amounts of exogenous and endogenous protein at three sites of the intestinal tract after a meal containing guanidinated casein. At the proximal jejunum more protein was found than was supplied in the meal, but only 60% of the exogenous protein was recovered. At the distal jejunum and ileum only 2–3% of the exogenous protein were found and only about 50% of the endogenous protein which originally had appeared at the proximal jejunum. That means that not only the exogenous protein but also a large part of the endogenous protein disappeared up to this site.

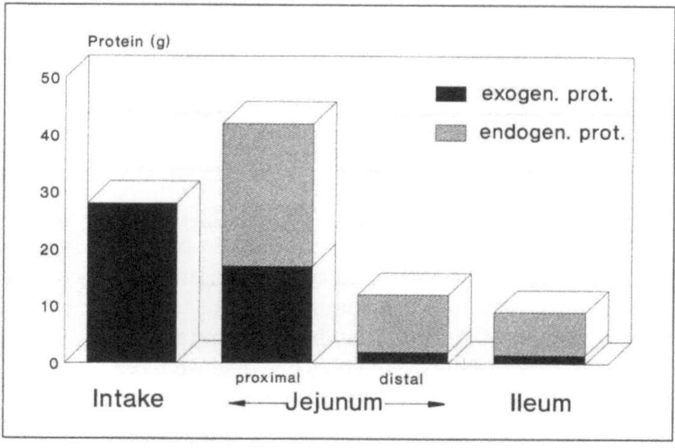

Fig. 1. Dietary casein and endogenous protein at different sites of the small intestine of Göttingen minipigs (trial 1)

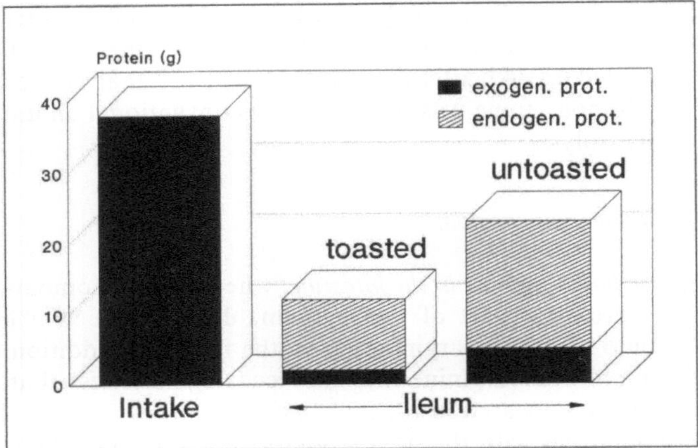

Fig. 2. Dietary and endogenous protein at the ileum of Göttingen minipigs depending on different heat treatments of soy beans (trial 2)

In Fig. 2 the amounts of endogenous and exogenous protein in the chyme at the ileum following the intake of a meal containing either toasted or untoasted soy beans are illustrated.

In comparison to toasted soybeans the untreated beans led both to a higher amount of endogenous protein and to a higher amount of food protein at the ileum. But the lower prececal apparent digestibility of untoasted protein is more due to the enhanced endogenous protein than to the rise of exogenous proteins.

Limitations of the homoarginine method

There are some limitations in the suitability of the homoarginine technique: if a considerable amount of the lysine cannot be transformed into homoarginine, because Maillard products have been formed in heat-treated protein, one must doubt whether the homoarginine method gives valid results. As shown in Table 1, severe heat-treatment of casein, particularly in the presence of glucose, leads to a low degree of guanidination (only 65% of the lysine could be transformed). In a rat assay this preparation showed only a true lysine digestibility of 67%.

On the other hand a high disappearance of homoarginine of 93% in the small intestine has been found in trials with pigs where the guanidinated derivatives of this severely heated protein were assayed. The homoarginine method obviously overestimates the protein absorption of severely heat-treated casein because of blockages by Maillard products which are not accessible for guanidination.

Another limitation of the homoarginine method may become apparent in cases of abnormal low activities of pancreatic proteolytic enzymes because we have observed in vitro that guanidinated casein was less rapidly cleaved by trypsin than native casein.

However, as shown in studies with guanidinated protein the high digestibility coefficients demonstrate that this effect on tryptic cleavage is not relevant in vivo.

70

Table 1. Degree of guanidination of heat treated casein and comparison between true lysine digestibility in rats and the homoarginine disappearance in pigs

Casein	Degree of guanidination	True lysine digestibility	Disappearance of homoarginine
		% of intake	
Native	97	98	99.4
Heat-treated	80	91	97.4
Glucose-supplemented, heat-treated	65	67	93.2

Conclusion

The suitability and reliability of the homoarginine method for evaluation of protein digestibility have been discussed. With the homoarginine method it becomes possible to distinguish between exogenous and endogenous protein in chyme and to measure the influence of anti-nutritive factors on protein absorption. However, some limitations of the homoarginine method may be caused by somewhat changed rates of tryptic and under some conditions chymotryptic cleavage of guanidinated substrates. A falsely high absorption value is found if a damaged protein does not allow a random guanidination of the lysine side-chains.

References

1. Bergner U, Uecker E, Rossow N, Simon O, Bergner H (1984) Untersuchungen zu endogenen N-Umsatzprozessen an ^{15}N markierten Schweinen. 4. Mitt Arch Tierernähr 34:593–605
2. Greenstein JP (1938) A synthesis of homoarginine. J Org Chem 2:480–483
3. Hagemeister H, Erbersdobler H (1985) Chemical labelling of dietary protein by transformation of lysine to homoarginine: a new technique to follow intestinal digestion and absorption. Proc Nutr Soc 44:133 A
4. Maga JA (1981) Measurement of available lysine using the guanidination reaction. J Food Sci 46:132–134

Lactoferrin Content in Feces in Ileostomy-operated Children Fed Human Milk

L. Hambraeus[1], G. Hjorth[1], B. Kristiansson[2], H. Hedlund[2], H. Andersson[3], B. Lönnerdal[4], L.-B. Sjöberg[5]

[1] Dept. of Nutrition, University of Uppsala, Sweden, [2] Dept. of Pediatrics, Östra Sjukhuset, Gothenburg, Sweden, [3] Dept. of Clinical Nutrition, Sahlgren's Hospital, Gothenburg, Sweden, [4] Dept. of Nutrition, University of California, Davis, USA, and [5] Semper AB, Stockholm, Sweden

Introduction

Lactoferrin (LF) is the major iron-binding protein in human milk, but it also occurs in other body fluids, e.g., tears, saliva and pancreatic juice. Human milk is unusually rich in lactoferrin which constitutes one of the three dominating whey proteins, the others being α-lactalbumin and secretory IgA [5]. Lactoferrin has been shown to inhibit the growth of microorganisms in vitro [2] but so far there is meager information regarding the bacteriostatic effect in vivo.

Lactoferrin has been postulated to play a physiological function in the breast-fed infant in two ways [1]. Firstly, it would function as a bacteriostatic component protecting the infant from gastrointestinal infections by binding the iron so strongly that it is not available for pathogens such as *Escherichia coli*. Secondly, it has been postulated that lactoferrin is responsible for the high bioavailability of iron from human milk [1], [6]. The role of lactoferrin in iron absorption is still under investigation. In both cases, however, the physiological role of lactoferrin can only be a reality if the lactoferrin molecule can survive intact during the passage of the gastrointestinal tract and not be digested by proteolytic enzymes.

Lactoferrin constitutes about 10−25% of the total protein content in human milk [5]. The role of lactoferrin as a source of protein from the nutritional point of view has however been questioned [4, 8]. Since its role in the defense against gastrointestinal infections requires that it is functionally intact throughout the intestinal tract, it may interfere with the nutritional role of lactoferrin. Significant amounts of lactoferrin and secretory IgA have been found in feces of breast-fed infants [3, 7].

The nutritionally available protein in human milk may have to be reduced by as much as 40% in case neither lactoferrin, secretory IgA, nor lysozyme are absorbed [4], all of them potentially playing a role in the defense against gastrointestinal infections. Further studies on the absorption and/or excretion in feces of lactoferrin, secretory IgA, and lysozyme are thus of great importance for the calculation of protein balance and requirement in the breast-fed infant. This should be taken into account when assessing protein requirement during infancy when human milk is used as reference for optimal protein intake.

Interestingly, although lactoferrin has been shown to occur in the stool of breast-fed infant only 2−6% of the lactoferrin consumed remains intact in feces during the first week of life and less than 2% after four months [3]. This means that the

72

abovementioned hypothesis that lactoferrin remains intact throughout the intestinal tract may not be entirely correct. On one hand, it means that the protective effect of lactoferrin against intestinal infections may not be valid in vivo. On the other hand, the role of lactoferrin in iron transport to the intestinal brush border membrane could still be valid.

In this study we analyzed the lactoferrin content in samples of stools which were originally collected for other studies on nitrogen and energy balance. The breastmilk consumption was measured but unfortunately no samples were collected. We had the opportunity, however, to analyze lyophilized stool samples with regard to the content of lactoferrin, secretory IgA, albumin, and α_1-antitrypsin by immunochemical methods.

Material and methods

Samples of feces were collected from 17 infants at the age of 12 to 260 days. Colostomy was performed in eight children and ileostomy in nine infants. The diagnoses are given in Table 1. Twelve of the patients were studied while they were consuming breastmilk in substantial amounts. Furthermore, in two patients samples were also obtained at a later stage (at 200 and 260 days of age) when they were no longer given any breastmilk. After collection of stools, samples were homogenized and lyophilized and kept frozen until analysis.

Intake of breastmilk was measured by weighing the child before and after feeding. The percent of lactoferrin excreted in the stools was calculated based on the assumption that the lactoferrin content of the breastmilk consumed was within a normal range. The lactoferrin content was analyzed by rocket immunoelectrophoresis as described in [3].

Results and discussion

A total of 67 samples were analyzed. Seventeen samples were obtained from four patients with necrotizing enterocolitis, nine samples from three patients with

Table 1. Diagnoses and ages of the patients

Diagnosis	No. pat.	No. spec.	Age Days	Ileostomy/ colostomy
Necrotising enterocolitis	4	17	45 (24– 59)	2/2
Hirschsprung's disease	2	8	25 (23– 27)	2/–
No breastmilk	3[a]	9	181 (158–213)	1/2
Ileum atresia	1	3	16 (15– 17)	1/–
No breastmilk	1	5	109 (107–111)	1/–
Anal atresia	2	7	15 (12– 19)	–/2
No breastmilk	3[a]	9	205 (154–260)	–/3
Misc.	3	9	31 (16– 56)	3/–

[a] 1 patient had received breastmilk earlier

Hirschsprung's disease, nine samples from three patients with anal atresia, and five samples from two patients with ileumatresia. Furthermore, nine samples obtained from three patients with miscellaneous diseases were analyzed. Ileostomy was performed in eight breast-fed patients and colostomy in four breast-fed patients.

Table 2 shows the lactoferrin content in feces expressed in percent of calculated lactoferrin intake. The lactoferrin concentration in breastmilk was assumed to be 2 mg/ml in milk obtained after three weeks of age and to 3 mg/ml before 21 days.

It is seen from Table 3 that the lactoferrin content in feces corresponded to 20% in patients with necrotizing enterocolitis (two of them with ileostomy and two with colostomy). The patients were between one- and two-months old. Interestingly, the two patients with Hirschsprung's disease showed as much as 32% intact lactoferrin in their feces although they were almost one-month old.

In the two groups of patients with atresia of ileum or rectum about 10% of the lactoferrin remained in feces although they were only two weeks old.

There was no real difference in the amount of breastmilk consumed which varied between 350 and 500 ml in the various groups of patients. The figures seem to indicate that the disease seems to be of more importance than the age or the amount of breastmilk consumed in the case of lactoferrin digestion. The fact that the high lactoferrin content in the feces of patients with necrotizing enterocolitis and Hirschsprung's disease might not be an effect of the inflammatory reaction, and local synthesis of lactoferrin in the intestine is indicated by the fact that in those patients with the same diagnosis which were not breast-fed no lactoferrin was found in the

Table 2. Breastmilk consumption and lactoferrin in stools in % of consumed LF

Diagnosis	No. pat.	No. spec.	Breastmilk consumed	Lactoferrin %	Age $\bar{x}$	Ileostomy/ colostomy
Necrotising enterocolitis	4	17	426 (165–620)	20 (5–36)	45 (24–59)	2/2
Hirschsprung's disease	2	8	370 (250–490)	32 (9–85)	25 (23–27)	2/–
Ileum atresia	1	3	580 (510–630)	9 (8–10)	16 (15–17)	1/–
Anal atresia	2	7	391 (340–540)	9 (6–12)	15 (12–19)	–/2
Misc.	3	9	451 (360–520)	10 (4–14)	31 (16–56)	3/–

Table 3. Lactoferrin in stools from patients who had ileostomy and colostomy (the values refer to % of consumed LF)

	No. of patients	No. of specimens	% Lactoferrin $\bar{x}$	Age $\bar{x}$
Ileostomy	8	30	20 (5–51)	32 (16–55)
Colostomy	4	14	15 (6–24)	32 (13–58)

stools. Furthermore, in two infants samples were collected also during a period when they had no breastmilk intake and in these cases no lactoferrin was found in the stools.

Conclusion

Analysis of the lactoferrin content in 67 stool specimens obtained from 17 patients which had been operated upon and had ileostomy (eight cases) or colostomy (three cases) were performed. In the 12 patients who consumed breastmilk, intact lactoferrin detectable by immunochemical methods could be found in the stools corresponding to between 9% and 32%. The underlying disease seems to be of greater importance than the age or the amount of breastmilk consumed. Thus 20% and 32%, respectively, of the calculated lactoferrin intake was found in the stools in patients with necrotizing enterocolitis (20%) and Hirschsprung's disease (32%). In patients with atresia of ileum or rectum 9% to 10% remained intact in the stools.

References

1. Brock JH (1980) Lactoferrin in human milk: its role in iron absorption and protection against enteric infection in the newborn infant. Arch Dis Child 55:417−421
2. Bullen JJ, Rogers HJ, Leigh L (1972) Iron binding protein in milk and resistance to Escherichia coli infection in infants. Br Med J I:69−75
3. Davidson LA, Lönnerdal B (1987) Persistence of human milk proteins in the breast-fed infant. Acta Paediatr Scand 76:733−740
4. Hambraeus L, Fransson G-B, Lönnerdal B (1984) Nutritional availability of breastmilk protein. Lancet II:167−168
5. Hambraeus L, Lönnerdal B, Forsum E, Gebre-Medhin M (1978) Nitrogen and protein components of human milk. Acta Paediatr Scand 67:561−565
6. Lönnerdal B (1984) Iron in milk. In: Stekel A (ed) Iron nutrition in infancy and childhood. Nestle Nutrition workshop series. Vol. 4. Raven Press, New York, pp 95−118
7. Spik G, Brunet B, Mazurier-Dehaine C, Fontaine G, Montreuil J (1982) Characterization and properties of the human and bovine lactotransferrine extracted from the faeces of newborn infants. Acta Paediatr Scand 71:979−985
8. Woelderen BF van (1987) Changing insight in human milk proteins: some implications. Nutr Abstr Rev 57:129−134

Specificity of the Intestinal Lactoferrin Receptor

L. A. Davidson and B. Lönnerdal

Department of Nutrition, University of California, Davis, USA

Introduction

In spite of a low iron content in human milk, breast-fed infants maintain adequate iron status significantly longer than formula-fed infants consuming similar amounts of iron [21, 22]. This indicates a high bioavailability of human milk iron. As the major iron-binding protein in human milk, lactoferrin has been postulated to be involved in the process of iron absorption in the suckling infant. An intestinal lactoferrin receptor was first described by Cox et al. following studies demonstrating delivery of iron from human lactoferrin (but not from transferrin), to human mucosal tissue [8]. Fransson et al. [14] found significantly faster uptake of radioiron into red blood cells of suckling piglets from a formula supplemented with bovine lactoferrin as compared with ferrous sulfate. Subsequent work has identified lactoferrin receptors in rabbit and mouse intestine [15, 20].

Our work focused on the Rhesus monkey as an animal model to study iron absorption. Not only is Rhesus milk similar in composition to human milk [18], but specifically we have shown Rhesus milk lactoferrin to be structurally and functionally similar to human lactoferrin [9], including similar molecular weights, amino acid composition, N-terminal amino acid sequence, and carbohydrate composition of the glycan chain. In fact, of the milks from the many species we tested, only human and Rhesus lactoferrin react with the human lactoferrin antibody. In addition, intestinal development of the infant Rhesus is similar to that of the human neonate. We therefore believe that the Rhesus monkey is an appropriate model for studying the involvement of lactoferrin in iron absorption from human or monkey milk. In this work, we further characterized the intestinal lactoferrin receptor.

Rhesus lactoferrin receptor

Our previous work [11] showed specific saturable binding of human and Rhesus lactoferrin to intestinal brush border membranes isolated from the Rhesus monkey. Intestinal lactoferrin receptors were found to occur in monkeys from all ages studied, including fetal tissue and through post-weaning ages. Bovine lactoferrin and human transferrin did not bind specifically to the brush border membrane, demonstrating specificity of the receptor for human and Rhesus lactoferrin. In addition, it appears that binding to the receptor is dependent upon the glycan chain of lactoferrin, as removal of fucose from lactoferrin by incubation with fucosidase resulted in decreased receptor binding. In addition, the presence of the fucose polymer, fucoidan, in the incubation caused a significant decrease in binding to the receptor.

In the present study we examined additional receptor characteristics in order to further understand the mechanism and activities of the intestinal lactoferrin receptor.

Delivery of other minerals by lactoferrin

In addition to iron, lactoferrin carries a large percentage of the manganese in human milk [19]; we therefore studied the binding of manganese-lactoferrin to the brush border membranes. While approximately 75% of human milk manganese is carried by lactoferrin, as compared to about 25% of human milk iron (Table 1), the ratio of iron to manganese bound to lactoferrin is about 1 000:1, i.e., less than 0.1% of the available metal binding sites of lactoferrin are occupied by manganese. Because some investigators have proposed that lactoferrin may be involved in zinc transport and delivery [2, 4, 5], zinc-lactoferrin was also studied.

Lactoferrin was labeled with ^{59}Fe, ^{54}Mn, or ^{65}Zn using an excess of citrate to keep the minerals in solution, followed by addition of apo-lactoferrin. Following labeling, the proteins were subjected to polyacrylamide gel electrophoresis followed by autoradiography to determine whether the protein had been labeled with the isotope. We found a high efficiency of labeling of lactoferrin with iron and manganese. However, apo-lactoferrin could not be labeled with zinc, either as a zinc citrate or zinc sulfate solution. Indeed, lactoferrin purified from human milk has been shown to contain no detectable zinc [17]. In previous studies examining "zinc-lactoferrin," no evidence for binding of zinc to lactoferrin was provided. Due to these observations, we believe lactoferrin is not involved in the binding or transport of zinc in human milk. This is consistent with our knowledge of the metal binding characteristics of lactoferrin [1].

Following labeling of lactoferrin with iron or manganese, binding studies were performed to quantify specific binding of lactoferrin to the Rhesus brush border membranes. Experiments were carried out as previously described [11]. Briefly, labeled human lactoferrin was incubated with brush border membrane vesicles at 37 °C for 5 min followed by vacuum filtration to separate free lactoferrin from that which is bound to the membranes. Correction was made for non-specific binding to the membranes. Figure 1 shows specific binding of iron- and manganese-lactoferrin of various concentrations to brush border membranes. It can be seen in these saturation curves that the magnitude of binding is similar for both minerals bound to lactoferrin. In a competition study, the binding of ^{54}Mn-lactoferrin or ^{59}Fe-lacto-

Table 1. Distribution of iron and manganese in human milk

	Iron	Manganese
Fat	30%	15%
Casein	10%	10%
Whey	60%	75%
Lactoferrin	25%	75%

77

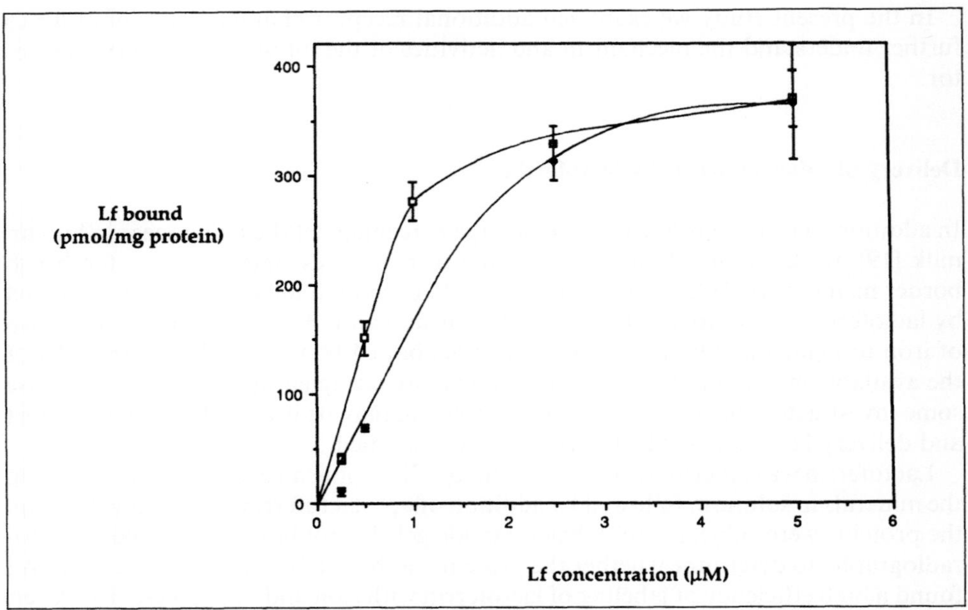

Fig. 1. Saturation curves for ^{59}Fe-lactoferrin ($\square$) and ^{54}Mn-lactoferrin ($\blacksquare$) binding to brush border membrane vesicles

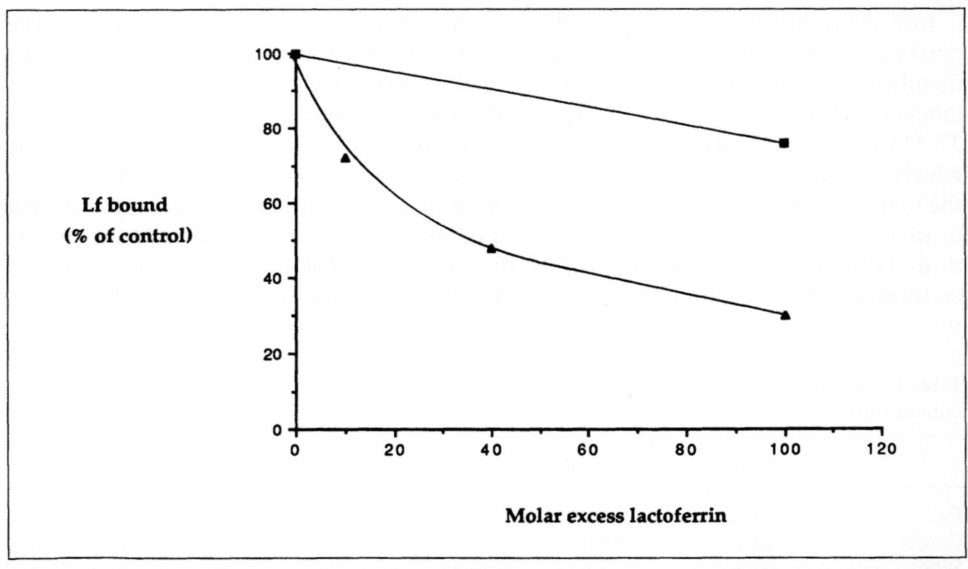

Fig. 2. Inhibition of ^{59}Fe-lactoferrin ($\blacksquare$) or ^{54}Mn-lactoferrin ($\blacktriangle$) binding by excess ferric lactoferrin

78

ferrin to the receptor was quantified following addition of increasing amounts of ferric lactoferrin. As shown in Fig. 2, addition of ferric lactoferrin results in a decrease in the binding of ^{54}Mn-lactoferrin to a greater extent than that of ^{59}Fe-lactoferrin, indicating a preference of the receptor for ferric lactoferrin over manganese lactoferrin.

Effect of degree of iron saturation on receptor binding

Lactoferrin in human milk is saturated with iron to a low degree [13]; only 3–5% of its binding capacity is utilized. However, during passage through the stomach and upper small intestine changes in pH and digestion of the protein occur and it is possible that the degree of iron saturation may vary. The interaction of lactoferrin with various degrees of iron saturation with the receptor was therefore investigated. Binding studies were performed with lactoferrin saturated with iron to 5, 20, 50, and 100%. We observed specific binding of all lactoferrin solutions to the brush border membrane, as shown in Fig. 3. However, in a competitive binding assay, an excess of 100% iron saturated lactoferrin inhibited binding of 5 or 20% saturated lactoferrin to a greater extent than lactoferrin solutions of 50 or 100% iron saturation. Similar inhibition occurred with an excess of 50% iron-saturated lactoferrin. This indicates a preference of the receptor for lactoferrin with a high degree of iron saturation.

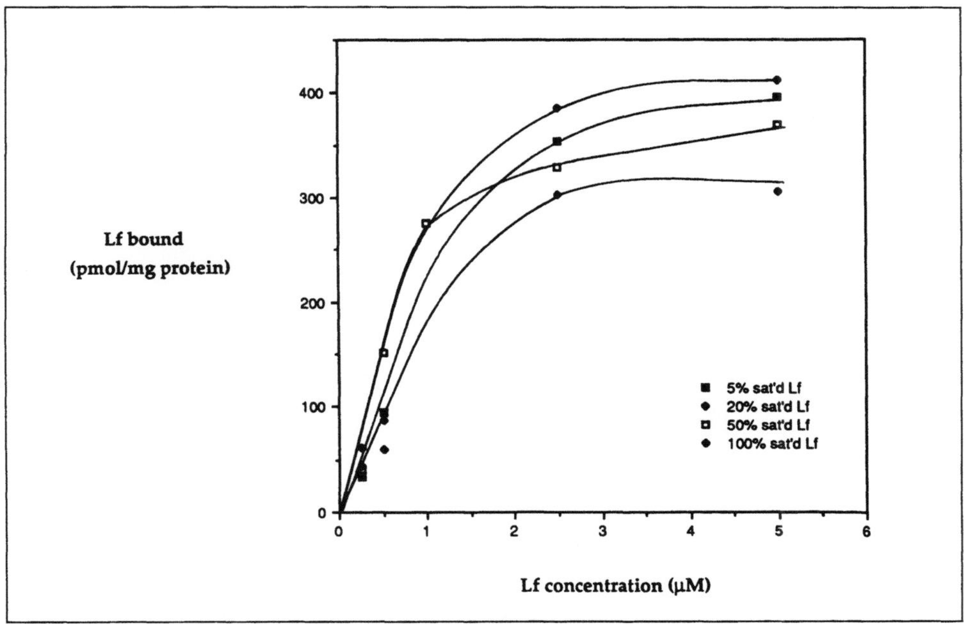

Fig. 3. Saturation curves for binding of 5, 20, 50, and 100% iron-saturated lactoferrin to brush border membranes

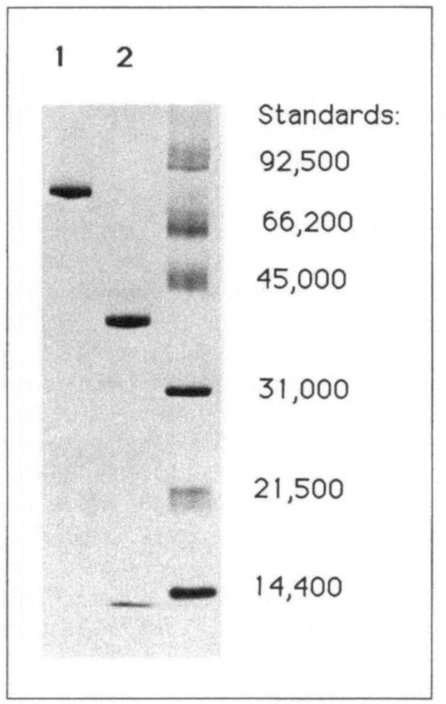

Fig. 4. SDS-polyacrylamide gel electrophoresis of intact lactoferrin (lane 1), half lactoferrin fragments (lane 2) and molecular weight standards

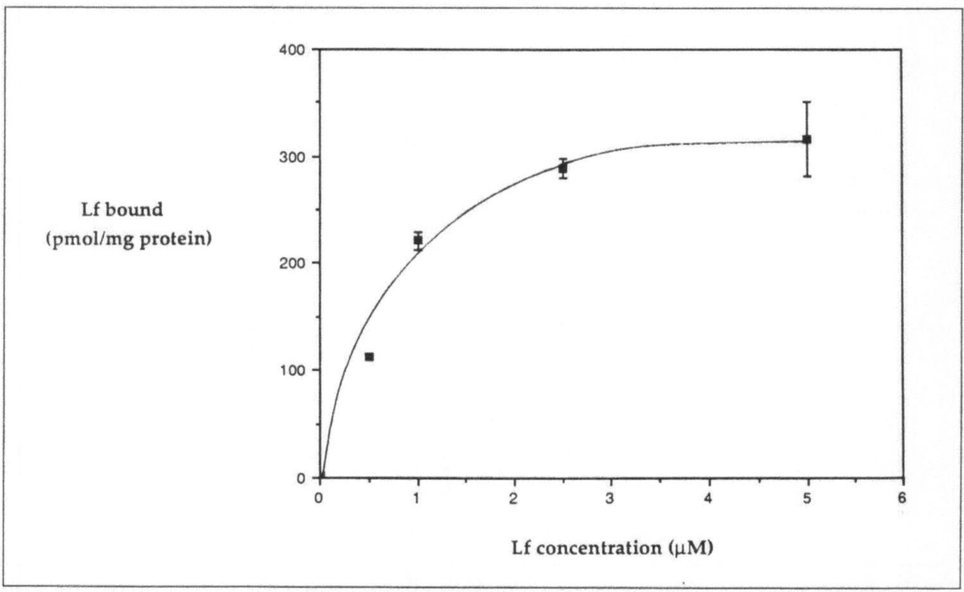

Fig. 5. Saturation curve for "half lactoferrin" binding to brush border membrane vesicles

80

Binding of lactoferrin digestion fragments

In vitro digestion of lactoferrin by various proteolytic enzymes has been examined
by several groups [6, 7, 16]. These studies have shown lactoferrin to be quite resistant
to proteolysis, and a number of large digestion fragments have been identified. In
vivo studies [10, 23] have shown a small percentage of ingested lactoferrin to survive
passage through the entire intestinal tract of the breast-fed infant, as immunologically intact lactoferrin can be detected in the feces of these infants. The tertiary structure
of human lactoferrin has recently been published [3]. Similar to transferrin, the
molecule consists of two lobes of almost identical size which are connected by an
extended polypeptide chain. We have separated these two half-molecules (Fig. 4) to
study their interaction with the receptor. The "half lactoferrin" molecules were
found to specifically bind to the receptor to a similar extent as intact lactoferrin
(Fig. 5). Again, competition with intact lactoferrin resulted in a greater inhibition
of half lactoferrin binding, suggesting higher affinity of the receptor for intact lactoferrin.

Conclusion

While the intestinal lactoferrin receptor from the Rhesus monkey is quite specific for
human or Rhesus lactoferrin, i.e., it does not recognize bovine lactoferrin or human
transferrin, it does recognize various forms of human lactoferrin. The receptor
readily accepts lactoferrin with a low level of iron saturation, as would occur in
human milk. However, when lactoferrin molecules with varying degrees of saturation are present, lactoferrin with a higher degree of iron saturation is more readily
bound.

Lactoferrin may provide a mechanism for manganese absorption from human
milk, as well as for iron absorption. A majority of human milk manganese is carried
on lactoferrin and this complex will interact with the intestinal lactoferrin receptor.
High bioavailability of manganese from human milk as compared to cow's milk and
infant formula was recently demonstrated in human adults [12]. In contrast, we were
unable to even label lactoferrin with zinc, presented either as zinc citrate or as zinc
sulfate. This is in agreement with the fact that no zinc can be detected bound to
lactoferrin isolated from human milk.

A small percentage of ingested lactoferrin has been shown to survive passage
through the gastrointestinal tract of the breast-fed infant. It is therefore probable
that significant amounts of intact and partially digested lactoferrin occur in the
intestine of the suckling infant. Even after partial digestion, i.e., cleavage into two
iron-binding fragments, lactoferrin retains enough structural integrity to interact
with the receptor. These observations provide additional insight into the functions
of the intestinal lactoferrin receptor.

References

1. Ainscough EW, Brodie AM, Plowman JE (1979) The chromium, manganese, cobalt and copper
   complexes of human lactoferrin. Inorg Chim Acta 33:149–153
2. Ainscough EW, Brodie AM, Plowman JE (1980) Zinc transport by lactoferrin in human milk.
   Am J Clin Nutr 33:1314–1315

3. Anderson BF, Baker HM, Dobson EJ, Norris GE, Rumball SV, Waters JM, Baker EN (1987) Structure of human lactoferrin at 3.2 Å resolution. Proc Natl Acad Sci USA 84:1769–773
4. Blakeborough P, Salter D, Gurr M (1983) Zinc binding in cow's milk and human milk. Biochem J 209:505–512
5. Blakeborough P, Salter DN (1987) The intestinal transport of zinc studied using brush-border-membrane vesicles from the piglet. Br J Nutr 57:45–55
6. Bluard-Deconinck J-M, Williams J, Evans RW, Van Snick J, Osinski PA, Masson PL (1978) Iron-binding fragments from the N-terminal and C-terminal region of human lactoferrin. Biochem J 171:321–327
7. Brines RD, Brock JH (1983) The effect of trypsin and chymotrypsin on the in vitro antimicrobial and iron-binding properties of lactoferrin in human milk and bovine colostrum. Biochim Biophys Acta 759:229–235
8. Cox TM, Mazurier J, Spik G, Montreuil J, Peters TJ (1979) Iron binding proteins and influx of iron across the duodenal brush border. Evidence for specific lactotransferrin receptors in the human intestine. Biochim Biophys Acta 588:120–128
9. Davidson LA, Lönnerdal B (1986) Isolation and characterization of Rhesus monkey milk lactoferrin. Pediatr Res 20:197–201
10. Davidson LA, Lönnerdal B (1987) Persistence of human milk proteins in the breast-fed infant. Acta Paediatr Scand 76:733–740
11. Davidson LA, Lönnerdal B (1988) Specific binding of lactoferrin to brush-border membrane: ontogeny and effect of glycan chain. Am J Physiol 254:G580–G585
12. Davidsson L, Cederblad Å, Lönnerdal B, Sandström B (1988) Manganese absorption from human milk, cow's milk and infant formulas. In: Hurley LS, Keen CL, Lönnerdal B, Rucker RB (eds) Trace Elements in Man and Animals – TEMA 6. Plenum Press, New York (in press)
13. Fransson G-B, Lönnerdal B (1980) Iron in human milk. J Pediatr 96:380–384
14. Fransson G-B, Thóren-Tolling K, Jones B, Hambraeus L, Lönnerdal B (1983) Absorption of lactoferrin-iron in suckling pigs. Nutr Res 3:373–384
15. Hu W-L, Mazurier J, Sawatzki G, Montreuil J, Spik G (1988) Lactotransferrin receptor for mouse small-intestinal brush border. Binding characteristics of membrane-bound and Triton X-100 solubilized forms. Biochem J 249:435–441
16. Line WF, Sly DA, Bezkorovainy A (1976) Limited cleavage of human lactoferrin with pepsin. Int J Biochem 7:203–208
17. Lönnerdal B, Hoffman B, Hurley LS (1982) Zinc and copper binding proteins in human milk. Am J Clin Nutr 36:1170–1176
18. Lönnerdal B, Keen CL, Glazier CE, Anderson J (1984) A longitudinal study of Rhesus monkey (Macaca mulatta) milk composition: trace elements, minerals, protein, carbohydrate, and fat. Pediatr Res 18:911–914
19. Lönnerdal B, Keen CL, Hurley LS (1985) Manganese binding proteins in human and cow's milk. Am J Clin Nutr 41:550–559
20. Mazurier J, Montreuil J, Spik G (1985) Visualization of lactotransferrin brush-border receptors by ligand-blotting. Biochim Biophys Acta 821:453–460
21. Saarinen UM (1978) Need for iron supplementation in infants on prolonged breast-feeding. J Pediatr 90:177–180
22. Siimes MA, Salmenpera L, Perheentupa J (1984) Exclusive breast-feeding for 9 months: risk of iron deficiency. J Pediatr 104:196–199
23. Spik G, Brunet B, Mazurier-Dehaine C, Fontaine G, Montreuil J (1982) Characterization and properties of the human and bovine lactotransferrins extracted from the feces of newborn infants. Acta Paediatr Scand 71:979–985

The Effects of Quantity and Type of Dietary Protein on the Rehabilitation after a Period of Energetic Undernourishment

H. de Waard[1] and P. L. M. Reyven[2]

[1] NIZO, Department of Nutrition, Ede, The Netherlands
[2] Present address: Biomedical Center, Department of General Surgery, State University Limburg, Maastricht, The Netherlands

Forty-two male Wistar rats of 3.5 months of age were subjected to a 50% food restriction of their ad lib. consumption, until their body weights had decreased by, on the average, 31%. One group of six rats then was sacrificed and served as a reference group. The remaining rats were divided into six groups of six animals and given different diets, ad lib., containing 5, 10, or 15% protein (on weight base) being either milk protein (soluble lactic protein, SLP, complete milk protein, not denatured) or soy protein (soy isolate) for a period of three weeks.

In the last week of food restriction the rats were placed in metabolic cages for another subsequent three weeks (the rehabilitation period) in order to determine nitrogen balance. At the end of the experiment blood samples were drawn and the livers removed.

There were significant differences between the groups in body weight, food consumption and efficiency, total protein in blood plasma, liver weight, N-balance (retention), creatinine clearance, blood plasma urea and urea excretion. Although most differences were level effects, the latter four parameters also showed significant effects attributable to the type of protein used (in all cases $P < 0.025$, with the exception of the N-balance, in which $P < 0.005$).

These things taken together in this respect, it can be noticed that at the low (5%) and moderate (10%) levels of dietary protein, the results with soy isolate stayed behind those observed with SLP, whereas at the highest protein level (15%) these differences were not observed anymore.

From this study it can be concluded that at suboptimal levels of dietary protein compared with soy protein, milk protein is of a superior quality.

During slimming the losses of body protein (on diets with low or moderate protein levels), were smaller with casein than with soy isolate; at high levels of dietary protein this effect of protein quality was not observed [1].

During rehabilitation after undernourishment there is a favorable effect of milk protein compared with soy isolate at low or moderate protein levels of intake; at high levels of dietary protein this difference was not found.

Under these functional conditions a high level or a high quality of the protein is recommended.

Reference

1. De Waard H (1983) Influence of the quantity and quality of dietary protein on the nitrogen losses during a negative energy balance in rats. Kieler Milchwirtschaftliche Forschungsberichte 35: 277–279

General Discussion:
Milk Protein and Nitrogen Equilibrium

P. Fürst

Institut für Biologische Chemie und Ernährungswissenschaft, Universität Hohenheim, Stuttgart, FRG

A novel method, the so-called homoarginine technique has been proposed by Dr. Hagemeister and colleagues, although yet only applicable in experimental animals, it assesses proteolysis and absorption of dietary proteins during passage through the gastrointestinal tract. Interesting new information obtained from his presentation is that a low digestibility (e.g., following TI-feeding) may emerge due to a reduction of proteolysis of exogenously ingested protein, but presumably this is quantitatively more important owing to the rise of endogenous secretion.

Two short communications contained in this volume deal with lactoferrin. Dr. Hambraeus and colleagues from Uppsala, Sweden, investigated the content of lactoferrin in feces obtained from ileostoma- or cholestoma-operated children fed human milk; they could demonstrate that substantial amounts (between 9–30%), of lactoferrin ingested remained. Obviously age did not influence the quantity remaining, but rather, it was dependent on the underlying disease. Some questions which are yet to be answered are the nutritional availability of lactoferrin and the extent of physiological activity relative to the low amounts utilized.

Dr. Lönnerdal discusses the specificity of the intestinal lactoferrin receptor. There is obviously a fucose-dependent receptor for lactoferrin in the brush-border membrane of the infant rhesus monkey. This recognizes human and monkey lactoferrin but not bovine lactoferrin or human transferrin. Lactoferrin may deliver both iron and manganese to the receptor; the affinity being certainly higher for iron-lactoferrin. Notable is that iron-saturated lactoferrin has a higher affinity to the receptor than less saturated forms. Importantly, in infants a specific receptor for lactoferrin facilitates iron and possibly, to some extent, manganese uptake. Although the receptor prefers the iron-saturated form of lactoferrin, it also accepts proteolytic fragments.

Dr. Reeds offers an insightful evaluation concerning milk protein and tissue nitrogen equilibrium on events occurring during postnatal development, such as changes in chemical composition of the body, the distribution of the body mass between cellular and extracellular components, and the contributions of various tissues to body protein mass. Considering growth in light of protein deposition and amino acid requirements two factors might be kept in mind: 1) The relative requirements of indispensible and dispensible amino acids, and 2) The division of these needs between those that support the accretion of new tissue and those required for the replacement of the basal or endogenous losses. Thus, preterm or newborn infants have no or limited ability to synthesize histidine, cystine, tyrosine, and presumably, glycine.

Although extensive information on amino acid composition is available in young animals, no directly comparable information exists for human infants. In one available study there is a remarkable similarity between the human fetus and that of weaning rat, except that percentages of glycine and proline are considerably higher in human fetus, suggesting a substantial higher contribution of collagen to fetal body protein.

The relative poor net utilization of milk proteins by the infants emphasizes an interesting aspect of human growth. Since basal nitrogen losses are very similar in the various species at least 50% of the infant's minimal requirement of nitrogen appears to be expended on the replacement of basal losses. It is also conceivable that the removal of some specific amino acids into non-protein pathways of metabolism may play a role in restricting the reutilization of amino acids released from protein breakdown. Losses of nitrogen in the feces can account for up to 50% of the total. In this respect it is important to distinguish between dietary or endogenous origin. Specific milk proteins appear in the feces of infants who are fed human milk. These infants obviously receive a negative supply of glycine, thus glycine deposition must be ensued via de novo synthesis. Dr. Reeds subsequently discusses tissue protein synthesis and its relationship to protein deposition. Protein turnover represents the sum of tissue protein turnover and each tissue has a characteristic relationship between the rate of protein synthesis and deposition. Importantly, the immediate postnatal period appears to be marked by accelerated growth of the gastrointestinal tract and increase in skin protein. These observations may explain, at least in part, why high rates of protein turnover per unit of protein intake are generally observed in human neonates.

Finally, Dr. Reeds paid attention to the possible role of trophic factors stimulating mucosal growth and development. Important factors like the stage of lactation, interspecific differences, and the degree to which the gastrointestinal tract and perhaps other tissues respond to these trophic factors may be critically dependent on the stage of maturity of the newborn.

Dr. Young directed his evaluation to deal with the capacity of dietary proteins to meet the needs for indispensible amino acids. According to his research, amino acid composition of the diet considered in relation to amino acid requirements of humans may provide a fair basis to evaluate various proteins with reference to practical aspects of human protein amino acid nutrition. Dr. Young critically evaluated the report from 1985 by FAO/WHO/UNO concerning the proposed energy and protein requirements. Especially the very low adult amino acid requirement values recommended by the above boards have been scrutinized. Dr. Young emphasized that by using the FAO/WHO/UNO scoring pattern almost any diet provides indispensible amino acids two to three times in excess. Consequently, he proposes a new tentative amino acid scoring pattern for application for the age range from pre-school children through adults. The new pattern has been developed on the basis of factorial prediction of minimum amino acid requirements and these values acquired are very consistent with estimations derived from ^{13}C tracer studies.

High amino acid requirement values have been questioned by Dr. Millward. He states that the requirements are complex and can only be defined under specific artificial conditions due to the fact that consumption of protein usually results in oxidative losses of indispensible amino acids variable according to dietary composition. An age-related fall in the values for requirement may thus simply reflect different dietary designs in the original balance studies, which then induce different rates of oxidative losses.

The new scoring pattern as proposed by Dr. Young leads to some important practical deductions. Undoubtedly, diets based mainly on cereals, especially in such areas as in Africa, the Near East, and Far East are likely limited by lysine, while diets typical for developed regions are more than adequate. Special care, however, might be directed to the new amino acid pattern when animal protein contributes less than about one-third of total dietary protein.

These considerations certainly underscore the high nutritional value of milk proteins.

Dr. Millward's presentation is devoted to evaluating hormonal response to dietary proteins. This approach was reviewed in the context of *anabolic drive*. The first important concept in this respect might be that the regulatory influence of dietary amino acids must be clearly separated from the function of amino acids as substrates. A second essential consideration made by Dr. Millward is that the anabolic drive may account in part for the inefficiency of amino acid utilization due to enhanced oxidation rates associated with increased intracellular concentrations of amino acids. Actually, Dr. Millward renders not less than 60% of human amino acid requirements as obligatory, while he contemplates the remaining 40% as an unaccountable part relative to the anabolic drive. The nature of the anabolic drive was characterized by central and peripheral regulations mediated by GH, T_4, IGF, T_3, and insulin, respectively. It is argued that insulin is particularly important.

In this connection it is pertinent to refer to Dr. Barth's presentation concerning endocrine response to animal and vegetable proteins. Obviously, two different proteins exhibit different endocrine responses as exemplified in variations in the plasma thyroxin concentrations. This finding conforms with the general picture: ingestion of two different vegetable proteins, like soya-protein and gluten, results in a considerably higher plasma thyroid hormone level than that seen after administration of casein. An important biological consequence of these findings is that higher plasma thyroxine levels indeed influence lipid metabolism resulting in an enhanced HMG-CoA reductase activity, increased synthesis, and excretion of bile acids, while VLDL secretion is diminished. The overall results will be a lower serum cholesterol concentration.

In Dr. Millward's presentation of his so-called "bag" theory, his hypothesis is an interesting speculation on the nature of muscle growth regulation. In malnutrition and muscle wasting diseases, factors like the increased amount of extracellular water associated with Na^+ accumulation, as well as changes in the intracellular pH, might be seriously considered in the frame of the presented hypothesis. However, these factors are known to influence the apparent creatinekinase equilibrium and, consequently, ATP resynthesis and cellular energy charge and may thus well "suit" the hypothesis proposed by Dr. Millward.

86

Milk Proteins and Metabolic Requirements of Trace Elements, Minerals, and Vitamins

B. Lönnerdal

Department of Nutrition University of California, Davis, CA, USA

Introduction

For most age groups, milk is considered an excellent source of many nutrients. Even those nutrients which do not occur in high concentrations in milk appear to be well taken up from milk and dairy products. When examining the reasons for the high bioavailability of nutrients from milk, a distinction must be made between the newborn infant with a not yet fully developed gastrointestinal tract and the adult with a much higher capacity for digestion. It is likely that the mechanisms explaining the high bioavailability will be quite different for these two age groups. In this paper, I will therefore attempt to give this developmental perspective to nutrient absorption from milk. Since many nutrients are bound to proteins, this discussion will primarily focus on milk proteins and how they differ from other sources of protein.

Protein Digestion

The efficiency of protein digestion is likely to have a pronounced effect on the absorption of minerals, trace elements, and vitamins. Depending on this efficiency, proteins will be present in the intestinal lumen either as amino acids and dipeptides or as larger peptides or protein fragments. This is of particular relevance for milk proteins, since several of these proteins are relatively resistant to degradation by proteolytic enzymes (see below). Because trace elements, minerals, and several vitamins are bound to proteins, the extent of digestion will determine if they will be released, and in some cases loosely complexed to amino acids and dipeptides (which may occur for elements such as zinc), or more firmly bound to larger polypeptides and protein fragments. This binding, or lack thereof, will then determine how well a nutrient is absorbed by the intestinal mucosa.

During early life, secretion of gastric acid and pepsin is lower than later in life [24]. As pepsin is only active at very low pH, peptic digestion in infancy may be quite low. We have found in both infant monkeys and a few human infants that gastric pH after a meal of milk or formula is often around pH 4−5 [27]; thus, pepsin digestion is minimal. This was supported by analysis of proteins in the gastric aspirates which showed that most milk proteins were still intact. Furthermore, the secretion of pancreatic enzymes such as trypsin, chymotrypsin, and elastase is lower in newborn infants than in adults. Thus, efficiency of proteolysis may be further compromised. Duodenal aspirates from Rhesus infants showed that several proteins in milk were, at least in part, still intact in the upper small intestine [27]. This may be a consequence

of low digestion capacity combined with resistance of some proteins against proteolysis. These in vivo studies are supported by our in vitro studies showing that under
conditions (pH, enzyme level, incubation times) similar to those occurring in the
infant, significant proportions of trace elements and minerals are bound to insoluble
material (casein) and high molecular weight proteins in milk [30]. As stated earlier,
this distribution is likely to affect the uptake process.

Milk Protein and Trace Element Absorption

Zinc Absorption

Zinc deficiency in healthy, well-nourished infants was first described by Hambidge
et al. [20]. These investigators found that infants fed a cow milk based formula had
significantly lower plasma zinc levels than breast-fed infants at six months of age.
Furthermore, male formula-fed infants, with a higher requirement of zinc because of
more rapid growth, were significantly shorter than the breast-fed male infants. Since
the level of zinc in the milk formula was equal to or higher than in breast milk, a
lower bioavailability of zinc from cow milk than from breast milk was implied.
Lower initial uptake of zinc from cow milk than from human milk was subsequently
shown by plasma uptake studies in human adults [5]. Whole body retention studies
in human adults, using ^{65}Zn and a whole body counter, demonstrated lower bioavailability of zinc from cow milk and cow milk-based infant formula than from
human milk [51]. On the other hand, zinc absorption from milk and milk products
was considerably higher than from soy formula, and also at the higher end of the
range observed for most composite diets [50]. These results demonstrate that zinc
absorption from milk and milk proteins is high compared to that from most other
protein sources and also that zinc absorption from human milk is exceptionally high,
particularly when considering its low protein level. A positive correlation between
the protein level of a meal and zinc absorption has been demonstrated [50]. Human
milk contains only 0.8−0.9% protein, while cow milk contains about 3% protein.
  The individual components in milk that affect zinc absorption have received some
attention [29, 32]. Absorption of zinc from whey proteins appears to be higher than
from casein [33]. This is supported by the findings of higher zinc bioavailability from
whey-predominant formula than from casein-predominant formula in infant monkeys [34] and also in adult human subjects [32]. This suggests that it is not a matter
of insufficient digestive capacity, but that zinc bound to casein is poorly absorbed.
It may be that phosphopeptides formed during incomplete casein digestion [38] bind
zinc and have a negative effect on zinc uptake [29]. It is also possible that zinc may
be "trapped" within the colloidal calcium phosphate of cow casein, thereby being
insoluble and poorly absorbed. This is supported by findings of higher zinc absorption from cow milk that had colloidal calcium phosphate removed [23]. Human
casein does not have a negative effect on zinc absorption [33], which may in part be
due to its low concentration but also to the absence of colloidal calcium phosphate
in human milk. It is also possible that phosphopeptides formed from human casein
may have a different effect on zinc absorption than that of bovine phosphopeptides.
  The higher uptake of zinc from human milk compared to cow milk may also be
explained by the larger proportion of zinc bound to citrate in human milk [31]. While

both these milks contain high levels of citrate, 3–5 mM, the high degree of binding of zinc to casein in cow milk leads to a smaller proportion of zinc bound to citrate. Zinc is well absorbed from citrate [56], which is an excellent chelator of zinc and other cations. It is also possible that citrate exerts a general positive effect on fluid and cation transport across the mucosa. It has been shown that addition of citrate to oral rehydration solutions increases water and electrolyte transport and that this effect is optimal at a citrate level of 5 mM [43]. Since zinc transport is enhanced by increased sodium and water transport [16], there may be an effect of citrate on zinc absorption independent of its chelating properties.

While lactose appears to have a positive effect on calcium and iron absorption (see below), it does not affect zinc absorption. Substitution of the carbohydrate source in soy formula from glucose polymer/maltrin to lactose did not affect zinc absorption in human subjects [32]. The high level of calcium in cow milk does not appear to have a negative effect on zinc absorption. Addition of high levels of calcium to cow milk-based infant formula did not change zinc uptake in human subjects [32]. Iron supplementation of cow milk formula may also affect zinc absorption. It has been shown that a large oral dose of iron reduces plasma uptake of zinc and that this effect is increased with the iron/zinc ratio [59]. A study on formula-fed infants showed lower plasma zinc levels in infants fed non-fortified formula than those fed unfortified formula [9]. Other studies have not found such an interaction and it has been shown that the competitive interaction between iron and zinc is minimal when they are given together in a meal [53]. Thus, this potential interaction needs further evaluation.

It should be emphasized that the difference in zinc absorption between human milk and cow milk or cow milk formula is, although significant, not of large magnitude. In fact, absorption of zinc from dairy sources often is at the higher end of the range exhibited for various foods [50]. This is in strong contrast to soy formula, from which zinc is poorly absorbed in humans [51], infant monkeys [34] and suckling rats [52]. This negative effect of soy on zinc absorption is likely due to the presence of phytate in soy protein isolate. Addition of phytate to cow milk formula to a level similar to that found in soy formula reduced zinc absorption by 50% and to a value similar to that for soy formula [32]. Furthermore, removal of phytate by either precipitation methods or phytase treatment caused a significant increase of zinc bioavailability to a level similar to cow milk formula [34]. This presence of phytate in some food items may have a negative effect on the absorption of zinc from milk if they are consumed together. For example, zinc absorption from a mixed milk/ cereal formula was significantly lower than from milk alone [51]. Thus, the high bioavailability of zinc from dairy products may be impaired by phytate-containing foods such as soy and cereals. This is supported by the observation that addition of cereals to breast milk or milk formula during the weaning period had a pronounced negative effect on the amount of zinc absorbed [3].

Iron Absorption

The iron content of milk and dairy products is usually low. Products intended for infants are therefore usually fortified with iron, while adults with high iron requirements, such as pregnant and lactating women, are given iron supplements. The effect of various dietary components on iron absorption therefore becomes more crucial for infants.

Iron absorption from human milk has been shown to be significantly higher than from cow milk and infant formula [36, 37, 45], although some methodological problems have been noticed in these studies [28], and the degree of individual variation is very large [45]. Again, however, iron absorption from soy formula and soy products appears to be considerably lower than from either milk or milk formula [18, 39]. The high bioavailability of iron from human milk has received some attention. Several studies have focused on the potential role of lactoferrin in this process (see other chapters in this volume). Lactoferrin is high in human milk but not in cow milk, and the possibility of a receptor for lactoferrin in the small intestine of humans was suggested by Cox et al. [8]. A study of iron absorption in suckling piglets fed formula with iron (either bound to lactoferrin or as ferrous sulfate) demonstrated higher and more rapid uptake of iron from lactoferrin than from the iron salt [15]. Studies in suckling mice supported these findings [14], while a study in human adults failed to show an effect [37]. In this latter study, however, a very high ratio of iron to lactoferrin was used and it was not tested whether lactoferrin actually was specifically labeled with ^{59}Fe [28]. Therefore, the labeled iron may have been absorbed as any other non-lactoferrin-bound iron. We have recently documented the presence of a receptor for lactoferrin on the mucosal surface of infant monkey small intestine [10]. This brush-border membrane receptor facilitates iron uptake and is specific for human and monkey lactoferrin, while bovine lactoferrin and human transferrin cannot deliver iron. These proteins contain glycans with different terminal carbohydrates [60] an it is possible that the difference in binding is due to differences in side-chain glycosylation. The lack of binding of bovine lactoferrin to the receptor may also explain the lack of effect of bovine lactoferrin supplementation of formula on iron absorption [11].

Similar to the discussion on zinc (above), the lower bioavailability of iron from cow milk than from human milk may also be due to the higher level of casein in this milk and its different binding properties. It has been shown that α-casein in cow milk binds iron efficiently [22], which may have a negative effect on iron absorption. In addition, the high calcium level of cow milk may have a negative effect on iron absorption. It has been shown in experimental animals that iron uptake from cow milk is lower than from human milk and that addition of calcium to human milk significantly reduced iron uptake [2]. Some of these negative factors in milk may, however, be overcome in milk formula; the calcium level is reduced and the level of citrate is usually high. Citrate has been shown to have a positive effect on iron absorption [17], possibly by keeping more iron in solution [21]. In addition, supplementation of milk formula with ascorbic acid has a pronounced beneficial effect on iron absorption [18].

The low bioavailability of iron from soy formula may be caused by several dietary factors. Phytate has been shown to have a negative effect on iron absorption [19]. Removal of phytate, however, increases iron bioavailability only to a limited extent, suggesting that other inhibitory factors are still present. Removal of phosphate, produced by degradation of phytic acid with phytase, further increases iron absorption [34]. Although this implies that phosphate inhibits iron absorption, it should be noted that the phytase removal of phosphorus also releases calcium bound to phytate and that the dialysis used to remove phosphate also removes calcium. Thus, the positive effect on iron absorption may also have been caused by removal of calcium. As described previously, calcium can affect iron absorption negatively [2]. It is also possible that the protein in soy formula can affect iron absorption. Further studies are needed to clarify the roles of protein, phytate, phosphate, and calcium on iron absorption. This is particularly important since calcium and phosphate are also high in milk. Their effect, or possibly lack of effect, needs to be known when evaluating iron absorption from combinations of cereals, which often are high in phytate and milk. This combination is one of the most commonly used weaning foods in industrialized countries and often a significant component of the diet throughout childhood.

Milk Protein and Mineral Absorption

Calcium Absorption

The bioavailability of calcium from milk has been shown to be high in both infants and adults [42, 44, 57, 58]. While some reports state that milk "inhibits" calcium absorption as compared to a radioisotope given in water solution, it should be noted that this is true for *any* food item given and that the absorption of radiolabeled calcium is an unphysiological situation with little relevance in human nutrition. Instead, when comparing the absorption of calcium from milk to that from several commercial sources of calcium salts given at physiological levels, it is found to be similarly high [42]. When evaluating calcium absorption from various foods, however, it is also important to consider the *amount* of calcium provided by the diet. Dietary surveys in the U.S. show that dairy products constitute about 70% of the total calcium intake.

Several factors in milk may be responsible for the high bioavailability of calcium. Milk protein, and in particular casein, was suggested more than two decades ago to have a positive effect on calcium absorption [38]; the formation of casein phospho-peptides would help in keeping calcium in solution and thus facilitate calcium uptake. Several reports support this concept. Lee et al. [25] showed that a phosphopeptide is formed from bovine casein during digestion in the rat and that this peptide aids in keeping calcium in solution. Phosphopeptides formed in vitro and in vivo have subsequently been demonstrated to have a positive effect on calcium absorption in chickens [40] and rats [54]. This effect appeared most pronounced in the distal part of the small intestine, the major site for calcium absorption. While it is possible that casein phosphopeptides may also be formed from α-casein, it has been shown that they can be obtained from β-casein [26]. This phosphopeptide exhibited a solubilizing effect on calcium as well as enhanced calcium absorption by ligated rat small intestine. In contrast, peptides formed from soy isolate did not have these properties [54].

Lactose in milk and dairy products is known to enhance calcium absorption [6], which can be particularly important during infancy [65]. It has been shown that lactose can increase calcium absorption by more than 30% in human adults [6]. This effect, however, is not always found and may be dependent on several factors such as the level of lactose given [58]. High individual variability in calcium absorption may also contribute to some of the differences found. The effect of citrate in milk on calcium absorption should also be considered (as described for zinc). Calcium has been demonstrated to be better absorbed when gives as calcium citrate as compared to calcium carbonate [41].

Calcium absorption from human milk appears to be higher than from cow milk [44, 57]. This may be explained by the different level of casein in these milks as well as its composition. It is possible that during early life, the digestive capacity of the infant's gut may be inadequate for casein phosphopeptide formation from bovine casein or that calcium is even better absorbed from phosphopeptides derived from human casein. Also, in human milk only 6% of the total calcium content is associated with casein, compared to 41% in cow milk [13]. It is also possible that differences in levels of vitamin D and its metabolites in the two milks may contribute to the differences observed. However, it is known that calcium absorption in the very young is less dependent on vitamin D.

Milk Protein and Vitamin Absorption

While the role of milk in providing fat-soluble vitamins is well-recognized, the importance of milk in absorption of water-soluble vitamins is less known. In adults, milk can most likely be considered as a source of water-soluble vitamins which are well absorbed. During infancy, however, there may be specific roles for milk proteins in facilitating vitamin uptake under conditions of limited digestion.

Folate-binding Protein

A specific binding protein for folate has been found in milk from several species. The folate-binding protein (FBP) in human milk has been isolated and characterized [12] and appears both in the whey and in a membrane-bound form [1]. The role of FBP in folate absorption during early life has been investigated in the newborn goat [48]. By aspirating intestinal contents, it was shown that purified goat milk FBP can resist low pH and proteolytic degradation and be found in intact form.

Higher uptake of folate from FBP than from the free form has been described when using isolated rat intestinal cells in vivo [7]. However, another study [46] did not find any enhancing effect of FBP on folate transport in everted sacs from rat intestine. In this study the transport of free and FBP-bound folate was found to be similar in jejunum and ileum. It was proposed that the prior observations [7] may have been a result of binding of protein-bound folate to the mucosal surface and not transport, since temperature (4° vs 37°C) did not affect uptake. Said et al. [46] proposed that FBP-bound folate is transported more slowly than free folate in the jejunum, which is the major site for absorption of free folate. Thus, folate absorption in the proximal intestine would be slowed down and a more gradual uptake of folate would occur, thereby avoiding rapid excretion of folate into urine and increasing tissue utilization of folate.

92

Recently, Mason and Selhub [35] pointed out that previous studies had been performed in adult rats and not suckling animals and that species-specific FBP had not been used. When using rat FBP and suckling rat pups, these investigators found significantly higher absorption of folate from FBP than free form in ileum but not in jejunum. In addition, Salter and Blakeborough [47] found that goat milk FBP increased folate transport in brush border membrane vesicles from neonatal goats. Thus, there may be a specific role for FBP to facilitate folate absorption in the distal part of the small intestine at young age.

Vitamin B_{12}-binding Protein

Similar to folate, vitamin B_{12} is carried by a specific binding protein in milk [4]. In human milk, this cobalamin-binding protein is predominantly of the R-type [49], and there is little transcobalamin. The concentration of vitamin B_{12}-binding protein in human milk is considerably higher than that of vitamin B_{12}, resulting in a high binding capacity for the vitamin. It is possible that the vitamin B_{12}-binding protein may be involved in the absorption of this vitamin from milk.

Several aspects of the involvement of vitamin B_{12}-binding proteins in the uptake of the vitamin have been studied recently [61–63]. Using suckling pigs as an animal model, it was found that suckled animals younger than seven days retained a higher proportion of vitamin B_{12} than artificially reared animals [61]. In 15-day-old animals no difference in vitamin B_{12} retention was found. In vivo studies also demonstrated that vitamin B_{12}-binding protein could be found intact in all parts of the small intestine [62]. This resistance against proteolysis appeared higher when the protein was saturated with vitamin B_{12} than when it was unsaturated; this is similar to what has been found for intrinsic factor (IF). Additional support for a role of vitamin B_{12}-binding protein in withholding vitamin B_{12} was also found; neither free nor bound radiolabeled cobalamin was bound to intestinal bacteria in the presence of the binding protein.

Brush border membrane vesicles from suckling pigs were also used to study vitamin B_{12} uptake [63]. The presence of vitamin B_{12}-binding protein was found to strongly promote vitamin B_{12} uptake in vesicles from seven- and 28-day-old piglets, while uptake in the absence of the binding protein was negligible. While the milk vitamin B_{12}-binding protein bound equally to all segments of the small intestine, IF only bound to the ileum. The vitamin B_{12}-binding protein was found to mediate uptake of the vitamin by a specific, rapid, and saturable mechanism.

Conclusion

Milk proteins have been shown to facilitate uptake of several nutrients such as trace elements, minerals, and vitamins. The effect appears to be most pronounced for the species-specific milk proteins. This suggests that specific mechanisms may have evolved to assure efficient nutrient utilization during a period of rapid growth but with limited digestive capacity.

References

1. Antony AC, Utley CS, Marcell PD, Kolhouse JF (1982) Isolation, characterization, and comparison of the solubilized particulate and soluble folate binding proteins from human milk. J Biol Chem 257:10081–10089
2. Barton JC, Conrad ME, Parmley RT (1983) Calcium inhibition of inorganic iron absorption in rats. Gastroenterology 84:90–91
3. Bell JG, Keen CL, Lönnerdal B (1987) Effect of infant cereals on zinc and copper absorption during weaning. Am J Dis Child 141:1128–1132
4. Burger RL, Allen RH (1974) Characterization of vitamin B_{12}-binding proteins from human milk and saliva by affinity chromatography. J Biol Chem 249:7220–7227
5. Casey CE, Walravens PA, Hambidge KM (1981) Availability of zinc: loading tests with human milk, cow's milk, and infant formulas. Pediatrics 68:394–396
6. Cochet BA, Juny M, Griessen P, Bartholdi P, Schaller P, Donath A (1983) Effects of lactose on intestinal calcium absorption in normal and lactase-deficient subjects. Gastroenterology 84:935–940
7. Colman N, Hettiarachchy N, Herbert V (1981) Detection of a milk factor that facilitates folate uptake by intestinal cells. Science 211:1427–1429
8. Cox TM, Mazurier J, Spik G, Montreuil J, Peters TJ (1979) Iron-binding proteins and influx of iron across the duodenal brush border. Biochim Biophys Acta 588:120–128
9. Craig WJ, Balbach L, Harris S, Vyhmeister N (1984) Plasma zinc and copper levels of infants fed different milk formulas. J Am Coll Nutr 3:183–186
10. Davidson LA, Lönnerdal B (1988) Specific binding of lactoferrin to brush border membrane: ontogeny and effect of glycan chain. Am J Physiol 254:G580–G585
11. Fairweather-Tait SJ, Balmer SE, Scott PH, Minski MJ (1988) Lactoferrin and iron absorption in newborn infants. Pediatr Res 22:651–654
12. Ford JE, Salter DN, Scott KJ (1969) The folate-binding protein in human milk. J Dairy Res 36:435–446
13. Fransson G-B, Lönnerdal B (1983) Distribution of trace elements and minerals in human and cow's milk. Pediatr Res 17:912–915
14. Fransson G-B, Keen CL, Lönnerdal B (1983) Supplementation of milk with iron bound to lactoferrin using weanling mice. I. Effects on hematology and tissue iron. J Pediatr Gastroenterol Nutr 2:693–700
15. Fransson G-B, Thoren-Tolling K, Jones B, Hambraeus L, Lönnerdal B (1983) Absorption of lactoferrin-iron in suckling pigs. Nutr Res 3:373–384
16. Ghishan FK (1984) Transport of electrolytes, water and glucose in zinc deficiency. J Pediatr Gastroenterol Nutr 3:608–612
17. Gillooly M, Bothwell TH, Torrance JD, MacPhail AP, Derman DP et al. (1983) The effects of organic acids, phytates, and polyphenols on the absorption of iron from vegetables. Br J Nutr 49:331–342
18. Gillooly M, Torrance JD, Bothwell TH, MacPhail AP, Derman D et al. (1984) The relative effect of ascorbic acid on iron absorption from soy-based and milk-based infant formulas. Am J Clin Nutr 40:522–527
19. Hallberg L, Rossander L, Skanberg A-B (1987) Phytates and the inhibitory effect of bran on iron absorption in man. Am J Clin Nutr 45:988–996
20. Hambidge KM, Walravens PA, Casey CE, Brown RM, Bender C (1979) Plasma zinc concentrations of breast-fed infants. J Pediatr 94:607–608
21. Hazell T, Johnson IT (1987) In vitro estimation of iron availability from a range of plant foods: influence of phytate, ascorbate and citrate. Br J Nutr 57:223–233
22. Hegenauer J, Saltman P, Ludwig D, Ripley L, Ley A (1979) Iron-supplemented cow milk. Identification and spectral properties of iron bound to casein micelles. J Agric Food Chem 27:1294–1301
23. Kiely J, Flynn A, Singh H, Fox PF (1988) Improved zinc bioavailability from colloidal calcium phosphate-free cow's milk. In: Hurley LS, Keen CL, Lönnerdal B, Rucker RB (eds) Trace Elements in Man and Animals – 6, Plenum Press, New York, in press
24. Lebenthal E, Lee PC, Heitlinger LA (1983) Impact of development of the gastrointestinal tract on infant feeding. J Pediatr 102:1–9

25. Lee YS, Noguchi T, Naito H (1980) Phosphopeptides and soluble calcium in the small intestine of rats given a casein diet. Br J Nutr 43:457–467
26. Lee YS, Noguchi T, Naito H (1983) Absorption of calcium in rats given diets containing casein or amino acid mixture: the role of casein phosphopeptides. Br J Nutr 49:67–76
27. Llanes TE, Lönnerdal B (1987) Digestibility of human, bovine and Rhesus monkey milk. Fed Proc 46:895
28. Lönnerdal B (1984) Iron in breast milk. In: Stekel A (ed) Iron Nutrition in Infancy and Childhood. Nestle Nutrition Workshop Series 4:95–118. Vevey/Raven Press, New York
29. Lönnerdal B (1985) Dietary factors affecting trace element bioavailability from human milk, cow's milk and infant formulas. Progress in Food and Nutrition Science 9:35–62. Pergamon Press, Oxford
30. Lönnerdal B, Glazier C (1988) An approach to assessing trace element bioavailability from milk in vitro: extrinsic labeling and proteolytic degradation. Biol Trace Element Res (in press)
31. Lönnerdal B, Stanislowski AG, Hurley LS (1980) Isolation of a low molecular weight zinc binding ligand from human milk. J Inorg Biochem 12:71–78
32. Lönnerdal B, Cederblad Å, Davidsson L, Sandström B (1984) The effect of individual components of soy formula and cow's milk formula on zinc bioavailability. Am J Clin Nutr 40:1064–1070
33. Lönnerdal B, Keen CL, Bell JG, Hurley LS (1985) Zinc uptake and retention from chelates and milk fractions. In: Mills CF, Bremner I, Chesters JK (eds) Trace elements in Man and Animals – TEMA 5. Commonwealth Agricultural Bureaux, Farnham Royal, UK
34. Lönnerdal B, Bell JG, Hendrickx AG, Burns RA, Keen CL (1988) Effect of phytate removal on zinc absorption from soy formula. Am J Clin Nutr (in press)
35. Mason JB, Selhub J (1988) Folate-binding protein and the absorption of folic acid in the small intestine of the suckling rat. Am J Clin Nutr 48:620–625
36. McMillan JA, Landaw SA, Oski FA (1976) Iron sufficiency in breast-fed infants and the availability of iron from human milk. Pediatrics 58:686–691
37. McMillan JA, Oski FA, Lourie G, Tomarelli RM, Landaw SA (1977) Iron absorption from human milk, simulated human milk, and proprietary formulas. Pediatrics 60:896–900
38. Mellander O (1950) The physiologic importance of the casein phosphopeptide calcium salt. II. Peroral calcium dosage of infants. Acta Soc Med Upsal 55:247–255
39. Morck TA, Lynch SR, Skikne BS, Cook JD (1981) Iron availability from infant food supplements. Am J Clin Nutr 34:2630–2634
40. Mykkänen HM, Wasserman RH (1980) Enhanced absorption of calcium by casein phosphopeptides in rachitic and normal chicks. J Nutr 110:2141–2148
41. Nicar MJ, Pak CYC (1985) Calcium bioavailability from calcium carbonate and calcium citrate. J Clin Endocrinol Metab 61:391–393
42. Recker RR, Bammi A, Barger-Lux J, Heaney RP (1988) Calcium absorbability from milk products, an imitation milk, and calcium carbonate. Am J Clin Nutr 47:93–95
43. Rolston DDK, Moriarty KJ, Farthing MJG, Kelly MJ, Clark ML, Dawson AM (1986) Acetate and citrate stimulate water and sodium absorption in the human jejunum. Digestion 34:101–104
44. Rowe J, Rowe D, Horak D et al. (1984) Hypophosphatemia and hypercalciuria in small premature infants fed human milk. J Pediatr 104:112–117
45. Saarinen UM, Siimes MA, Dallman PR (1977) Iron absorption in infants: High bioavailability of breast milk iron as indicated by the extrinsic tag method of iron absorption and by the concentration of serum ferritin. J Pediatr 91:36–39
46. Said HM, Horne DW, Wagner C (1985) Effect of human milk folate binding protein on folate intestinal transport. Arch Biochem Biophys 251:114–120
47. Salter DN, Blakeborough P (1988) Influence of goat's-milk folate-binding protein on transport of 5-methyltetrahydrofolate in neonatal-goat small intestinal brush-border-membrane vesicles. Br J Nutr 59:497–507
48. Salter DN, Mowlem A (1983) Neonatal role of milk folate-binding protein: studies on the course of digestion of goat's milk folate binder in the 6-d-old kid. Br J Nutr 50:589–596
49. Sandberg DP, Begley JA, Hall CA (1981) The content, binding, and forms of vitamin B_{12} in milk. Am J Clin Nutr 34:1717–1724

50. Sandström B, Arvidsson B, Cederblad Å, Björn-Rasmussen E (1980) Zinc absorption from composite meals. I. The significance of wheat extraction rate, zinc, calcium, and protein content in meals based on bread. Am J Clin Nutr 33:739–745
51. Sandström B, Cederblad Å, Lönnerdal B (1983) Zinc absorption from human milk, cow's milk and infant formulas. Am J Dis Child 137:726–729
52. Sandström B, Keen CL, Lönnerdal B (1983) An experimental model for studies of zinc bio-availability from milk and infant formulas using extrinsic labelling. Am J Clin Nutr 38:420–428
53. Sandström B, Davidsson L, Cederblad Å, Lönnerdal B (1985) Oral iron, dietary ligands and zinc absorption. J Nutr 115:411–414
54. Sato R, Noguchi T, Naito H (1986) Casein phosphopeptide (CPP) enhances calcium absorption from the ligated segment of rat small intestine. J Nutr Sci Vitaminol 32:67–76
55. Sczekan SR, Joshi JG (1987) Isolation and characterization of ferritin from soyabeans (Glycine max). J Biol Chem 262:13780–13788
56. Seal CJ, Heaton FW (1983) Chemical factors affecting the intestinal absorption of zinc in vitro and in vivo. Br J Nutr 50:317–324
57. Senterre J, Salle B (1982) Calcium and phosphorus economy of the preterm infant and its interaction with vitamin D and its metabolites. Acta Paediatr Scand Suppl 296:85–92
58. Smith TM, Kolars JC, Savino DA, Levitt MD (1985) Absorption of calcium from milk and yogurt. Am J Clin Nutr 42:1197–1200
59. Solomons NW, Jacob RA (1981) Studies on the bioavailability of zinc in humans: effect of heme and nonheme iron on the absorption of zinc. Am J Clin Nutr 34:475–482
60. Spik G, Strecker G, Fournet B, Bouquelet S, Montreuil J (1982) Primary structure of the glycans from human lactotransferrin. Eur J Biochem 121:413–419
61. Trugo NMF, Newport MJ (1985) Vitamin B_{12} absorption in the neonatal piglet. 2. Resistance of the vitamin B_{12}-binding protein in sows' milk to proteolysis in vivo. Br J Nutr 54:257–267
62. Trugo NMF, Ford JE, Sansom BF (1985) Vitamin B_{12} absorption in the neonatal piglet. 1. Studies in vivo on the influence of the vitamin B_{12}-binding protein from sow's milk on the absorption of vitamin B_{12} and related compounds. Br J Nutr 54:245–255
63. Trugo NMF, Ford JE, Salter DN (1985) Vitamin B_{12} absorption in the neonatal piglet. 3. Influence of vitamin B_{12}-binding protein from sows' milk on uptake of vitamin B_{12} by microvillous membrane vesicles prepared from small intestine of the piglet. Br J Nutr 54:269–283
64. Ziegler EE, Figueroa-Colón R, Serfass RE, Nelson SE (1987) Effect of low dietary zinc on zinc metabolism in infancy: stable isotope studies. Am J Clin Nutr 45:849
65. Ziegler EE, Fomon SJ (1983) Lactose enhances mineral absorption in infancy. J Pediatr Gastroenterol Nutr 2:288–294

Manganese Absorption from Human Milk, Cow Milk, and Infant Formulas

L. Davidsson, Å. Cederblad, B. Lönnerdal and B. Sandström

Departments of Clinical Nutrition and Radiation Physics,
University of Gothenburg, Sahlgrenska Hospital, Göteborg, Sweden,
and Department of Nutrition, University of California, Davis, USA

Introduction

Our knowledge of manganese absorption and metabolism in man is limited. Consequently, the amounts of manganese required for humans of different ages remain largely unknown. Most diets consumed by adults are relatively high in manganese; thus, a pronounced deficiency is unlikely. In early life, however, there is a greater risk for manganese deficiency as well as for toxicity. Human milk and cow milk formulas are low in manganese concentration, while soy formulas may contain high amounts of manganese. In addition, other factors in infant formulas, e.g., phytate and supplemental iron could interfere with manganese absorption. Manganese nutrition in the neonatal period is poorly understood due in part to the lack of information of manganese content in infant foods and its bioavailability.

In this study manganese absorption from human milk, cow milk, and infant formulas was studied in healthy adults using a recently developed radionuclide method [1]. The method involves feeding an extrinsically labelled diet, using the gamma-emitter ^{54}Mn or ^{52}Mn, and monitoring the whole-body retention with a sensitive whole-body counter. The use of an extrinsic label was recently validated by using a double-isotope technique and intrinsic labelling [2].

Materials and methods

Twenty-six women and six men volunteered for the study. They were all healthy with normal levels of manganese in whole blood and normal iron status indices. Fourteen of the subjects participated twice; eight were first served human milk and approximately two months later, cow milk; six subjects were fed whey-predominant cow milk formula with or without iron fortification on two separate occasions.

Test meals included:
1) Human milk, pooled and pasteurized, obtained from the milk bank at Östra Hospital, Göteborg;
2) Cow milk, 3% fat, purchased from a local vendor;
3) Whey-predominant (60/40) cow milk formula (Baby Semp 1, Semper AB, Stockholm, Sweden), iron-fortified to 7 mg Fe/L;
4) Whey-predominant (60/40) cow milk formula (Enfamil, Mead Johnson, Evansville, Indiana, USA);
   a) iron fortified to 12 mg Fe/L;
   b) without iron fortification (2 mg/L);
5) Soy formula (Prosobee, Mead Johnson).

Each serving consisted of 450 g. Milks and formulas were extrinsically labelled with 0.2 MBq ^{54}Mn or ^{52}Mn and 1.5 MBq ^{51}Cr before serving. ^{51}Cr was used to establish the time point when the non-absorbed manganese isotope left the body and "true" whole-body retention was measured.

The subjects arrived in the morning after 12 h of fasting. Each subject's background radioactivity was measured in the whole body counter and blood samples were drawn for determination of whole blood manganese and iron status indices. The extrinsically labelled test meal was served and a 100% value of administered activity was obtained by measurement in the whole body counter immediately after intake. No other food or drink was allowed during the next 3 h. The subjects maintained their normal food intake during the study. Whole-body retention was measured during approximately 30 days after initiating the study. Manganese absorption was calculated by extrapolation from whole-body retention measurements during days 10–30.

Results and discussion

The fractional manganese absorption from human milk was significantly different ($p < 0.01$) from the absorption of manganese from cow milk, iron-fortified cow milk formula (12 mg Fe/L), and soy formula (Table 1). However, when total amount of absorbed manganese was calculated only minor differences were observed between the milks and the whey-predominant cow milk formulas. The fractional absorption of manganese from soy formula was very low, but due to the high manganese content the total amount of absorbed manganese was significantly higher from soy formula as compared to manganese absorption from human milk.

An effect of iron level on manganese absorption is indicated by the results from the two identical formulas which differed only in iron content.

Table 1. Manganese and iron content and manganese absorption from 450 g of diet

Diet	Mn μg	Fe mg	n	Mn-absorption % ($\bar{X} \pm SD$)	μg ($\bar{X} \pm SD$)
1. Human milk	7.2	0.3	8	8.2 ± 2.9	0.59 ± 0.21
2. Cow milk	44	0.1	8	2.4 ± 1.7 [b]	1.06 ± 0.75
3. Whey-predominant cow milk formula (Baby Semp)	23	3.0	14	6.1 ± 4.7	1.40 ± 1.08
4. Whey-predominant cow milk formula (Enfamil)					
a) iron-fortified	59	5.6	6	1.6 ± 0.8 [b]	0.94 ± 0.47
b) non-iron-fortified	59	1.0	6	3.6 ± 3.2	2.12 ± 1.89 [a]
5. Soy formula	154	7.2	4	1.0 ± 0.4 [b]	1.54 ± 0.62 [b]

[a] Significantly different ($p < 0.05$) from human milk
[b] Significantly different ($p < 0.01$) from human milk

The optimum amount of manganese to be provided in infant formulas needs to be determined.

References

1. Davidsson L, Cederblad Å, Lönnerdal B, Sandström B (1988) Manganese retention in man: A method for estimating manganese absorption in man. Am J Clin Nutr (in press)
2. Davidsson L, Cederblad Å, Hagebö E, Lönnerdal B, Sandström B (1988) Assessment of the validity of using extrinsic labeling for studies of manganese absorption in man. J Nutr (in press)

Intrinsic Labelling of Iron in Milk

J. Gislason*, B. Jones**, B. Lönnerdal***, L. Hambraeus*

* Department of Nutrition, University of Uppsala, Sweden,
** Department of Clinical Chemistry, Swedish University of Agricultural Sciences,
Uppsala, Sweden, and
*** Department of Nutrition, University of California, Davis, USA

Introduction

Despite its high affinity to iron, lactoferrin (LF) is only saturated to 2–4% of its iron-binding capacity [1] and LF-bound iron constitutes 25–30% of the total iron content in human milk. It has been shown that when an iron isotope is added to a milk sample in vitro more than 80% is bound to lactoferrin. The reported high bioavailability of *extrinsically* labeled iron in human milk [3] may therefore be valid essentially for only the *lactoferrin-bound* iron. The bioavailability of the other iron complexes in human milk is thus still unknown.

The purpose of this study was to 1) analyze the possibility to label the various iron complexes in milk *intrinsically;* and 2) study the effect of iron status of the lactating animal (mother) on the recovery of an intravenously administered iron isotope.

Materials and methods

The recovery of an *intravenously* administered iron isotope (^{59}Fe) was studied in five adult, apparently healthy, lactating goats. The animals were studied in two experiments under anemic conditions during which they were bled and in one experiment after iron supplementation. At the end of each period of treatment (bleeding or supplementation), blood samples were collected for hematologic analysis, the mammary glands were emptied, and the background radioactivity in the milk was measured. Subsequently, a dose of 3.7 MBq ^{59}Fe-citrate was injected intravenously. Milk samples were collected during the next five days, at first, every hour until eight hours after injection, then after longer intervals. Some samples were fractionated by ultracentrifugation (140 000 g, 1 h and 4 °C). Radioactivity was measured in fat, casein, and whey fractions in 5 ml milk samples in a gamma counter (Nuclear Chicago 1186 with a well-type NaI detector).

Result and discussion

Isotope recovery in 1 ml of milk was found to be as low as 0.0001–0.0007% of the injected dose. As illustrated in Fig. 1 the process of iron excretion is slow, with the maximum concentration of ^{59}Fe reached 10 to 15 hours after intravenous injection, and then declining slowly to a constant minimum 40 to 70 hrs after injection.

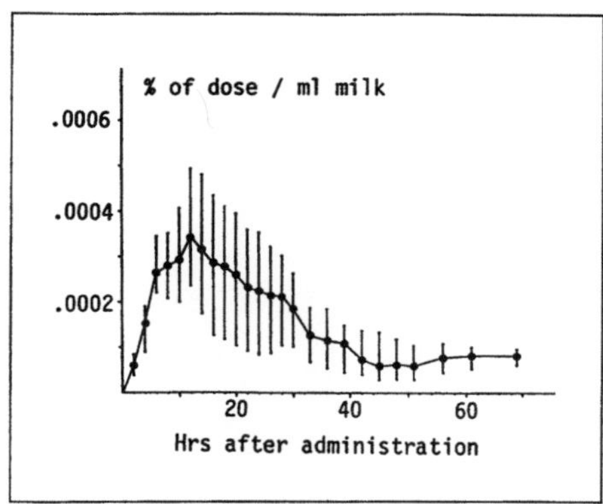

Fig. 1. Distribution of ^{59}Fe excreted in milk after intravenous injection of the isotope. The dotted curve represents the mean of the three experiments and vertical lines represent the range

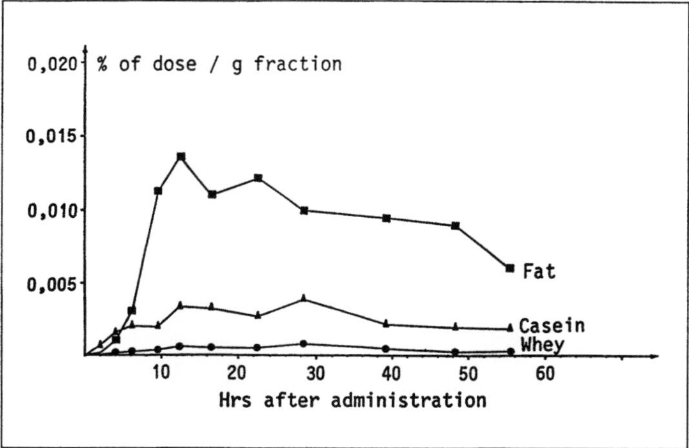

Fig. 2. Distribution of ^{59}Fe in the various iron compartments in milk after intravenous administration of the isotope

Figure 2 shows the distribution of the isotope in the fat, whey, and casein fractions, respectively. It is evident that all iron compartments in milk are labeled by the intrinsic labeling technique.

Figure 3 shows the correlation between iron status of the lactating animal and recovery of iron isotope in milk. In three goats the amount of blood loss and the duration of supplementation (number of days after the period of bleeding) is plotted against the cumulative percent recovery of injected dose in 1 ml milk portions during 100 h. It is evident that with increasing blood loss, recovery of isotope in milk is decreased.

Goat 1 shows the most rapid return to higher isotope recovery although she was not supplemented with iron after the bleeding period. As she was one of two goats that lost the most blood in the first two experiments the findings indicate that the iron excretion in milk is highly prioritized during anemia.

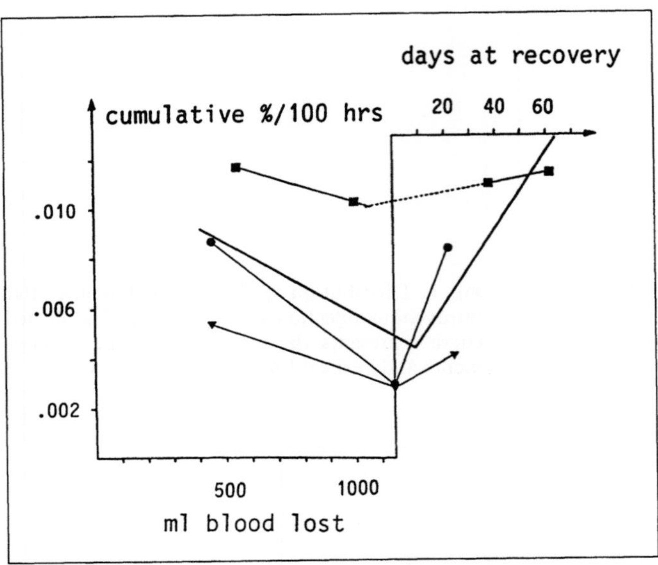

Fig. 3. Correlation between iron status and recovery of iron isotope in milk in three goats (1, 2, and 3). The amount of blood loss and the duration of supplementation (number of days after the period of bleeding) is plotted against the cumulative percent recovery of injected dose in 1 ml milk portions during 100 h; ● Goat 1, ■ Goat 2, ▼ Goat 3

References

1. Fransson G-B, Lönnerdal B (1980) Iron in human milk. J Pediatr 96:380–384
2. Lönnerdal B (1985) Biochemistry and physiological function of human milk proteins. Am J Clin Nutr 42:1299–1317
3. Saarinen UM, Siimes MA, Dallman PR (1977) Iron absorption in infants: high bioavailability of breast milk iron as indicated by the extrinsic tag method of iron absorption and by the concentration of serum ferritin. J Pediatr 91:36–39.

Does Bovine Lactoferrin Resist Absorption in the Small Intestine of Neonatal and Adult Pigs?

M. Schmitz[1], H. Hagemeister[1], Iris Görtler[2], J. G. Bindels[2], and C. A. Barth[1]

[1] Institut für Physiologie und Biochemie der Ernährung,
Bundesanstalt für Milchforschung, Kiel, FRG
[2] Milupa AG, Friedrichsdorf/Ts., FRG

Introduction

The iron binding protein lactoferrin (LF) is found in the milk of different species. Its concentration is very high in the first period of lactation.

Three functions of LF are discussed in the literature. It is proposed that LF enhances the absorption of its ligand and that it supports the host resistance of the newborn [3]. Furthermore, a stimulating effect on proliferation of enterocytes has been demonstrated [4].

The biological importance of the role as a bacteriostatic agent is difficult to assess without the demonstration of its survival in the intestinal tract.

In our investigation in pigs we wanted to answer the following questions:
1) How much LF resists degradation up to the ileum?
2) Is there an age-dependent change in degradation of LF?

Methods

The degradation of LF was examined in six adult Göttingen minipigs and in three piglets, 21-days old. The adult pigs were kept on a semi-synthetic diet meeting the maintenance requirement. They were fed 1 wt.% of bovine LF in a test meal of 180 g dry matter. Six h postprandially the animals were sacrificed and digesta were collected as completely as possible from different segments of the intestine (see abscissa of Fig. 1).

The suckling piglets consumed 75 mg LF in 50 ml fresh sow's milk and intestinal samples were obtained 3 h postprandially. LF was determined by Rocket Immunoelectrophoresis.

Results

Less than 0.25% of the LF consumed were detected in total digesta of adult pigs (data not shown). On the other hand in piglets there was a considerable amount of LF in the distal part of the small intestine (Fig. 1).

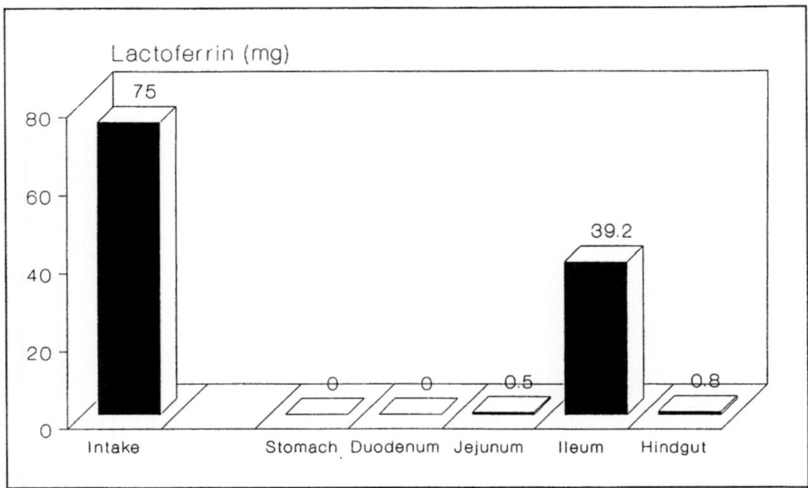

Fig. 1. Amount of bovine lactoferrin in different segments of the gastrointestinal tract of piglets, 3 h postprandially. The averages of three piglets are given. In the ileum the values ranged from 32–51 mg

Conclusion

Supplementation of the diet with bovine LF caused up to 20% undegraded LF in the feces of the human newborn [1, 2, 5, 6]. In piglets we found a higher amount amounting to 52% in the prececal segment of the intestine. The discrepancy of results between the human newborn and piglets may be related to the species and the fact that heterologous lactoferrin was used in our experiments. Furthermore, it can be expected that a high percentage of LF is degraded in the hindgut by microbial activity which was excluded in our investigations.

The results are in accordance with the concept of the bacteriostatic activity of LF in the intestinal tract.

References

1. Davidson LA, Lönnerdal B (1987) Persistence of human milk proteins in the breast-fed infant. Acta Paediatr Scand 76:733–740
2. Görtler J, Bindels J, Sawatzki G (1988) In-vitro- und in-vivo-Verdauung von Lactoferrin. Ernährungsumschau 35:159 (Abstr.)
3. Hambraeus L, Fransson GB, Lönnerdal B (1984) Nutritional availability of breast milk protein. Lancet II: 21:167–168
4. Nichols BL, McKee KS, Henry JF, Putman M (1987) Human lactoferrin stimulates thymidine incorporation into DNA of rat crypt cells. Pediatr Res 21:563–567
5. Prentice A, Ewing G, Roberts SB, Lucas A, MacCarthy A, Jarjou LMA, Whitehead RG (1987) The nutritional role of breast-milk IgA and lactoferrin. Acta Paediatr Scand 76:592–598
6. Spik G, Brunet B, Mazurier-Dehaine C, Fontaine G, Montreuil J (1982) Characterization and properties of the human and bovine lactotransferrins extracted from the faeces of newborn infants. Acta Paediatr Scand 71:979–985

Can Lactoferrin Supplementation Improve the Availability of Iron from Milk?

G. Schulz-Lell, H.-D. Oldigs, K. Dörner, and J. Schaub

Department of Pediatrics, Christian-Albrechts University, Kiel, FRG

Introduction

The availability of iron from human milk is better than from cow milk or cow milk formulas. Lactoferrin, an important iron-binding protein in human milk has been considered responsible for the high bioavailability of breast-milk iron. To evaluate the possible role of lactoferrin on the availability of iron, we measured iron uptake from human milk and two cow-milk formulas; one of the latter was supplemented with bovine lactoferrin (100 mg/100 ml).

Methods

Iron balance studies were performed in 26 healthy term infants from their 3rd until their 17th week of life. The balance studies were done at home and comprised up to five periods of three subsequent 24 h collections with intervals of three to four weeks. Altogether 97 balance periods could be analyzed in the three study groups. Ten infants were breast-fed, seven received an adapted infant formula supplemented with 100 mg bovine lactoferrin per 100 ml milk, and nine received the same adapted formula without lactoferrin. Median iron content in human milk was 0.44 mg/l. Iron concentration in the supplemented formula was 1.06 mg/l, and 45% was bound to lactoferrin. The non-supplemented formula contained 0.77 mg/l.

The iron concentration in the milk and stool samples was measured after dry ashing by automated flameless atomic absorption spectrophotometry.

Results

The mean iron intake of the breast-fed infants was 0.2 mg/kg b.w. × 3 d from their 3rd to 17th week of life. They retained 0.09 mg/kg b.w. × 3 d. The lactoferrin supplemented group received 0.5 mg iron/kg b.w. × 3 d and retained 0.2 mg/kg b.w. × 3 d. The mean iron intake of the infants fed the adapted formula without supplementation of lactoferrin was 0.36 mg/kg b.w. × 3 d. The retention of iron was 0.13 mg/kg b.w. × 3 d.

In Table 1 mean iron intake, excretion, and retention in the breastfed and formula-fed infants is given.

Table 1. Iron balances in the breast-fed (HM) and formula-fed (LF$^+$ = supplemented, and LF$^-$ = non-supplemented formula) infants is given in mg/kg b.w. × 3 d

Type of feeding	Balance period				
	1	2	3	4	5
HM	n = 6 [a]	n = 9	n = 8	n = 9	n = 7
Intake	0.29	0.24	0.18	0.15	0.15
	(0.16–0.55)	(0.1 –0.33)	(0.12–0.22)	(0.09–0.24)	(0.09–0.28)
Excretion	0.16	0.17	0.08	0.04	0.08
	(0.04–0.29)	(0.1 –0.36)	(0.01–0.32)	(0 –0.1)	(0.01–0.15)
Retention	0.13	0.07	0.1	0.11	0.07
	(−0.05–0.44)	(−0.08–0.17)	(−0.1 –0.21)	(0.03–0.21)	(0.01–0.18)
LF$^+$	n = 4	n = 6	n = 6	n = 7	n = 7
Intake	0.55	0.53	0.53	0.43	0.5
	(0.44–0.63)	(0.43–0.61)	(0.38–0.84)	(0.28–0.65)	(0.35–0.64)
Excretion	0.42	0.35	0.28	0.29	0.25
	(0.39–0.46)	(0.13–0.52)	(0.19–0.48)	(0.19–0.47)	(0.08–0.42)
Retention	0.13	0.18	0.25	0.14	0.25
	(−0.02–0.21)	(−0.02–0.39)	(−0.02–0.55)	(0.01–0.21)	(0 –0.53)
LF$^-$	n = 7	n = 8	n = 5	n = 4	n = 4
Intake	0.39	0.37	0.37	0.3	0.35
	(0.24–0.67)	(0.21–0.63)	(0.26–0.47)	(0.18–0.46)	(0.24–0.47)
Excretion	0.27	0.24	0.25	0.18	0.19
	(0.16–0.52)	(0.14–0.32)	(0.12–0.31)	(0.12–0.24)	(0.15–0.26)
Retention	0.12	0.13	0.12	0.12	0.16
	(−0.07–0.51)	(−0.01–0.48)	(0.02–0.34)	(−0.01–0.34)	(0.03–0.26)

[a] n = number of balances in each period
[b] numbers in parentheses are ranges

Discussion

No significant differences in the ratio iron intake/iron retention were observed among the three groups. In this study the supplementation of the adapted infant formula with lactoferrin did not improve iron absorption.

Two reasons for these findings may be discussed:

1) Lactoferrin *regulates* iron absorption as postulated by Fransson et al. [2]: in iron deficient mice iron absorption from a lactoferrin supplemented meal was significantly higher than in iron saturated mice, whereas iron absorption from an iron salt was always high, regardless whether iron status was sufficient or not.

Our studies were carried out in the first four months of life when most infants still have sufficient body iron, thus additional iron absorption from lactoferrin seems unnecessary;

2) Lactoferrin promotes iron absorption *species specifically* as was seen by Cox et al. [1]: iron uptake in human enterocytes from human lactoferrin was several times higher than from bovine lactoferrin.

Acknowledgements:

We wish to thank all the participating mothers without whose efforts this study would not have been possible. This study was supported by Milupa AG, Friedrichsdorf, FRG, and their aid is gratefully acknowledged.

References

1. Cox TM, Mazurier J, Spik G, Montreuil J, Peters TJ (1979) Iron binding proteins and influx of iron across the duodenal brush border. Evidence for specific lactotransferrin receptors in the human intestine. Biochim Biophys Acta 558:129–141
2. Fransson G-B, Keen CL, Lönnerdal B (1983) Supplementation of milk with iron bound to lactoferrin using weanling mice: I. Effects on hematology and tissue iron. J Pediatr Gastroenterol Nutr 2:693–700

The Nutritive Value of Bovine Lactoferrin

I. Schoppe, C. A. Barth, and H. Hagemeister

Institut für Physiologie und Biochemie der Ernährung,
Bundesanstalt für Milchforschung, Kiel, FRG

Introduction

A low degradation and absorption of the iron-binding milk protein lactoferrin has
been associated with its proposed nutritional role of enhancing absorption of iron
and supporting host resistance of the newborn [2]. Such a low degradation and
absorption can be expected to affect parameters of the nutritive value such as
apparent net protein utilization (NPU) and apparent digestibility (AD). Therefore
NPU and AD of bovine lactoferrin were studied in the rat.

Materials and methods

Male Wistar rats (40 g) were obtained from Winkelmann, Borchen, FRG. Before the
metabolic experiment the animals were kept on a standard diet for seven days up to
a body weight of 65–70 g. The balance experiment, including the design of the
metabolic cages, followed the procedure of Eggum [1]. During the five days balance
period each animal was supplied with 750 mg nitrogen from bovine lactoferrin (test
protein) or casein (reference protein) in a nitrogen-free mixture. During the balance
period urine and feces were collected separately and nitrogen was determined using
the Kjeldahl method. Plasma iron was determined at the end of the balance period
by standard procedure.

The diets were based on:
- lactoferrin as test protein (experiment 1);
- lactoferrin plus 0.5% anise-oil as test protein or casein plus 0.5% anise-oil as
  reference protein (experiment 2);
- lactoferrin plus 0.34% isoleucine as test protein or casein plus 0.21% methionine
  as reference protein (experiment 3);
- lactoferrin plus parenteral iron; 12.5 mg Fe^{3+} were applied i.p. in total over the
  four days adaptation and five days balance period (experiment 4).

Results and discussion

The results from the different experiments are given in Table 1. Experiment 1 shows
that lactoferrin is highly digestible and that 37% of the lactoferrin nitrogen consumed was retained. Because the animals refused to completely consume the diet we
flavored lactoferrin and casein diets in experiment 2, but despite flavoring, the food

Table 1. Nitrogen intake, apparent NPU, AD, and plasma-iron levels following different lactoferrin- and casein-diets. LF = Lactoferrin. Plasma iron concentrations of experiments 3 were significantly different at the p < 0.01 level (Student's t-test). $\bar{X} \pm$ SEM

Experiment	1	2	2	3	3	4	4
Diet	LF	LF + anise-oil	Casein + anise-oil	LF + iso-leucine	Casein + methionine	LF[a]	LF[b]
N intake (mg) per rat and balance period	439 ±17	464 ±6	509 ±76	450 ±17	715 ±41	506 ±23	469 ±25
AD	0.93±0.1	0.90±0.1	0.89±0.1	0.95 ±0.4	0.93 ±0.4	0.94 ±0.4	0.93 ±0.1
Apparent NPU	0.37±0.1	0.37±0.2	0.42±0.2	0.42 ±0.2	0.80 ±0.1	0.43 ±0.2	0.44 ±0.2
Plasma iron (µg/dl)				108.4±6.5	167.9±18.2	117.1±9.6	115.2±10.2

[a] Parenteral iron treatment
[b] Parenteral solvent treatment

intake remained low in both dietary groups. We wondered whether the food intake could be raised by supplementing lactoferrin and casein with their first limiting amino acids and therefore experiment 3 was performed. While food consumption and apparent NPU in casein-fed animals was raised significantly, this was not the case in lactoferrin-fed animals.

We conclude that the bovine lactoferrin tested was digestible and utilized as a food protein. Perhaps an even better nitrogen retention would result if a complete food intake of lactoferrin diets could be achieved. This could be done either by feeding lower doses of lactoferrin or possibly by reducing the high tryptophan content of lactoferrin. The AD was better than 0.97 and the NPU of 0.37 was not much lower than the value obtained with casein (0.42 in experiment 2). Insofar, bovine lactoferrin does not seem to contain major parts which are not digested by the rat.

Because of the surprisingly low plasma iron levels in lactoferrin-fed rats, we tried to raise the food acceptance by supplying iron parenterally (experiment 4). However, neither food intake nor plasma iron, nor apparent NPU were raised considerably. We conclude that the low food intake was not due to the as yet unexplained low plasma iron levels in lactoferrin-fed animals.

References

1. Eggum BO (1973) A study of certain factors influencing protein utilization in rats and pigs. 406. Beretning fra forsogslaboratoriet, Kopenhagen
2. Hambraeus L, Fransson G-B, Lönnerdal B (1984) Nutritional availability of breast milk protein. Lancet 21:167−168

General Discussion: Milk Proteins and Ligands

L. Hambraeus

Dept. of Nutrition, University of Uppsala, Sweden

The section on milk proteins and ligands comprises three outstanding papers in which the physiological role of lactoferrin in relation to its iron saturation and fragmentation by digestive proteolysis could be said to be the common focus of interest.

The role of lactoferrin is presented from various aspects, which can be summarized under the following headings:

1) the *antimicrobial* effect;
2) the *nutritional* effect with respect to *macronutrients,* e.g., the role of lactoferrin as a dietary protein, and to *micronutrients,* e.g., the role of lactoferrin in iron transport;
3) the *mitogenic* or trophic activity on the intestinal mucosa.

It is obvious that the antimicrobial effect is still essentially based on in vitro studies. In her paper, however, Dr. Spik states that an in vivo effect has been observed. When 1 g of lactoferrin was administered for seven days, a decrease in the total bacterial amounts was found in the gastrointestinal tract and some streptococci and staphylococci strains disappeared.

It seems to be most accurate to stimulate further studies on the effect of lactoferrin using in vivo studies. It is of special interest to study the effect of various iron saturation of lactoferrin on its antimicrobial effect as well as when the lactoferrin molecule loses its physiological effect during digestive proteolysis. This is of special concern since Dr. Goldblum has shown the occurrence of lactoferrin fragments both in stools and urine in infants fed enriched human milk.

The question of the need of homologue lactoferrin for a bacteriostatic effect in man also needs further studies. As discussed during the conference, bovine lactoferrin seems to have little antimicrobial effect in man.

A special question remains as to whether we should perform studies in other populations which have a higher incidence of infections, i.e., developing countries. As suggested by Dr. Reiter, this might make it easier to get evidence for a suggested antimicrobial effect of lactoferrin during infancy and childhood in vivo.

The nutritional role of lactoferrin seems to be more challenging. The problem can be divided into two sections:

1) a *macronutrient* aspect regarding lactoferrin as a dietary protein source;
2) a *micronutrient* aspect regarding the role of lactoferrin for iron absorption and transport.

Dr. Spik originally showed the occurrence of lactoferrin in the stools of breast-fed infants. Her findings have been verified in the USA by Davidson and Lönnerdal, in the UK by Prentice and collaborators, as well as in Sweden by ourselves. Thus,

110

4–6% of consumed lactoferrin seems to remain in the stools in healthy newborns during the first weeks of life, but under certain circumstances, up to 20–30% may remain, as indicated by our own observations. Of particular interest are the observations by Dr. Goldblum and his collaborators of lactoferrin fragments in stools and urine that occur after ingestion of fortified human milk.

These findings may indicate that we cannot calculate the amount of lactoferrin (which constitutes 10–15% of the total protein content in human milk) that is nutritionally available without further studies. This calls for the use of isotopes in order to evaluate the origin of lactoferrin and its fragment found in stools and urine, but we also need studies to analyze the lactoferrin exchange over the intestinal mucosa.

The role of lactoferrin in the transport of iron additionally requires further studies. Notable in this respect is the discussion regarding lactoferrin receptors in the intestinal brush border mucosa cells, offered by Drs. Lönnerdal and Spik. Not only is the model presented by Dr. Lönnerdal of great interest for further studies, but also its localization in the brush border and the specificity of the lactoferrin receptor and its implication for the possible role of bovine lactoferrin as an iron vehicle deserves our attention.

The pH-dependency and the interaction between various minerals, e.g., trace elements zinc and manganese, calcium, and iron absorption must also be studied further.

An additional and partly new aspect of the role of lactoferrin is the *mitogenic* or *trophic* activity. This has especially been studied by Nichols and collaborators. It represents a challenging aspect on the role of lactoferrin for the differentiation of human intestinal mucosa cells. It also includes the hypothesis that the consumption of human milk may stimulate the local production of lactoferrin in various mucosa cells, e.g., the urogenital tract as proposed by Dr. Goldblum.

In conclusion it could be said that the physiological role of lactoferrin, which may constitute a good example of the milk protein and ligand complex, gets more and more intriguing. Obviously, as more knowledge is accumulated, the primary role of lactoferrin as an antimicrobial unit has been supplemented and perhaps surpassed by its role in iron transport and as a mitogenic agent. Of special concern are questions regarding the digestion and fragmentation of the lactoferrin molecule in the gastrointestinal tract; it remains for us to find out when it loses its physiological role, and we must further study the lactoferrin receptor in the intestinal mucosa. The antimicrobial effect which originally was considered as *the* role of lactoferrin must be better verified in vivo before we understand the mechanism and the possible use of heterologues of lactoferrin molecule.

Structure and Variability of Milk Proteins

B. Ribadeau-Dumas

Dairy Research Laboratory, Institut National de la Recherche Agronomique,
Jouy-en-Josas, France

Introduction

A protein molecule is best characterized when its tertiary structure has been elucidated by high resolution x-ray crystallography. As far as the main milk proteins are
concerned, this was achieved in 1986 for two of them, baboon α-lactalbumin [1] and
bovine β-lactoglobulin [2]. No one casein has ever been crystallized and, owing to
their self-aggregation properties, it may be that these proteins will never be obtained
in crystalline form. However, as for the other milk proteins, a great deal of information can be deduced from their primary structures. Indeed, for any biologically active
protein, genetic variability does not usually affect the tertiary structure to a large
extent even if the activity is strongly modified. A good example of this statement is
the phylogenic relationship between α-lactalbumin and lysozyme that we shall discuss later on.

In what follows the known structures of the main cow milk proteins will be
discussed, as well as that of a rodent milk protein, whey acidic protein (WAP), which
has not been detected in dairy species. The structure and features of an ubiquitous
milk protein, lactoferrin, will be discussed elsewhere.

Another part of this review will be devoted to intra- and inter-species variability
of these proteins due either to genetic considerations or to post-translational modifications, neglecting variations (mainly quantitative) resulting from lactation stage,
age, feeding, mammary status, and milk storage.

Structure of milk proteins

The primary structures of the main cow milk proteins were elucidated using conventional methodology between 1972 and 1977. From this period a number of milk
proteins from many species were characterized using either the same techniques or,
more recently, recombinant DNA technology. The list of the milk proteins of known
primary structure is given in Table 1. It must be remembered that a protein sequence
deduced from that of the corresponding cDNA does not give any information on
post-translational modifications such as phosphorylation, glycosylation, proteolysis, etc.

Data given in this chapter will concern the overall features of the main milk
proteins, cow milk proteins being taken as examples when information is available.

Whey proteins: these proteins have different cellular origin. α-Lactalbumin,
β-lactoglobulin, xanthine oxidase, possibly lactoferrin, lactoperoxidase, and lyso-

112

Table 1. Main milk proteins of known primary structures

α-Lactalbumin
Cow (p, c, g), sheep (p, c), goat (p), horse (p), camel (p), rabbit (p), guinea-pig (p, c), rat (p, c, g), red-necked wallaby (p), man (p, c, g)

β-Lactoglobulin
Cow (p, c), water buffalo (p), sheep (p, c, g), goat (p), mouflon (p), horse I and II (p), donkey I (p), pig (p), Eastern grey kangaroo (p)

α_{s1}-Casein
Cow (p, c), sheep (c), goat (p), guinea-pig (c), rat (c)

α_{s2}-Casein
Cow (p, c), sheep (c), guinea-pig (c), rat (c), mouse (c)

β-Casein
Cow (p, c), sheep (p), rat (c, g), mouse (c), human (p)

$\varkappa$-Casein
Cow (p, c), sheep (p), goat (p), rat (c), mouse (c), human (p)

Other milk proteins
Human lactoferrin (p), rat and mouse WAPs (c), cow folate-binding protein (p), cow colostrum serine proteinase inhibitor (p), cow colostrum low-Mr cysteine proteinase inhibitor (p)

---

p, protein; c, cDNA; g, gene

zyme, are secreted by the mammary epithelium. IgGs, IgMs, serum albumin, plasminogen and a number of minor proteins originate from blood. IgAs are synthesized by lymphocytes located near the mammary epithelium [3].

Whey proteins have tight, globular, tertiary structures which make them fairly resistant to proteinases. Each one has a specific biological activity, either in the mother or in the newborn, or both. Most of them are easily denatured, in particular by heat treatments which make them insoluble.

α-Lactalbumin: this protein was crystallized from cow whey just before World War II and was continuously studied from that time. However, the discovery of its biological role in 1966 [4] led to a burst of investigations. α-Lactalbumin is one of the two subunits of lactose synthetase, the enzyme involved in the last step of lactose biosynthesis. The primary structure of the bovine molecule was elucidated in 1970 [5]. It is a 123-residue single peptide chain with four disulphide bridges. Some corrections were brought in 1984 to the initially published sequence [6]. The amino acid sequence was also deduced in 1987 from that of the corresponding cDNA [7]. This brought other corrections to the previous results. Finally the whole nucleotide sequence of bovine α-lactalbumin gene was published in 1987 [8]. It confirms the amino acid sequence established in [7]. α-Lactalbumin displays approximately 40% homology with either mammalian or avian c-type lysozymes [6], enzymes which have a quite different catalytic activity. This homology was also observed at the structural gene level [9]. Only in 1980 was α-lactalbumin found to contain a calcium ion [10].

As indicated earlier the tertiary structure of baboon α-lactalbumin was recently resolved by x-ray crystallography at 1.7 Å resolution [1]. The sequence of this protein was not known, but preliminary experiments suggested very few differences between the sequences of human and baboon α-lactalbumins, the former being known. It appeared that α-lactalbumin possesses a Ca^{2+} binding loop that resembles the "EF-hand" present in calcium-modulated proteins. A clear homology exists with the corresponding loop in c-type lysozymes which do not bind calcium. The calcium ion in α-lactalbumin is bound to the carbonyl groups of Lys 79 and Asp 84, to carboxylate of Asp 82, 87, and 88, and to two water molecules [1]. The biological role of metal ion binding is unclear. It seems to stabilize the molecular conformation. Below pH 4 α-lactalbumin loses its calcium and undergoes a conformational change [11]. It was recently shown in our laboratory (Ivon M., Miranda G., to be published) that, both in vivo and in vitro, α-lactalbumin is digested by pepsin only below pH 3.5.

Bovine β-lactoglobulin: this protein was obtained in crystalline form at the same time as α-lactalbumin and an important amount of data concerning its properties and structure have been accumulated over the years. Its primary structure was determined in 1972 [12]. It contains 162 amino acid residues, two disulphide bridges and a free SH group at position 121 [2]. An identical sequence lacking the first 11 residues, was recently deduced from that of the corresponding cDNA [13]. Bovine β-lactoglobulin exists as a dimer in milk.

An indication concerning its biological role was obtained only in 1985 when two groups of researchers observed a limited sequence homology to plasma retinol-binding protein (RBP) [14, 15]. The recent elucidation of β-lactoglobulin tertiary structure at a resolution of 2.8 Å strongly confirmed this suggestion [2]. RBP is the protein that transports retinol in plasma from its storage in the liver. Both proteins have quite similar tertiary structures. The β-lactoglobulin molecule consists of an anti-parallel β-sheet, formed by nine strands wrapped around to form a flattened cone. The core of the molecule is an eight-stranded, anti-parallel β-barrel. Strand I is involved in the formation of the dimer by causing anti-parallel interactions with its counterpart. The interface between monomers also involves hydrophobic interactions and stacking of imidazole rings. The free sulphydryl, at position 121, is buried [2]. For many years it has been known that β-lactoglobulin strongly binds two retinol molecules, one on each protomer, with a binding constant of $2 \cdot 10^{-8}$ M. This binding is not pH dependent in the range 2–7.5 [16]. It was recently found that β-lactoglobulin belongs to a family of proteins able to transport small hydrophobic molecules; this family includes human plasma apolipoprotein D, human plasma retinol-binding protein, human urinary protein HC, human α1 microglobulin, ungulate β-microglobulin, rodent α2u-globulin, tobacco hornworm insecticyanin, and frog BG protein [14, 17, 18]. ^{125}I-labelled β-lactoglobulin-retinol complex binds specifically to purified microvilli prepared from one-week old calf intestine. This specific binding is not present in six-month old animals [2]. It was shown more than 30 years ago that a proteinuria develops during the first 30–40 h of life in calves fed on colostrum and that the main protein present in urine was β-lactoglobulin. The protein was crystallized from urine [19, 20]. These results strongly suggest a role of β-lactoglobulin in vitamin A uptake and transport in the calf.

114

Other whey proteins: lactoferrin, an important human milk protein, will be considered elsewhere. Although its structure is known, milk serum albumin will not be considered here because of its minor significance. The primary structures of lactoperoxidase and xanthine oxidase are not known. Milk plasminogen was unambiguously identified with its blood counterpart. Although it is of economic significance, only its action on caseins will be considered in the next section.

Cow colostrum often contains more than 10 g immunoglobulins per 100 ml, in which IgGs are predominant and belong to two different classes, IgG1 and IgG2, with slightly different heavy chains. These IgGs, together with IgMs (pentamers of IgG-like molecules joined by disulphide bridges and a secretory component), are of blood origin, while IgAs are synthesized by lymphocytes in the mammary gland as dimers of two IgG-like molecules linked by a joining piece, a 15 kD glycoprotein. A secretory component (SC,71 kD glycoprotein), synthesized by secretory epithelial cells, attaches to them for secretion of sIgAs in milk. Some free SC occurs in milk [21]. IgGs are essential for the newborn calf which does not acquire passive immunity pre-partum via placental transport as it occurs in some other species; they are able to be transferred from the intestinal lumen to the blood. sIgAs are found in all exocrine secretions and play an important role by providing local immunity. We shall not give any detail on the structure of immunoglobulins which is now well known. A human IgA was sequenced in 1979 [22].

A rat whey glycoprotein, which was present at three levels of phosphorylation (0, 1, 2 P/mole), was characterized in 1978 [23]. A similar protein, which did not appear to be phosphorylated, was similarly characterized in mouse milk in 1981 and called WAP (whey acid protein) [24]. The latter was shown to be synthesized in the mammary gland. The amino acid sequences of both proteins were deduced from that of their corresponding cDNAs [25, 26]. WAP is the major protein in mouse whey. Among other species it has been only characterized in the rabbit (E. Devinoy, personal communication). Mouse and rat WAPs (134 and 137 amino acids, respectively) are homologous, acidic, cysteine-rich proteins with 66% homology. Most of the cysteines are located in two clusters containing six cysteines each, arranged in an identical pattern. The occurrence of sequences 19–24 Ser-Ser-Ser-Glu-Asp and Ser-Pro-Ile-Glu-Gly in the rat and mouse WAPS, respectively, may explain why only the former is phosphorylated (see below).

The cysteine pattern of WAPs resembles that of a family of proteins which contain a "four disulfide core" (snake venom neurotoxins, wheat germ agglutinin, ragweed pollen allergen Ra5, and neurophysin). However, although neurophysin and WAPs have an almost identical cysteine pattern, there is no homology between them at the amino acid level [26]. On the other hand, the WAPs seem to have a close evolutionary relationship with red sea turtle protease inhibitor. However no protease inhibitor activity was found for rat WAPs [25].

Caseins

The four caseins, α_{s1}, α_{s2}, β, and $\varkappa$, are all synthesized by the mammary secretory epithelium. They do not have any known biological activity. However their structure is especially loose, owing likely in part to their high proline content. This makes them highly susceptible to digestion by proteinases, namely those of the newborn's digestive tract. In milk they occur as micelles formed of the four proteins and inorganic calcium phosphate, whose occurrence is due to the presence of phosphate groups on each of them, especially on α_{s1}-, α_{s2}-, and β-caseins. These groups seem to be included in a semi-crystalline array of brushite, a form of calcium phosphate. The caseins are soluble at neutral pH, but give polydisperse self- or hetero-aggregates of high molecular weights. In the presence of a low level of calcium only $\varkappa$-casein is soluble. This protein is responsible for the stability of casein micelles in milk. Most $\varkappa$-casein appears to be located at or near the micelle surface. In the stomach of young ruminants $\varkappa$-casein is rapidly split into two parts, caseinomacropeptide (CMP) and para-$\varkappa$-casein, by chymosin and pepsin. CMP is soluble while the whole micellar system, which now includes para-$\varkappa$-casein, coagulates. This leads to a slow release of peptides by the stomach proteases [27, 28].

The study of a number of different caseins from different species or corresponding to different genetic variants (see next section) has shown that phosphorylation of caseins in the mammary gland only occurs on serine or threonine residues which are included in the sequence Ser/Thr-X-A, X being any amino acid and A an acidic residue (Glu, SerP, Asp). The occurrence of Thr in first position, or Asp in the third position, leads to a less efficient phosphorylation. The presence of the code sequence shown above appears to be an absolute requirement for phosphorylation of caseins. However this is not sufficient since some such sequences are not at all, or incompletely phosphorylated [29].

α_{s1}-*Casein:* the primary structure of the bovine protein was elucidated in 1971 [30]. α_{s1}-Casein has a single peptide chain of 199 residues which does not contain any cysteine or cystine. Its major form in cow milk bears eight phosphate groups attached to serine residues.

Three hydrophobic regions are discernible in the molecule, roughly including residues 1–44, 90–113 and 132–199. Sequence 41–77, containing 7 SerP, 8 Glu, 3 Asp, carries a high net negative charge at the pH of milk, while the remainder of the molecule has essentially no net charge at this pH.

In 1984 two groups independently determined the sequence of α_{s1}-casein cDNA [31, 32]. The corresponding protein sequences were identical to that determined earlier, with one exception: residue 30, formerly determined as Gln, is Glu. Pre-α_{s1}-casein was recently cloned and expressed within *E. coli* cells [33].

α_{s2}-*Casein:* this was the last bovine casein characterized and sequenced. It occurs in milk in several forms, formerly called α_{s2}, α_{s3}, α_{s4}, and α_{s6} (α_{s5} being a dimer α_{s3}–α_{s4}), differing in their levels of phosphorylation. Its primary structure was determined in 1977 [34]. The peptide chain contains 207 residues, including two close cysteines at positions 36 and 40. The number of proline residues is lower and that of lysine higher than those found in the other caseins. Thirteen phosphate groups were localized, 12 on serines, one on threonine [35]. They are grouped in three areas of the molecule (7–31, 55–66, 129–143). A clear sequence homology was observed between segments 50–123 and 132–207. The cDNA corresponding to bovine α_{s2}-casein was sequenced in 1987 [36]. The corresponding protein sequence showed one difference with that published earlier. Gln was found at position 87, instead of Glu; there was likely an error in the former sequence. This cDNA study extended the internal homology observed earlier to give a tandem repeat of codons 33–125 and 126–205.

β-Casein: this casein has a single chain of 209 residues without any cysteine and cystine. It is especially rich in proline. Five serines located in the first 35 residues are phosphorylated. *β*-Casein is the most hydrophobic of all the caseins. At the pH of milk the N-terminal 21-residue segment is highly negatively charged, while the rest of the molecule, which is highly hydrophobic, has no net charge. This is why *β*-casein forms micellar aggregates in solution. *β*-Casein was sequenced in 1972 [37]. The protein was fully re-sequenced in 1988 [38]. Four errors were found in the first report, involving three times Glu↔Gln. The fourth error was an inversion of two residues.

Three separate sequence determinations of *β*-casein, deduced from those of cDNAS, were made available in 1987 [36, 39, 40]. The first two fully agree with the corrected sequence obtained from the protein in 1988. Although the third corrects the original sequence like the other three determinations at positions 117 and 175, it gives the same results in the other two places mentioned above. Furthermore, Leu was found instead of Met at position 93. Taking also into consideration a number of differences at the nucleotide level between the sequence shown in [40] and those obtained in [36] and [39], we conclude that the sequence reported in [40] is not accurate.

x-Casein: being the target of rennet, this protein has been most intensively studied. It is, in milk, glycosylated at various levels. The primary structure of the carbohydrate-free fraction was determined in 1973 [41]. Its chymosin sensitive bond had been identified earlier [42] from the characterization of a tryptic peptide overlapping para-*x*-casein and CMP. Carbohydrates are bound to CMP by O-glycosidic linkages with serine or threonine residues. It appears that Thr 131, 133, 135 (or 136), 142 and Ser 141 are potential glycosylation sites [43]. The various carbohydrate chains, the structures of which have been determined by several groups, are formed of three to six units (N-acetylgalactosamine, N-acetylglucosamine, galactose, N-acetyl neuraminic acid). They have in common the disaccharide Gal *β*-1-3-Gal Nac, directly linked to the peptide chain [44]. *x*-Casein has two cysteine residues in the para-*x* region (1→105). This region is highly hydrophobic and insoluble. One (or two) phosphate groups are linked to serines in the CMP (106→169). CMP is highly polar, negatively charged, and soluble. Its amphipathic character leads *x*-casein to form micelles in solution. The structure surrounding the chymosin-sensitive bond appears to be essential for an efficient cleavage by pepsin and, above all, chymosin (e.g., see [45]).

The *x*-casein sequence was confirmed in 1984 by the study of the corresponding cDNA [32]. Asn was found at position 81 instead of Asp. A similar result was obtained by another team, however they found His instead of Tyr at position 35. These last authors cloned and expressed in *E. coli* a DNA fragment corresponding to the whole cDNA coding region with the addition of Met-Ala at the N-terminal end [46].

Variability of milk proteins

We think that the most interesting variability of milk proteins is the one regarding genetic events. However, colostrum composition is extremely different from that of mature milk as far as the concentrations and proportions of its various protein components are concerned.

From what we know about milk proteins, intra- or inter-species variability from genetic origin can affect their primary structure, their concentration or proportion in milk, and the extent of their post-translational modifications. The intra-species variability of milk proteins has only been studied in the cow, and to a lesser extent, in the goat.

Cow milk: an extensive review dealing with the genetic polymorphism of bovine milk proteins and its relationships with milk quantity, composition, and "cheesability" was recently published [47]. Most information given below is derived from it.

Genetic variants have been found, mostly by electrophoresis, for all main milk proteins (the four caseins, α-lactalbumin, β-lactoglobulin). Only two were detected for α-lactalbumin and α_{s2}-casein. Most of them derive from point mutations leading to the substitution of a single residue. Only two display a deletion. The study of polymorphism in cattle was extended to two other species of the *Bos* genus, Zebu, and Yak.

It has been known for years that α_{s1}-, β-, and $\varkappa$-caseins are genetically closely linked and that almost undissociable combinations of alleles (haplotypes) at the three corresponding loci are transmitted from parents to offspring. The allelic frequencies at the six main lactoprotein loci in 11 French breeds and the haplotype frequencies for $\alpha_{s}1$-, β- and $\varkappa$-caseins in five French breeds are shown in [47].

A large survey of literature has found, among contradictory data, some quantitative and qualitative relationships associated with some genetic variants [47]. β-Lactoglobulin polymorphism has a major effect on its own concentration (β-Lg$^{A/A}$ > β-Lg$^{B/B}$) which indirectly leads to the following relation between β-lactoglobulin phenotypes on the one hand, and whole casein content and casein index on the other hand: β-Lg$^{B/B}$ > β-Lg$^{A/B}$ > β-Lg$^{A/A}$. The casein index is higher by 2.5–3%, on the average, in milk from β-Lg$^{B/B}$ cows than in that from β-Lg$^{A/A}$ cows [47].

From a qualitative point of view a highly significant effect of $\varkappa$-casein polymorphism was found on natural and maximum milk heat stability ($\varkappa$-CN$^{B/B}$ > $\varkappa$-CN$^{A/B}$ > $\varkappa$-CN$^{A/A}$) [48]. Furthermore, $\varkappa$-CN B and β-CN B give milk better "cheesability" [47].

As far as post-translational modifications are concerned in bovine milk protein variability, let us note that $\varkappa$-casein from colostrum is richer in carbohydrates than its counterpart in mature milk. Furthermore the carbohydrate moieties of the former are somewhat different [43]. Proteolysis of $\alpha_{s}2$- and $\varkappa$-caseins by plasmin can be considered as a post translational modification. We have observed long ago that individual fresh milks display on electrophoresis quite different patterns of $\varkappa$-caseins at the quantitative point of view. Whether this is due to genetic traits or other individual features is not known.

Goat milk: a study of the genetic polymorphism of goat α_{s1}- and α_{s2}-caseins was published in 1984 [49]. While α_{s2}-CN polymorphism was fairly simple, that of α_{s1}-CN was uncommonly complex. By electrophoresis α_{s1}-CN showed both quantitative variations and differences in electrophoretic mobility. The same authors showed in 1987 that this polymorphism was under control of a minimum of six alleles of locus α_{s1}-CN, termed A, B, C, B-, F, and 0. Alleles A, B, and C were found to be associated with a "high" content of α_{s1}-casein in milk (approximate contribution of each allele being 3.6 g/l), allele F with a low content (0.6 g/l) and B- with an intermediate content (1.6 g/l), 0 (zero) being probably a true null allele (no production of α_{s1}-casein) [50].

The A variant was recently sequenced. It was also shown that A and C differ from B each by two amino acid substitutions [51]. Other studies are in progress in connection with our group. Its seems that some low-expressed variants underwent severe structural modifications when compared to the three known variants. The correlation observed between the polymorphism of α_{s1}-casein and the total casein content of milk suggests the promotion, in milking goat populations, of a selection favoring the three alleles associated with a high α_{s1}-casein content [50, 51].

Inter-species variability: it has been known for many years that there are large qualitative and quantitative differences between milks according to the species. A number of data concerning the milk proteins in many species can be found in [21]. In general all the milks studied so far contain proteins homologous to their bovine counterparts, but their concentrations and/or proportions are often quite different. It would be too long to discuss these variations here. In what follows we will first talk of species in which some of the six main milk proteins are absent. Some examples of strongly different post-translational modifications among species will then be given. Finally a short review will be made of the milk protein phylogeny.

Occurrence of the six main milk proteins among species: α-lactalbumin has been found in all milks that have been checked except that of the California sea lion which contains no lactose [21]. However the obvious correlation between the occurrence of lactose and that of α-lactalbumin in milk was not found during a recent study of the milk of a marsupial, the Tammer wallaby. In this species the α-lactalbumin concentration remained almost constant through lactation (40 weeks) while the concentration of total lactose (free lactose plus lactose contained in oligosaccharides) fell to zero after 34 weeks post-partum [52].

β-Lactoglobulin was once thought to occur only in milk from ruminants. It has now been found in the milk of seven non-ruminant species: pig, horse, donkey, dog, dolphin, manatee, and kangaroo [53]. However no trace of β-lactoglobulin was ever found in milk from mouse, rat, rabbit, guinea pig, and human. We recently thoroughly reinvestigated human milk proteins and did not find this protein [54]. Similarly, human milk appears to be devoid of α_{s1}- and α_{s2}-caseins [55]. Finally, let us recall that whey acidic protein (WAP) has been found only in rat and mouse milk.

Inter-species variability due to post-translational modifications: milk proteins are submitted to three main post-translational modifications: enzymic removal of signal peptide (which affects all of them and will not be considered here), phosphorylation (caseins and WAP), and glycosylation ($\varkappa$-casein, immunoglobulins, α-lactalbumin, lactoferrin). Only the extent of phosphorylation and glycosylation appears to vary to a large extent between species.

Whereas there is only a minor proportion of bovine α_{s1}-casein (formerly called α_{s0}-casein) with an additional phosphate group, two α_{s1}-caseins occur in similar proportions in goat, sheep, and water buffalo. They only differ by a phosphate group [51, 56, 57].

Similarly there are two β-caseins, with the same difference, in goat and sheep milks [56, 58]. Finally human β-casein consists of six proteins with identical peptide chains and zero to five phosphate groups [21]. In any case the occurrence of such multiple forms among caseins originates from the partial phosphorylation of some potential phosphorylation sites.

As far as differences in glycosylation among species are concerned, we will just mention two proteins, α-lactalbumin, which appears to be more or less glycosylated according to species [21], and $\varkappa$-casein. The latter, which contains ca. 10% carbohydrates in cow milk, is highly glycosylated in human [59] and porcine [60] milks in which the carbohydrate content amounts to ca. 50%. This makes them soluble in 12% trichloroacetic acid.

Phylogeny of milk proteins: it does not seem that there is any evolutionary relationship between α-lactalbumin, β-lactoglobulin, and caseins. Even in the last groups of proteins, homologies between $\varkappa$-casein and the other three caseins have not been found.

The conservation of the amino acid sequence of the leader peptide of the ovine Ca-sensitive caseins, which led originally to the proposal of a casein gene family [61], has been strikingly substantiated by additional sequences determined on DNA which also show conservation of the 5′ untranslated sequences. However, available data show that the evolution of each of the caseins was extremely rapid. These proteins accumulated a number of point mutations, additions, and deletions. For example, rat α_{s1}-casein has 31% homology with its bovine counterpart at the amino acid level, but the alignment requires a total of one insertion and eight deletions. A 60-residue-insertion, corresponding to a repetition of a 10-residue segment, starts at position 123 in the rat sequence [62]. The homology of rat and bovine $\varkappa$-caseins is somewhat higher (38%). Comparison of five $\varkappa$-casein sequences (from goat, ewe, cow, human, and rat) shows that the rate of evolution of para-$\varkappa$-casein is higher than that of CMP. This is mainly due to the high homology which can be observed between residues 104 and 126 (cow numbering): 55 and 22% identity in the five species for this fragment, and for the whole molecule, respectively [59]. This region contains bond 105–106 whose cleavage by pepsin or chymosin leads to milk coagulation in the stomach.

As indicated above there is now clear evidence of a common origin for the Ca-sensitive caseins. However, at the level of the mature proteins, only polyphosphorylated regions have clear homologies. The proposal has been made [63] that the ancestral casein gene resulted from the recruitment into a functional gene of a minimum of four exons corresponding to the 5′ non-coding region, the leader peptide, a hydrophilic sequence containing a major phosphorylation site, and a hydrophobic sequence responsible for aggregation behavior.

120

Indeed it seems that, during evolution, caseins have only kept features that allow them to be secreted in the mammary gland, to make micelles, and to be easily degraded in the digestive tract.

The two main whey proteins each have a specific biological function. This is likely the reason for an evolution slower than that of caseins. They each are part of a protein family, that of lysozyme for α-lactalbumin, and that of a protein family able to transport small hydrophic ligands. α-Lactalbumin has lost characteristic residues involved in catalysis in lysozyme, but has acquired new binding sites for galactosyltransferase and probably Ca^{2+} ions. If it is assumed that β-lactoglobulin only occurs in milk, it must derive from an ancient transport protein. If it is also assumed that the function of β-lactoglobulin is to transport retinol, it must have kept, during evolution, both amino acid residues involved in the binding of retinol, and those involved in interactions with a receptor. Its dimeric state in the milk of ruminants and kangaroo does not appear to be essential, as it occurs as a monomer in the other species studied so far. All β-lactoglobulins seem to have two disulphide bridges. All but the horse I protein have a free thiol group [64]. In horse and donkey two forms of β-lactoglobulin were identified. β-Lg I and II from horse only have 70% homology. This means that the two proteins are synthesized from two different genes [65].

References

1. Stuart DI, Acharya KR, Walker NPC, Smith SG, Lewis M, Phillips DC (1986) α-Lactalbumin possesses a novel calcium binding loop. Nature 324:84–87
2. Papiz MZ, Sawyer L, Eliopoulos EE, North ACT, Findlay JBC, Sivaprasadarao R, Jones TA, Newcomer ME, Kraulis PJ (1986) The structure of β-lactoglobulin and its similarity to plasma retinol-binding protein. Nature 324:383–385
3. Mepham TB, Gaye P, Mercier JC (1982) Biosynthesis of milk proteins. In: Fox PE (ed) Developments in Dairy Chemistry, vol 1. Applied Science Publishers, London New York, pp 115–156
4. Ebner KE, Brodbeck U (1968) Biological role of α-lactalbumin: a review. J Dairy Sci 51: 317–322
5. Brew K, Castellino FJ, Vanaman TC, Hill RL (1970) The complete amino acid sequence of bovine α-lactalbumin. J Biol Chem 245:4570–4582
6. Shewale JG, Sinha SK, Brew K (1984) Evolution of α-lactalbumins. The complete amino acid sequence of the α-lactalbumin from a marsupial (Macropus rufogriseus) and corrections to regions of sequence in bovine and goat α-lactalbumins. J Biol Chem 259:4947–4956
7. Hurley WL, Schuler LA (1987) Molecular cloning and nucleotide sequence of a bovine α-lactalbumin cDNA. Gene 61:119–122
8. Vilotte JL, Soulier S, Mercier JC, Gaye P, Hue-Delahaye D, Furet JP (1987) Complete nucleotide sequence of bovine α-lactalbumin gene: comparison with its rat counterpart. Biochimie 69:609–620
9. Qasba PK, Surinder KS (1984) Similarity of the nucleotide sequence of rat α-lactalbumin and chicken lysozyme genes. Nature 308:377–380
10. Hiraoka Y, Segawa T, Kuwajima K, Sugai S, Murai N (1980) α-Lactalbumin: a calcium metallo-protein. Biochem Biophys Res Commun 95:1098–1104
11. Kronman MJ, Sinha SK, Brew K (1981) Characteristics of the binding of Ca^{2+} and other divalent metal ions to bovine α-lactalbumin. J Biol Chem 256:8582–8587
12. Braunitzer G, Chen R, Schrank B, Stangl A (1972) Automatische Sequenzanalyse eines Proteins (β-lactoglobulin AB). Hoppe-Seyler's Z Physiol Chem 353:832–834
13. Jamieson AC, Vandeyar MA, Kang YC, Kinsella JE, Batt CA (1987) Cloning and nucleotide sequence of the bovine β-lactoglobulin gene. Gene 61:85–90
14. Pervaiz S, Brew K (1985) Homology of β-lactoglobulin, serum retinol-binding protein, and protein HC. Science 228:335–337

15. Godovac-Zimmermann J, Conti A, Liberatori J, Braunitzer G (1985) Homology between the primary structures of β-lactoglobulin and human retinol-binding protein: evidence for a similar biological function? Hoppe-Seyler's Z Physiol Chem 366:431–434

16. Fugate RD, Song PS (1980) Spectroscopic characterization of β-lactoglobulin-retinol complex. Biochim Biophys Acta 625:28–42

17. Drayna D, Fielding C, McLean J, Bear B, Castro G, Chem E, Comstock L, Henzel W, Kohr W, Rhee L, Wion K, Lawn R (1986) Cloning and expression of human apolipoprotein D cDNA. J Biol Chem 261:16535–16539

18. Lee KH, Wells RG, Reed RR (1987) Isolation of an olfactory cDNA: similarity to retinol-binding protein suggests a role in olfaction. Science 235:1053–1056

19. Deutsch HF, Smith VR (1957) Intestinal permeability to proteins in the newborn herbivore. Am J Physiol 191:271–276

20. Pierce AE (1960) β-Lactoglobulin in the urine of the new-born suckled calf. Nature 188:940–941

21. Jenness R (1982) Interspecies comparison of milk proteins. In: Fox PF (ed) Developments in Dairy Chemistry, vol 1. Applied Science Publishers, London New York, pp 87–114

22. Lice YSV, Low TLK, Putnam FW (1979) Primary structure of a human IgA1 immunoglobulin, I, II, III, IV. J Biol Chem 254:2839–2874

23. Mc Kenzie RM, Larson BL (1978) Purification and partial characterization of a unique group of phosphoproteins from rat milk whey. J Dairy Sci 61:723–728

24. Piletz JE, Heinlen M, Ganschow RE (1981) Biochemical characterization of a novel whey protein from murine milk. J Biol Chem 256:11509–11516

25. Dandekar AM, Robinson EA, Appella E, Qasba PK (1982) Complete sequence analysis of cDNA clones encoding rat whey phosphoproteins: homology to a protease inhibitor. Proc Natl Acad Sci USA 79:3987–3991

26. Hennighausen LG, Sippel AE (1982) Mouse whey acidic protein is a novel member of the family of "four-disulfide core" proteins. Nucleic Acids Res 10:2677–2684

27. Schmidt DG (1982) Association of caseins and casein micelle structure. In: Fox PF (ed) Developments in dairy chemistry, vol 1. Applied Science Publishers, London New York, pp 61–86

28. Dalgleish DG (1982) The enzymatic coagulation of milk. In: Fox PF (ed) Developments in dairy chemistry, vol 1. Applied Science Publishers, London New York, pp 157–187

29. Mercier JC (1981) Phosphorylation of caseins, present evidence for an amino acid triplet code post-translationally recognized by specific kinases. Biochimie 63:1–17

30. Mercier JC, Grosclaude F, Ribadeau-Dumas B (1971) Structure primaire de la caseine αS1 bovine. Séquence complète. Eur J Biochem 23:41–51

31. Nagao M, Maki M, Sasaki R, Chiba H (1984) Isolation and sequence analysis of bovine αS1-casein cDNA clone. Agric Biol Chem 48:1663–1667

32. Stewart AF, Willis IM, Mc Kinlay AG (1984) Nucleotide sequences of bovine αS1- and $\varkappa$-casein cDNAs. Nucleic Acids Res 12:3895–3907

33. Nagao M, Nakagawa Y, Ishii A, Sasaki R, Harutaka H, Chiba H (1988) Expression of bovine αS1-casein cDNA in Escherichia coli. Agric Biol Chem 52:191–200

34. Brignon G, Ribadeau-Dumas B, Mercier JC, Pelissier JP, Das BC (1977) Complete amino acid sequence of bovine αS2-casein. FEBS Lett 76:274–279

35. Brignon G (1979) Determination de la structure primaire de la caseine αS2 bovine. Nouvelles évidences, concernant le code de phosphorylation des caséines. Thesis, Paris VII University

36. Stewart AF, Bonsing J, Beattie CW, Shah F, Willis IM, Mackinlay AG (1987) Complete nucleotide sequences of bovine αS2- and β-casein cDNAs: comparisons with related sequences in other species. Mol Biol Evol 4:231–241

37. Ribadeau-Dumas B, Brignon G, Grosclaude F, Mercier JC (1972) Structure primaire de la caseine β bovine. Sequence complète. Eur J Biochem 25:505–514

38. Carles C, Huet JC, Ribadeau-Dumas B (1988) A new strategy for primary structure determination of proteins: application to bovine β-casein. FEBS Lett 229:265–272

39. Bayev AA, Smirnov IK, Gorodestsky (1987) The primary structure of bovine β-casein cDNA. Mol Biol 21:255–265

40. Jimenez-Florez R, Kang YC, Richardson T (1987) Cloning and sequence analysis of bovine β-casein cDNA. Biochem Biophys Res Commun 142:617–621

122

41. Mercier JC, Brignon G, Ribadeau-Dumas B (1973) Structure primaire de la caseine $\varkappa$ bovine. Séquence complète. Eur J Biochem 35:222–235
42. Jollès J, Alais C, Jollès P (1968) The tryptic peptide with the rennin sensitive linkage of cow's $\varkappa$-casein. Biochim Biophys Acta 168:591–593
43. Zevaco C, Ribadeau-Dumas B (1984) A study of the carbohydrate binding sites of bovine $\varkappa$-casein using high performance liquid chromatography. Milchwissenschaft 39:206–210
44. Saito T, Itoh T, Adachi S, Susuki T, Usui T (1982) A new hexasaccharide chain isolated from bovine colostrum $\varkappa$-casein taken at the time of parturition. Biochim Biophys Acta 719:309–317
45. Carles C, Martin P (1985) Kinetic study of the action of bovine chymosin and pepsin A on bovine $\varkappa$-casein. Arch Biochem Biophys 242:411–416
46. Kang YC, Richardson T (1988) Molecular cloning and expression of bovine $\varkappa$-casein in Escherichia coli. J Dairy Sci 71:29–40
47. Grosclaude F (1988) Le polymorphisme génétique des principales lactoproteines bovines. Relations avec la quantité, la composition et les aptitudes fromagères du lait. Prod Anim 1:5–17
48. Mc Lean D, Graham ERB, Ponzoni RW (1987) Effects of milk protein variants and composition on heat stability of milk. J Dairy Res 54:219–235
49. Boulanger A, Grosclaude F, Mahé MF (1984) Polymorphisme des caséines αS1 et αS2 de la chèvre (Capra hircus). Genet Sel Evol 16:157–176
50. Grosclaude F, Mahé MF, Brignon G, Di Stasio L, Jeunet R (1987) A Mendelian polymorphism underlying quantitative variations of goat αS1-casein. Genet Sel Evol 19:399–412
51. Brignon G, Mahé MF, Grosclaude F, Ribadeau-Dumas B (1988) submitted for publication
52. Messer M, Elliott C (1987) Changes in α-lactalbumin, total lactose, UDP-galactose hydrolase and other factors in tammar wallaby (Macropus eugenii) milk during lactation. Aust J Biol Sci 40:37–46
53. Godovac-Zimmermann J, Conti A, James L, Napolitano L (1988) Microanalysis of the amino acid sequence of monomeric β-lactoglobulin I from donkey (Equus asinus) milk. Biol Chem Hoppe-Seyler 369:171–179
54. Brignon G, Chtourou A, Ribadeau-Dumas B (1985) Does β-lactoglobulin occur in human milk? J Dairy Res 52:249–254
55. Chtourou A, Brignon G, Ribadeau-Dumas B (1985) Quantification of β-casein in human milk. J Dairy Res 52:239–247
56. Richardson BC, Creamer LK (1976) Comparative micelle structure. V. The isolation and characterization of the major bovine caseins. NZJ Dairy Sci Technol 11:46–53
57. Addeo F, Mercier JC, Ribadeau-Dumas B (1977) The caseins of buffalo milk. J Dairy Res 44:455–468
58. Richardson BC, Creamer LK (1974) Comparative micelle structure III. The isolation and chemical characterization of caprine β1-casein and β2-casein. Biochim Biophys Acta 365:133–137
59. Brignon G, Chtourou A, Ribadeau-Dumas B (1985) Preparation and amino acid sequence of human $\varkappa$-casein. FEBS Lett 188:48–54
60. Cerning-Beroard J, Zevaco C (1984) Purification and characterization of porcine $\varkappa$-casein. J Dairy Res 51:259–266
61. Gaye P, Gautron JP, Mercier JC, Hazé G (1977) Amino terminal sequences of the precursors of ovine caseins. Biochem Biophys Res Commun 79:903–911
62. Hobbs AA, Rosen JM (1982) Sequence of rat α- and $\varkappa$-casein mRNAs: evolutionary comparison of the calcium dependent rat casein multigene family. Nucleic Acids Res 10:8079–8097
63. Jones WK, Yu-Lee LY, Clift SM, Brown TL, Rosen JM (1985) The rat casein multigene family. Fine structure and evolution of the β-casein gene. J Biol Chem 260:7042–7050
64. Godovac-Zimmermann J, Braunitzer G (1987) Modern aspects of the primary structure and function of β-lactoglobulin. Milchwissenschaft 42:294–297
65. Godovac-Zimmermann, Conti A, Liberatori J, Braunitzer G (1985) The amino acid sequence of β-lactoglobulin II from horse colostrum: β-lactoglobulins are retinol-binding proteins. Hoppe-Seyler's Z Physiol Chem 366:601–608

Modification of Milk by Gene Transfer

J. Paul Simons

AFRC Institute of Animal Physiology and Genetics Research, Edinburgh Research Station, Kings Buildings, Edinburgh, Scotland

Introduction

Recombinant DNA technology allows the isolation and precise structural character-ization of genes. Reintroduction of genes into intact organisms is now possible, allowing functional analysis of cloned genes and of the physiological consequences of expression of cloned genes [18]. In addition, gene transfer offers the possibility of genetic manipulation of livestock, directing changes as opposed to selecting from existing populations [21].

Genes direct the synthesis of proteins; a large proportion of milk is protein, the properties of which are responsible for many of the characteristics of milk. Most of the protein in milk consists of a few species which are synthesized in the mammary epithelium, encoded by singly-copy genes which are expressed specifically in the mammary gland. The composition of milk should, therefore, be amenable to manip-ulation by gene transfer.

Transgenic animals

Animals into which genes have been transferred are called "transgenic animals." Several methods have been developed for gene transfer into mice: microinjection [4], infection with retroviral vectors [12, 25], and embryonic stem cell-mediated gene transfer [9, 19]. Of these, microinjection has been the most widely used technique in mice, and to date the only method successfully applied to other mammals.

For gene transfer by microinjection, newly fertilized eggs are collected, typically from superovulated animals, before the first cleavage. At this stage, two pronuclei are present, one derived from the sperm nucleus and one from the egg. In turn, eggs are held firmly by suction on a blunt pipette while a very fine pipette is inserted into one of the pronuclei and DNA injected. Approximately two picolitres of DNA solution are injected containing tens to hundreds of molecules of DNA, sufficient to cause visible swelling of the pronucleus. About two-thirds of eggs survive this proce-dure and are transferred into the oviducts of recipient female animals. About a quarter of the transferred eggs undergo normal gestation, and of the animals born, up to a third carry the injected DNA [4].

Transgenic animals generated by microinjection typically carry multiple copies of the injected DNA, integrated into a single site on one of the chromosomes. There appears to be no specificity of integration site. Most transgenic animals contain the transferred gene in every cell, including the germ cells; this means that the integrated

124

DNA (or "transgene") may be transmitted to offspring, allowing the establishment of transgenic lines. Because the transgene is integrated into a single chromosome, the animals are hemizygous and the frequency of transmission is 50%.

Many genes have been transferred into mice (see [18] for review), and have been shown to be expressed in the same cell types as the endogenous copy of the gene. Expression of transgenes is often regulated correctly during development and in response to hormonal and other environmental signals. In addition to the transfer of natural genes, it is possible to alter genes prior to transfer. In this way it is possible to produce novel proteins (if the coding sequence is altered), or to change the regulation of expression of naturally occurring proteins. Many studies have employed hybrid genes in which the regulatory elements of one gene have been joined with the coding region of a second gene; as a result, the second gene has been expressed in a manner characteristic of the first.

Gene transfer into livestock

While mice are excellent experimental organisms, practical application to the manipulation of milk requires gene transfer in dairy animals; scale-up from mice to livestock is not trivial. The production of transgenic mice by microinjection is relatively simple and efficient: a single day's microinjection can yield several transgenic mice. This is largely possible because of the availability of almost unlimited numbers of mouse eggs, and the ability to reimplant large numbers of eggs quickly. There are a number of factors which make gene transfer into large animals considerably more difficult. Dairy species (cows, sheep, and goats) have small litters, small numbers of eggs are produced, even after hormonally induced superovulation, and few can be implanted into each foster mother. Experimental control over the timing of ovulation, fertilization and early development is not precise; many of the eggs which are recovered are not at the correct stage for microinjection. Eggs of large animals are relatively opaque, making pronuclei more difficult to see: sheep and goat pronuclei are visible by careful DIC microscopy, pig and cow pronuclei are only visible after the eggs have been centrifuged to stratify the cytoplasm [26]. The eggs of livestock are more fragile than mouse eggs and more care in microinjection is required. Despite these problems, transgenic pigs and sheep have been produced [2, 10, 23]. We have chosen to work with sheep for a number of reasons: they are relatively cheap, the generation interval is significantly shorter than that of cows, and sheep may yield sufficient milk for commercial scale production of human proteins (see below), our main interest to date. We have produced eight transgenic sheep, carrying three different fusion genes (see Tables 1 and 2). So far we have bred from six of these sheep and have obtained transgenic progeny from four; in one case we have second generation transgenic offspring.

The methods described above allow the introduction of new genes into animals. The modification of expression of endogenous genes may prove to be equally important in practical applications of gene transfer. Techniques are being developed to allow this [6, 11, 13, 14, 24], but are not yet sufficiently efficient for general application.

Table 1. Production of transgenic sheep. Breakdown of sheep eggs microinjected over three seasons (1984/5, 1985/6 and 1986/7), including data presented in [23], where the methods are described

DNA injected[a]	Stage of injected eggs			Animals resulting		Transgenic
	1	2	4	lambs	fetuses[c]	
pMK	93	42	0	26	9	1
pMK[b]	15	5	0	3	0	0
BLG-FIX	252	48	7	52	5	4
BLG-α_1AT	298	40	5	47	11	3
Total	658	135	12	128	25	8[d]

[a] pMK: a plasmid containing a hybrid gene composed of the mouse metallothionein-1 promoter and the herpes simplex virus thymidine kinase gene [3]; BLG-FIX: a sheep β-lactoglobulin/human factor IX hybrid gene [23]; BLG-α_1AT: a sheep β-lactoglobulin/human α_1-antitrypsin hybrid gene [23]

[b] Eggs were centrifuged prior to microinjection

[c] Not all fetuses were tested for the presence of the injected DNA because of degradation of the tissue

[d] All transgenic individuals were derived from microinjection at the one-cell stage, and are living

Table 2. Characteristics of transgenic sheep

Sheep number	Transgene	Copy number	Sex	Transmission
5LL229	pMK	1	♀	a
6LL225	BLG-FIX	40	♂	Yes
6LL231	BLG-FIX	10	♀	Yes
6LL239	BLG-FIX	1	♂	Yes
6LL240	BLG-FIX	10	♀	Yes
6LL273	BLG-α_1AT	4	♀	b
7LL167	BLG-α_1AT	1	♀	b
7LL183	BLG-α_1AT	10	♀	b

[a] 1 progeny analyzed, not transgenic

[b] No data available yet

Sheep β-lactoglobulin in mouse milk

β-lactoglobulin is a major component of ruminant milk, but is absent from the milk of rodents. We chose to study expression of the sheep β-lactoglobulin gene in transgenic mice. The gene-encoding sheep β-lactoglobulin was isolated from a genomic library [1] and characterized. It is approximately 4.9 kb long and is split into seven exons. Clones were isolated which, from their nucleotide sequences, were predicted to encode A or B variants of β-lactoglobulin. Segments of DNA from clone SS1, containing the gene and flanking regions (Fig. 1a) were introduced into mice by microinjection into fertilized eggs [22], as described above. Of the seven transgenic mice we chose for breeding, five transmitted the β-lactoglobulin gene to their offspring, establishing lines of transgenic mice.

Table 3. Sheep β-lactoglobulin in transgenic mouse milk. Milk was collected from transgenic mice 11 days after parturition. Concentrations of β-lactoglobulin were determined by scanning densitometry of coomassie blue-stained sds polyacrylamide gels, with concentration standards of pure sheep β-lactoglobulin. (Data from [22].)

Transgenic line (DNA segment)	Mouse	β-lactoglobulin concentration (mg/ml)
7	7.1	23.2
(*Sal* I)	7.10	23.0
14	14.10	3.0
(*Sal* I)	14.11	7.0
45	45.18	14.1
(*Sal* I – *Xba* I)	45.20	21.6

One of the lines of mice has the gene incorporated on the Y chromosome; only male mice are transgenic and thus this line could not be analyzed for appropriate expression of β-lactoglobulin. Milk was collected from transgenic mice from the other four lines and analyzed for the presence of sheep β-lactoglobulin. Mice of three lines produce large amounts of sheep β-lactoglobulin, up to 23 mg/ml (Table 3), the fourth line produces the sheep protein at low levels. The β-lactoglobulin produced by these mice has the same molecular weight as β-lactoglobulin produced by sheep, and reacts with antiserum raised against β-lactoglobulin purified from sheep milk. The sequence of the gene in clone SS1 corresponds to that of the B variant of β-lactoglobulin [1]. The isoelectric point of the protein produced in the mice was indistinguishable from that of β-lactoglobulin B from sheep. We have also obtained apparently genuine β-lactoglobulin A from mice carrying another clone, SS12, the sequence of which corresponds to β-lactoglobulin A (Ali and McClenaghan, unpublished data).

In order to determine whether expression of the sheep β-lactoglobulin gene is correctly regulated in transgenic mice, we prepared RNA from a number of organs and found β-lactoglobulin RNA only in the mammary gland. Analysis of the lactating transgenic mouse mammary RNA has shown that the 5' and 3' ends of the β-lactoglobulin mRNA are identical with those of β-lactoglobulin mRNA from sheep. We have examined the time course of expression of the β-lactoglobulin gene through pregnancy and lactation: some sheep RNA is detectable in the mammary glands of virgin transgenic mice, the amount of β-lactoglobulin mRNA increases steadily throughout pregnancy and rises further after parturition, broadly following the pattern of expression of mouse β-casein (S. Harris, unpublished data), and reflecting the normal regulation of β-lactoglobulin expression in sheep (S. Ali, unpublished data).

Lee et al. [16] have produced transgenic mice which carry the rat β-casein gene: expression was observed in the mammary glands of lactating female mice (and in the brain of one line of mice). The expression was regulated through pregnancy and

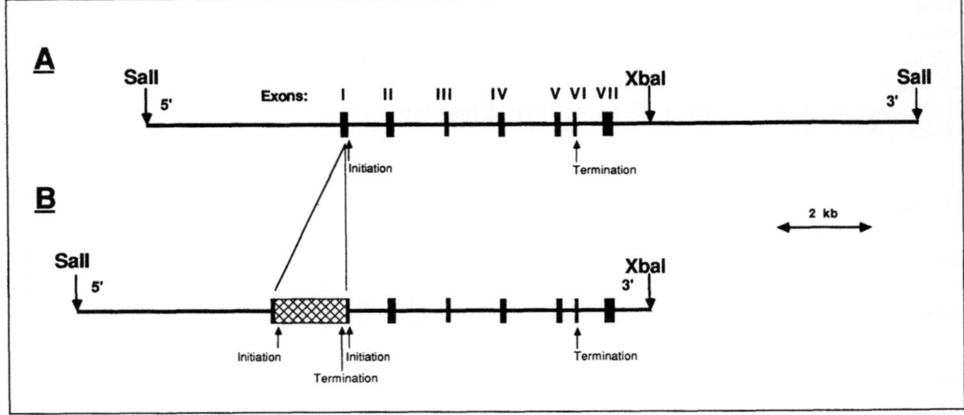

Fig. 1a. The structure of the sheep β-lactoglobulin gene transferred into mice. The entire 16.2 kb insert of clone SS1 (*Sal*I – *Sal*I) or a fragment with 5.7 kb of 3' flanking sequences removed (*Sal*I – *Xba*I) were used to produce transgenic mice. The positions and sizes of the exons are shown; the translation initiation codon is in exon 1, the termination codon is in exon 6. (See [1] and [22].)

Fig. 1b. The structure of hybrid genes transferred into sheep. The β-lactoglobulin-factor IX (BLG-FIX) and β-lactoglobulin-α_1-antitrypsin (BLG-AAT) hybrid genes were constructed by insertion of cDNAs encoding human factor IX or human α_1-antitrypsin (cross-hatched box) into exon 1 of the sheep β-lactoglobulin gene, upstream of the β-lactoglobulin translation initiation codon. The mature transcript predicted is a bicistronic mRNA with the potential to be translated into factor IX or α_1-antitrypsin and β-lactoglobulin. (See [23].)

lactation, but the levels of expression were low: 0.01 to 1% of endogenous mouse β-casein.

Production of human proteins in milk

There are many medical conditions which may be alleviated by treatment with particular human proteins. Hemophilias A and B, for example, are sex-linked disorders of blood coagulation which result from deficiencies of clotting factors VIII and IX, respectively. Hemophiliacs are currently treated with partially purified concentrates of factors VIII and IX prepared from donated human blood. In addition to the limited supply of human blood, problems have been encountered with viral contamination of clotting factor concentrates, notably with hepatitis viruses and HIV.

Numerous genes encoding proteins of potential therapeutic value have been cloned in recent years, and a great deal of effort has been expended on attempts to express these genes to produce large amounts of authentic proteins. While some proteins are produced satisfactorily in microorganisms such as *E. coli,* many proteins require modification after translation, necessitating their expression in mammalian cells. We have previously suggested [5, 15] that a viable alternative to expression in vitro, in cultured mammalian cells, may be expression in vivo, in transgenic animals. For this to be feasible, the rates of production of the proteins of interest would ideally be high, and they should be easily harvested. These criteria suggested the use

128

of milk protein gene regulatory elements to direct expression of human genes to the mammary gland during lactation, the proteins being secreted into milk. This may have the added advantage that the products would largely be isolated from the animal's circulation, minimizing any harmful effects of the presence of large amounts of biologically active human proteins.

We chose clotting human factor IX and human α_1-antitrypsin as candidates for expression in milk. Hybrid genes were constructed using the sheep β-lactoglobulin gene and cDNA clones encoding factor IX and α_1-antitrypsin: the cDNAs were inserted into the first exon, upstream of the β-lactoglobulin translation initiation codon (Fig. 1 b). The hybrid genes have been introduced into sheep by microinjection of DNA (see Tables 2 and 3). Four transgenic sheep carry the β-lactoglobulin-factor IX hybrid gene (BLG-FIX) and three carry the β-lactoglobulin-α_1-antitrypsin hybrid gene (BLG-α_1AT). So far we have been able to analyze expression of the transgene in the two ewes which carry BLG-FIX. Milk was collected, and RNA was prepared from biopsy samples. Both factor IX RNA and protein (Table 4) were detected in samples from both of these sheep, although at low levels. We are currently investigating how increased expression of human proteins in milk may be achieved. Gordon et al. [8] have recently reported the production of human tissue plasminogen activator (tPA) in mouse milk using the 5' end of the mouse whey acidic protein gene to direct expression of tPA cDNA.

Once transgenic animals have been obtained which secrete large amounts of therapeutic proteins, considerable effort will be required to characterize the protein produced, and to devise methods for purification of the protein free from all contaminants.

Modification of milk

We have shown that it is possible to modify the composition of milk by gene transfer, but how do we envisage this being applied to milk as a food? There are a number of possibilities, some of which I will describe below, along with some of the limitations.

Table 4. Human factor IX in transgenic sheep milk. 200 ml milk was defatted and acid fractionated. The whey fraction was applied to an anti-factor IX monoclonal antibody affinity column. Bound material was eluted, concentrated, and assayed for factor IX antigen and activity. Antigen was assayed by radioimmunoassay using a human-specific anti-factor IX polyclonal antiserum; activity was assayed by a single stage clotting assay. The recovery for factor IX from the affinity column was not quantitative.

Sheep number	Factor IX antigen[a]	Factor IX activity[a]
6LL231	0.015	0.031
6LL240	0.024	0.032
Control 1	< 0.0077	< 0.01
Control 2	N. D.	< 0.009

[a] Total amount of factor IX recovered by the above protocol, in international units

Protein content

At first sight, it may seem that the expression of extra genes in the mammary gland may result in an increased protein concentration. The levels of mRNA may not, however, be the limiting factor in determining the output of protein by the mammary gland. In these circumstances, any extra protein production would be at the expense of one or more other proteins. Preliminary data (M. McClenaghan, unpublished) on the effects of β-lactoglobulin expression on the total protein content of transgenic mouse milk indicate that there is little or no increase in total protein concentration. While transfer of milk protein genes may not allow increased protein production, it may be possible, for example, to increase the production of casein at the expense of the whey proteins.

Casein micelle structure

The caseins comprise the majority of milk protein, and are assembled into large micellar structures [see 20]. Many of the properties of milk are determined by the biochemical and biophysical properties of the casein micelles. α_{s1}-, α_{s2}-β-caseins are highly phosphorylated proteins which bind calcium ions and calcium phosphate. $\varkappa$-casein is predominantly located at the surface of the micelle and is important for micelle stability.

It would be possible to construct genes to produce modified α_{s1}-, α_{s2}-β-caseins, creating new phosphorylation sites which may result in increased calcium content and increased micelle stability. Another option is to increase the ratio of $\varkappa$-casein to α_{s1}-, α_{s2}- and β-caseins which would lead to reduced micelle size, and possibly increased stability. In cheesemaking, chymosin is used to cleave $\varkappa$-casein, destabilizing the micelle and leading to the formation of the curd. A reduced $\varkappa$-casein to α_{s1}-, α_{s2}- and β-caseins ratio may be desirable for cheese production.

Lactose

Lactose is present in cows milk at a level of 46 grams per litre, and after ingestion, is hydrolyzed by the action of lactase in the brush border of the small intestine. The products of lactose hydrolysis, glucose and galactose, are readily absorbed from the gut, lactose is not. In a large proportion of individuals of some human populations (African, American-Indian, and Asian), production of lactase ceases at weaning; ingestion of dairy products can result in symptoms of bloating, flatulence, abdominal cramps and diarrhea, due to the osmotic effects of the lactose and the metabolism of lactose by intestinal bacteria. Lactose is synthesized in the secretory epithelial cells of the mammary gland, by the action of galactosyl transferase in conjunction with α-lactalbumin. It may be possible to reduce or prevent the production of lactose [17] by reducing or eliminating the expression of α-lactalbumin, alternatively, production of lactase in the milk would result in hydrolysis in situ. The consequences of such manipulations may be expected to result in a dramatic alteration of milk volume, since lactose is a major osmotic component of cow's milk. The low-lactose milk produced may possibly be quite abnormal and prove not to be useful.

Fat

The production and secretion of fat globules is many steps removed from the expression of a gene, and thus a much more difficult process to envisage changing by gene transfer. It has been suggested that expression of $\Delta 12$ desaturase could increase linoleic acid at the expense of oleic acid [7]; in addition to the dietary benefits, this modification would be expected to result in softer butter. Milk fat and proteins are secreted by the same cells, thus any manipulation which has a profound effect on milk fat production may also be expected to influence the protein synthetic capacity of the mammary gland.

Antimicrobial proteins

Milk naturally contains antimicrobial proteins, most notably immunoglobulins. The immunoglobulins, however, are a diverse set of proteins. The genes encoding the immunoglobulins are assembled during development by processes of DNA rearrangement and mutation. It would be possible to introduce genes to produce a single immunoglobulin, but it is difficult to envisage the production of the populations of immunoglobulins which are important for immunity.

Conclusions

Techniques are now available for the genetic manipulation of animals, including dairy animals. In this paper I have outlined some of the possibilities offered by genetic manipulation for modification of milk. The production of therapeutic proteins in milk is in some respects the simplest, in that the desired product is a single protein; providing the product can be isolated from the milk in its native form, the overall characteristics of the milk are not important. Milk as a food, however, is complex. The quality of milk is a combination of a large number of factors, many of which we can envisage altering by gene transfer. The changes possible are very specific, and it is to be expected that different changes would be appropriate for the production of milk for different purposes. Now that the technology of gene transfer is becoming established, it is necessary for dairy scientists to collaborate with molecular biologists to design, produce and assess new milk products.

Acknowledgement:

I would like to thank my colleagues and collaborators who have all played an important role in the execution of the work presented here and in the formulation of ideas.

References

1. Ali S, Clark AJ (1988) Characterisation of the gene encoding β-lactoglobulin: similarity to the genes for retinol binding protein and other secretory proteins. J Mol Biol 199:415–426
2. Brem G, Brenig B, Goodman HM, Selden RC, Graf F, Kruff B, Springman K, Hondele J, Meyer J, Winnacker E-L, Kräußlich H (1985) Production of transgenic mice, rabbits and pigs by microinjection into pronuclei. Zuchthygiene 20:251–252

3. Brinster RL, Chen HY, Trumbauer M (1981) Somatic expression of herpes thymidine kinase in mice following injection of a fusion gene into eggs. Cell 27:223–231
4. Brinster RL, Chen HY, Trumbauer ME, Yagle MK, Palmiter RD (1985) Factors affecting the efficiency of introducing foreign DNA into mice by microinjecting eggs. Proc Natl Acad Sci USA 82:4438–4442
5. Clark AJ, Simons P, Wilmut I, Lathe R (1987) Pharmaceuticals from transgenic livestock. Trends Biotechnol 5:20–24
6. Doetschman T, Gregg RG, Maeda N, Hooper ML, Melton DW, Thompson S, Smithies O (1987) Targeted correction of a mutant HPRT gene in mouse embryonic stem cells. Nature 330:576–578
7. Gannon F (1986) Transgenics. In: Smith C, King JWB, McKay JC (eds) Exploiting new technologies in animal breeding: genetic developments. Oxford University Press, Oxford, pp 54–58
8. Gordon K, Lee E, Vitale JA, Smith AE, Westphal H, Hennighausen L (1987) Production of human tissue plasminogen activator in transgenic mouse milk. Bio Technology 5:1183–1187
9. Gossler A, Doetschman T, Korn R, Serfling E, Kemler R (1986) Transgenesis by means of blastocyst-derived stem cell lines. Proc Natl Acad Sci USA 83:9065–9069
10. Hammer RE, Pursel VG, Rexroad CE Jr, Wall RJ, Bolt DJ, Ebert KM, Palmiter RD, Brinster RL (1985) Production of transgenic rabbits, sheep and pigs by microinjection. Nature 315:680–683
11. Izant JG, Weintraub H (1984) Inhibition of thymidine kinase gene expression by anti-sense RNA: a molecular approach to genetic analysis. Cell 36:1007–1015
12. Jähner D, Kirsten H, Mulligan R, Jaenisch R (1985) Insertion of the bacterial gpt gene into the germ line of mice by retroviral infection. Proc Natl Acad Sci USA 82:6927–6931
13. Kim SK, Wold BJ (1985) Stable reduction of thymidine kinase activity in cells expressing high levels of anti-sense RNA. Cell 42:129–138
14. Kuehn MR, Bradley A, Robertson EJ, Evans MJ (1987) A potential animal model for Lesch-Nyhan syndrome through introduction of HPRT mutations into mice. Nature 326:295–298
15. Lathe R, Clark AJ, Archibald AL, Bishop JO, Simons P, Wilmut I (1986) Novel products from livestock. In: Smith C, King JWB, McKay JC (eds) Exploiting new technologies in animal breeding: genetic developments. Oxford University Press, Oxford, pp 91–102
16. Lee K-F, DeMayo FJ, Atiee SH, Rosen JM (1988) Tissue-specific expression of the rat β-casein gene in transgenic mice. Nucl Acids Res 16:1027–1041
17. Mercier J-C (1986) Genetic engineering applied to milk producing animals: some expectations. In: Smith C, King JWB, McKay JC (eds) Exploiting new technologies in animal breeding: genetic developments. Oxford University Press, Oxford, pp 122–131
18. Palmiter RD, Brinster RL (1986) Germline transformation of mice. Ann Rev Genet 20:465–499
19. Robertson E, Bradley A, Kuehn M, Evans M (1986) Germ-line transmission of genes introduced into cultured pluripotential cells by retroviral vector. Nature 323:445–448
20. Schmidt DG (1982) Association of caseins and casein micelle structure. In: Fox PF (ed) Developments in dairy chemistry, vol 1, proteins. Applied Science Publishers, London and New York, pp 61–86
21. Simons JP, Land RB (1987) Transgenic livestock. J Reprod Fertil [Suppl] 34:237–250
22. Simons JP, McClenaghan M, Clark AJ (1987) Alteration of the quality of milk by expression of sheep β-lactoglobulin in transgenic mice. Nature 328:530–532
23. Simons JP, Wilmut I, Clark AJ, Archibald AL, Bishop JO, Lathe R (1988) Gene transfer into sheep. Bio Technology 6:179–183
24. Thomas KR, Capecchi MR (1987) Site-directed mutagenesis by gene targeting in mouse embryo-derived stem cells. Cell 51:503–512
25. van der Putten H, Botteri FM, Miller AD, Rosenfeld MG, Fan H, Evans RM, Verma IM (1985) Efficient insertion of genes into the mouse germ line via retroviral vectors. Proc Natl Acad Sci USA 82:6148–6152
26. Wall RJ, Pursel VG, Hammer RE, Brinster RL (1985) Development of porcine ova that were centrifuged to permit visualization of pronuclei and nuclei. Biol Reprod 32:645–651
27. Walstra P, Jenness R (1984) Dairy chemistry and physics. Wiley, New York

Analysis of Protein Structure in Solution by Two-Dimensional NMR Spectroscopy: 2D-^{1}H NMR Investigation of Ribonuclease T_1 and Its Complexes with 2'- and 3'-Guanosine Monophosphates

H. Rüterjans, E. Hoffmann, J. Schmidt, and J. Simon

Institute of Biophysical Chemistry, Johann Wolfgang Goethe University, Frankfurt, FRG

Introduction

Two-dimensional (2 D) NMR spectroscopy may be used to study the tertiary structure of proteins and nucleic acids [28]. This NMR approach to determining the structure of biological macromolecules has some advantages over the conventional method using x-ray crystallography:
1) NMR spectroscopy can be applied to proteins that cannot be crystallized;
2) The tertiary structure of proteins or nucleic acids can be obtained under varying conditions such as pH, temperature, or inhibitor concentration;
3) NMR parameters may provide information on the dynamics of protein conformation, so that exchange processes and molecular flexibility may be studied.

Prior to an analysis of the tertiary structure a nearly complete assignment of proton resonances has to be achieved. In the following the assignment procedures, as well as the subsequent evaluation of the tertiary structure, are described for the enzyme ribonuclease T_1.

Ribonuclease T_1 (RNase T_1) cleaves the phosphodiester bond of RNA, specifically at the 3' end of guanosine; 2'- or 3'-guanosine monophosphates (2'- or 3'-GMP) as well as other purine nucleotides act as inhibitors of this reaction [25]. The tertiary structure of the RNase-T_1-2'-GMP complex was determined by x-ray crystallography [11, 24, 3]. The present study revealed not only the folding manner of the enzyme but also some interesting features of protein-nucleotide interactions. Using the distance parameters obtained from the 2D-NMR investigation, a preliminary molecular dynamics calculation was carried out to arrive at a tertiary structure of the complexes comparable to that obtained from crystals.

The immense complexity of ^{1}H-NMR spectra of RNase T_1 has hitherto restricted ^{1}H-NMR studies to the observation of the aromatic resonances or isolated methyl resonances of the protein spectrum [23, 1, 2, 10, 12–14, 21]. Interactions between histidines and tyrosines of the active site and of the base and phosphate moieties of nucleotides were already derived from observations of the pH dependence of some protein proton resonances and of ^{15}N, ^{31}P, and ^{1}H resonances of the nucleotide [19, 16, 20]. The high magnetic field strengths currently used as well as the application of 2D-NMR techniques make it possible to gain useful additional information about the structure of the free enzyme and about the changes in structure induced by nucleotide binding.

Materials and methods

On the basis of the 2D procedures developed by Wüthrich and others [4, 28] we were able to identify most of the amino acid spin systems and to assign about 90% of the side chain and backbone proton resonances to specific amino acid residues in this protein. We used the following 2D experiments for the sequential assignment of the ^{1}H resonances:
1) Correlated Spectroscopy (COSY; [22, 6, 18]);
2) Nuclear Overhauser enhancement and exchange spectroscopy (NOESY; [15, 17]);
3) Relayed coherence transfer spectroscopy (RCT; [9, 27, 7]).

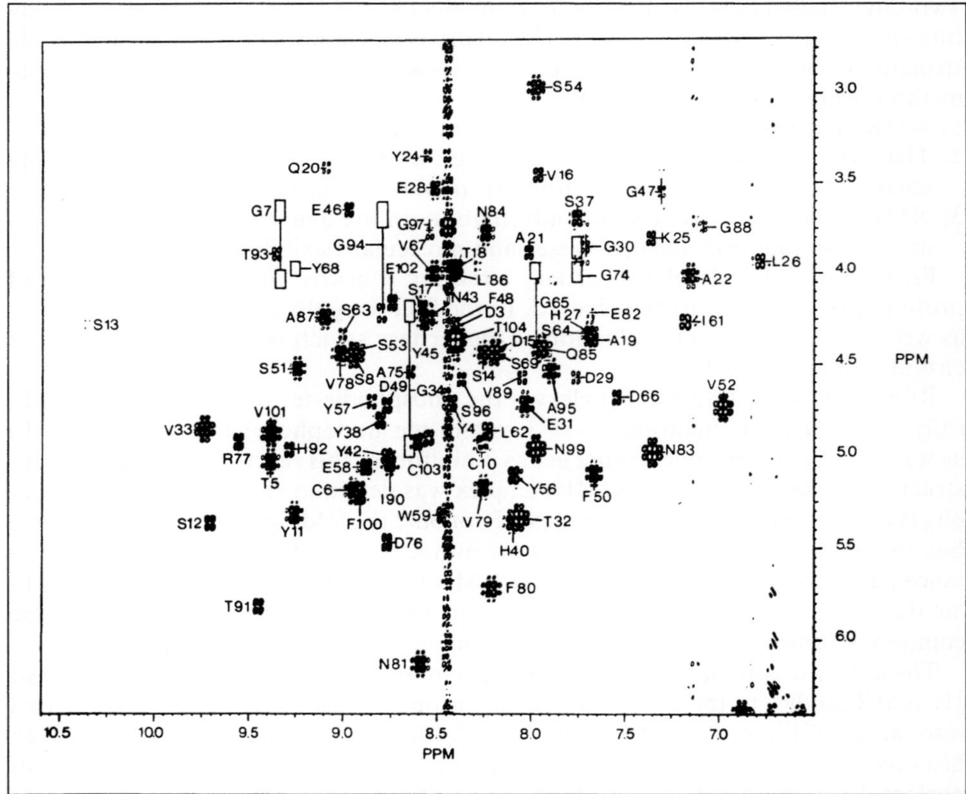

Fig. 1. Phase-sensitive COSY spectrum of RNase T_1. The region depicted contains cross peaks originating from the scalar coupling between $C_\alpha H$ and amide protons (fingerprint region). The assignments are indicated. The cross peaks of G 23 are outside the plotted region, approximately at 7.3 ppm and 2.0 ppm

134

All spectra were acquired on a Bruker AM-500 spectrometer. The pulse sequence $(T\text{-}90°\text{-}t_1\text{-}90°\text{-}t_2)_n$ was used for the acquisition of the COSY spectra. A typical COSY spectrum was recorded within approx. 36 h, with 144 scans accumulated per free induction decay (FID). The NOESY spectra were recorded with the pulse sequence $(T\text{-}90°\text{-}t_1\text{-}90°\text{-}t_m\text{-}90°\text{-}t_2)_n$. The mixing time t_m was set to 150 ms; in order to eliminate artifacts resulting from coherence transfer, t_m was varied randomly by 10%; 256 scans were accumulated per FID. A typical NOESY spectrum was obtained within approximately 68 h. RCT experiments were performed with the pulse train $(T\text{-}90°\text{-}t_1\text{-}90°\text{-}\tau_m/2\text{-}180°\text{-}\tau_m/2\text{-}90°\text{-}t_2)_n$. Undesired artifacts were eliminated by phase cycling as described by [8, 5]. As an important advantage, RCT spectra contain information on connectivities between two spins that exhibit negligible direct coupling, but do both exhibit significant coupling to a common third spin.

For experiments in H_2O, the solvent resonance was saturated by continuous selective irradiation at all times except t_1 and t_2.

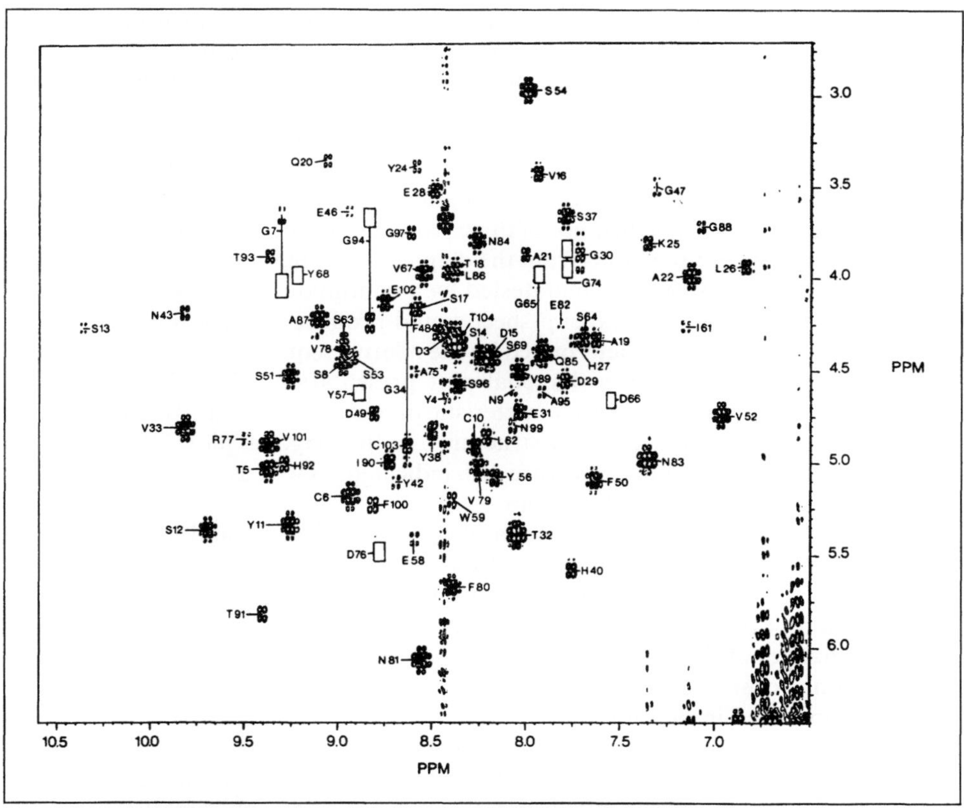

Fig. 2. COSY spectrum of the RNase-T_1-2′-GMP complex (fingerprint region)

135

The procedure for sequence-specific assignments of proton resonances of proteins was developed by Wüthrich [28]. The basic concept is to utilize scalar coupling phenomena to identify complete spin systems. The type of amino acid is then assigned via the characteristic chemical shifts of the coupled protons. After this identification of the spin system the $C_\alpha H$ and backbone amide protons of the amino acid in question have to be assigned. Through-space connectivities between $C_\alpha H$ and backbone amide protons are monitored via NOESY spectroscopy. In this manner successive residues can be identified step-by-step. With the help of the primary structure of the molecule, the sequence-specific assignment can be carried out.

Results and discussion

In Figs. 1 and 2 the so-called fingerprint peak region, i.e., the connectivities between the NH and the $C_\alpha H$ resonance of phase-sensitive COSY spectra, is shown for free RNase T_1 and for the RNase-T_1-2′-GMP complex. Approximately 100 cross peaks of this area could be assigned to specific amino acid residues.

From corresponding NOESY spectra, NOE values of adjacent protons in the molecule were derived. In Fig. 3 these NOE data are presented in a diagonal plot. Since the NOE values are related to distances between the corresponding protons, such a plot is particularly helpful for the determination of the secondary structure. The filled squares indicate NOEs between backbone protons (NH or $C_\alpha H$). The sequential NOEs appear as a nearly continuous line of squares immediately adjacent to the diagonal. The helical regions are characterized by lines of squares running parallel to the diagonal at a distance of three sequence positions, with some additional NOEs at two or four positions from the diagonal. The regular antiparallel β-sheet formed between four strands is manifested by a continuous line of squares perpendicular to the diagonal. A parallel β-sheet would produce a line of squares parallel to the diagonal at a distance determined by the relative sequence locations of the two neighboring strands. Hence, a diagonal plot presentation of the NOE data can provide a clear illustration of the characteristic patterns of short ^{1}H-^{1}H distances for the common polypeptide secondary structures. In the diagonal plot, NOEs between NH or $C_\alpha H$ and a side chain proton are also indicated, as well as NOEs between side chain protons of two different residues.

From a comparison of COSY spectra of the free RNase T_1 and its 2′-GMP and 3′-GMP complexes (Figs. 1 and 2; Figs. 4 and 5) two essential features for conformational changes induced by complex formation were derived:
1) Interaction of the inhibitor and the substrate with the active site leads to considerable changes of the side-chain and backbone positions. Correspondingly larger changes are observed in the proton resonance positions of amino acid residues directly involved in binding;
2) Smaller changes occur also inside the protein domain, in particular in the contact area between the α-helix and the β-sheet structure. Proton resonances of amino acids located in this area shift their respective positions only slightly.

T 38, H 40, Y 42, N 43, Y 45, Y 56, E 58, W 59, R 77, H 92, N 99, and F 100 proton resonances shift considerably (Figs. 1 and 2). The side-chain resonances of V 79 and I 90 (Figs. 4 and 5), which are located between the central β-sheet and the active

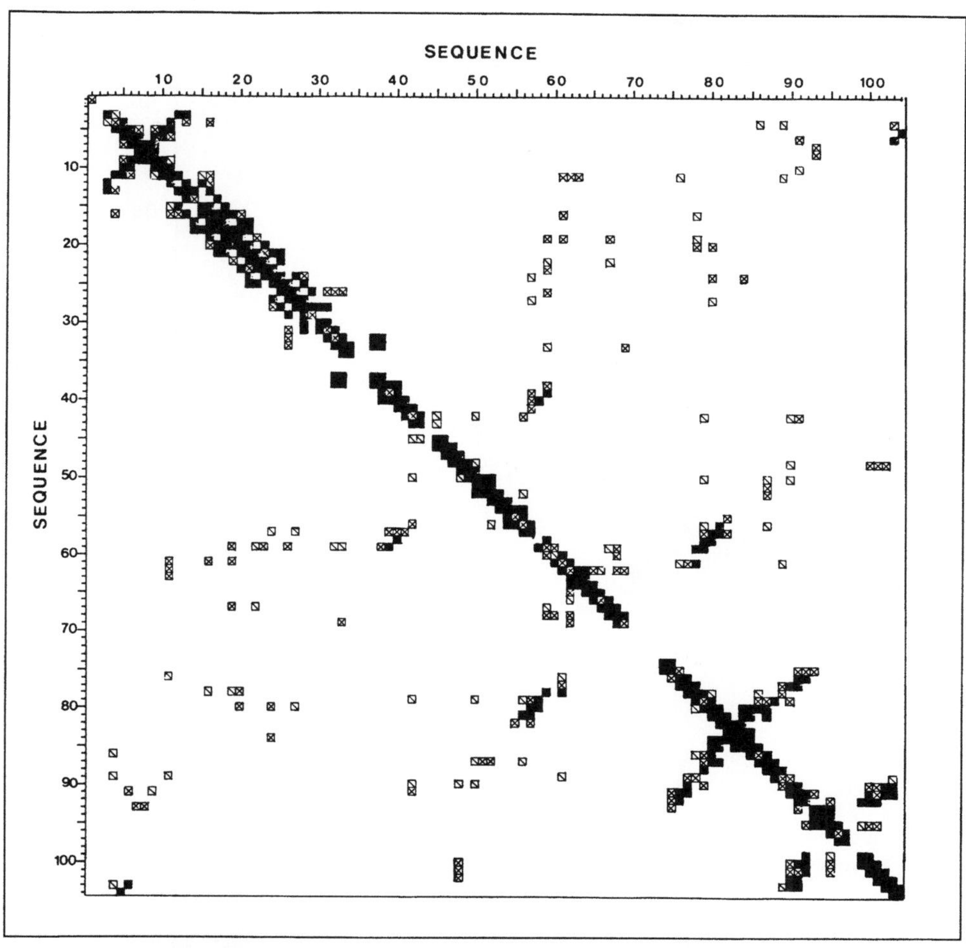

Fig. 3. Diagonal plot of the NOEs observed in RNase T$_1$. The two axes represent the amino acid sequence.

(■) NOEs between backbone protons (NH or C$_\alpha$H); (⊠) NOEs between NH or C$_\alpha$H and any side chain proton; (⊠) NOEs between the side chain protons

center, are also shifted. These residues are more or less directly involved in nucleotide binding or are near the active site.

Smaller high- or low-field shifts can be observed for the ^{1}H spin systems of A 19, Q 20, G 23, Y 24, L 26, H 27, T 32, V 33, P 39, Y 57, F 80, and N 81. These resonances are located in the contact area between the α helix and the β-sheet structure.

Distance changes between protons attached to aromatic side chains and protons of adjacent groups also lead to considerable shift changes.

From a comparison of the COSY spectra it is evident that the conformation of the RNase-T$_1$-3′-GMP complex is more similar to the conformation of the free enzyme than to that of the RNase-T$_1$-2′-GMP complex (spectra of the RNase-T$_1$3′-GMP complex not shown).

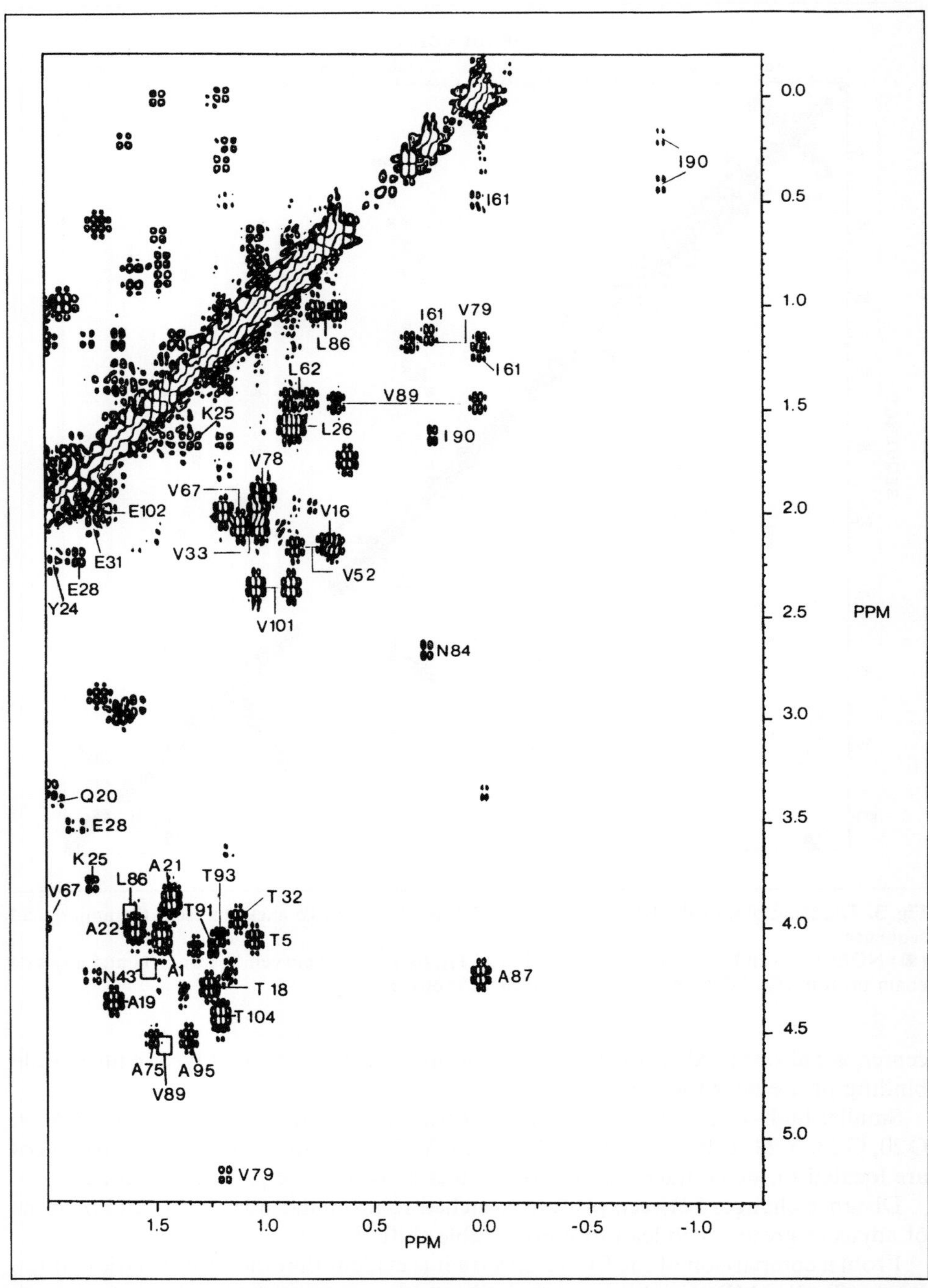

Fig. 4. Phase-sensitive COSY spectrum of RNase T_1. The cross peaks in this area originate from the scalar coupling between the methyl protons of alanine, threonine, valine, isoleucine, or leucine, and the C_α, C_β, or C_γ protons

138

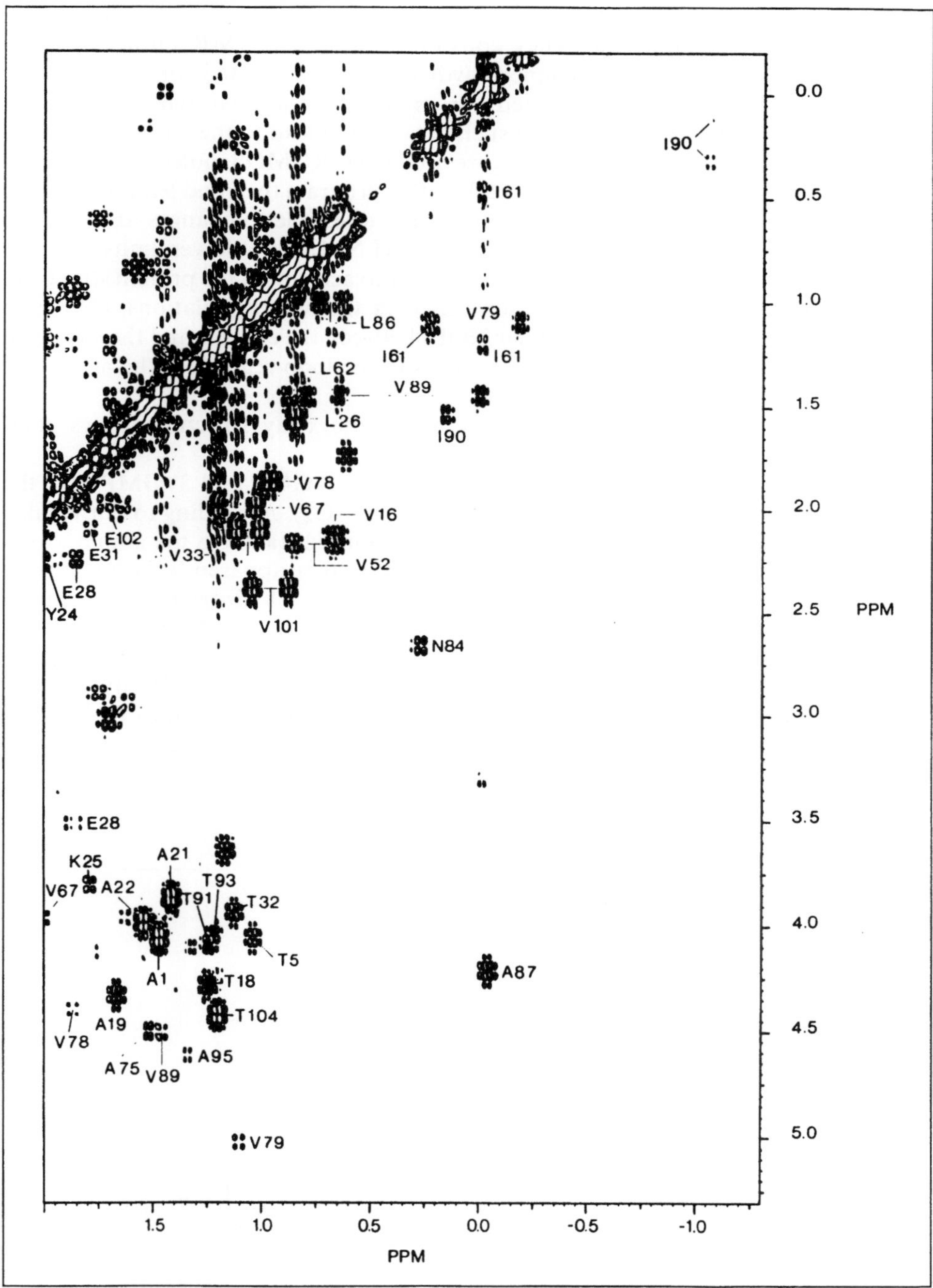

Fig. 5. Phase-sensitive COSY spectrum of the RNase-T$_1$-2′-GMP complex. Ala, Thr, Val, Ile, and Leu methyl cross peaks

The so-called long-range NOEs contain valuable information about distance constraints in the macromolecule. They constitute the major NMR information on spatial protein structures and connect hydrogen atoms located anywhere in the sequence. Therefore a distance geometry program or a molecular dynamics calculation for structure determination from such data should lead to the tertiary structure of a protein. For our study of RNase T_1 we used the GROMOS molecular dynamics program developed by van Gunsteren [26]. In this program the bond lengths are kept fixed, whereas all bond and torsional angles are accessible to changes at any time. Violations of upper distance limits obtained by ^{1}H NMR or of lower limits imposed by the atomic van der Waals' volumes are taken into account in the protein structure. 300 NOE long-distance constraints were used in the MD calculations. Owing to NMR-inherent reasons, intermolecular NOEs between the enzyme and the substrate are not observable, and intramolecular NOEs in the loop regions of the active site are rare.

The comparison of the refined structure and the x-ray structure shows good agreement with respect to the backbone folding.

As found in the crystal structure, the nucleotides 2'-GMP and 3'-GMP take the syn conformation for the glycosidic bond (C_8-N_9-C'_1-O'_4). According to our results, the peptide CO group of Asn 98 is located in H-bond distance to the amino group of the guanine base. The carboxyl group of Glu 46 is apparently bound to the Guo-N_1 position, whereas the Guo-O_6 position forms a hydrogen bond with the peptide NH of Asn 44. In contrast to the x-ray structure, the N_7 position is not hydrogen-bonded to the amide group and the peptide NH of Asn 43. In fact, the formation of this H-bond seems unlikely also according to ^{15}N-NMR studies [20]. In agreement with previous 1D-NMR studies, the phosphate group of 2'-GMP interacts with the imidazole ring of His 40. The guanidinium group of Arg 77 is located near the phosphate group, but too far away to form a hydrogen bond, and His 92 cannot reach the phosphate group either.

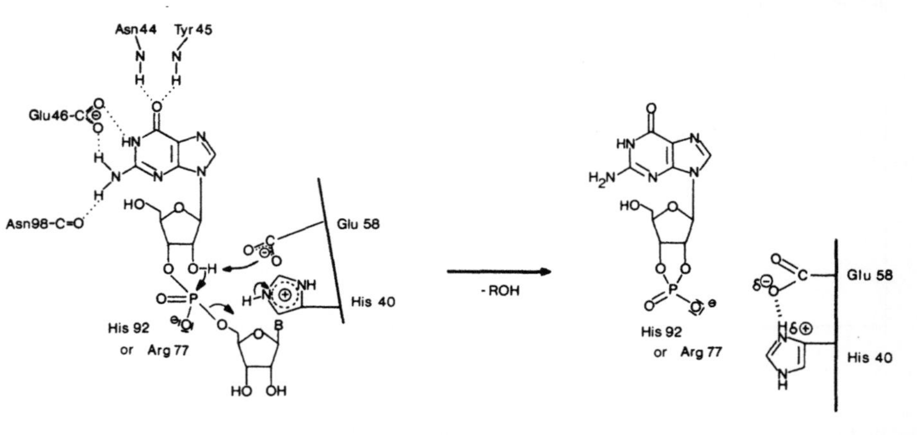

Fig. 6. Possible arrangement of hydrogen bonds for the binding of the base of 2'-GMP in the active site of RNase T_1, and presumable mechanism of RNase-T_1 action

140

Like in the crystal, the phenol ring of Tyr 45 is located on top of the guanine ring. Unfortunately, it was not possible to detect any NOE values for the interaction through space of the ring protons of Tyr 45 with any of the surrounding protons in the RNase-T$_1$-2'-GMP complex. Not even the COSY or NOESY cross peaks of the backbone NH and C$_\alpha$H and of the side chain proton resonances were observed in the complex. Apparently owing to exchange phenomena, the resonances are broadened.

For the initial transesterification step of RNase-T$_1$-catalyzed hydrolysis, a possible mechanism is proposed in Fig. 6. According to this mechanism, the carboxylate group of Glu 58 withdraws a proton from the 2'-hydroxyl group, which is then ready for a nucleophilic attack of the phosphate group, possibly activated by an increase in electrophilicity of the phosphate moiety via an interaction with His 92 or Arg 77. The O$_5'$ leaving group takes over a proton from the protonated imidazole group of His 40. It appears that the side chains of Glu 58 and His 40 are connected via a hydrogen bond in the free enzyme. The hydrolysis of the 2'−3' cyclic phosphate should occur in a similar way. Glu 58 transfers its proton to form the 2'-hydroxyl group, whereas His 40 activates the water molecule. Possibly, contrary to this reaction scheme, His 92 takes the role of His 40, since it is also located near the phosphate group. Further studies with mutants of RNase T$_1$, presently in progress, may clarify this point.

Acknowledgements

We thank the Deutsche Forschungsgemeinschaft for a grant (Ru 145/8-2). Thanks are to due Miss H. Tietz for checking the English text and typing the manuscript.

References

1. Arata Y, Kimura S, Matsuo H, Narita K (1976) Biochem Biophys Res Commun 73:133−140
2. Arata Y, Kimura S, Matsuo H, Narita K (1979) Biochemistry 18:18−24
3. Arni R, Heinemann U, Maslowska M, Tokuoka R, Saenger W (in press)
4. Aue WP, Bartholdi E, Ernst RR (1976) J Chem Phys 64:2229−2246
5. Bain AD (1984) J Magn Reson 56:418−427
6. Bax A, Freeman R (1981) J Magn Reson 44:542−561
7. Bax A, Drobny G (1985) J Magn Reson 61:306−320
8. Bodenhausen G, Kogler H, Ernst RR (1984) J Magn Reson 58:370−388
9. Eich G, Bodenhausen G, Ernst RR (1982) J Am Chem Soc 104:3731−3732
10. Fülling R, Rüterjans H (1978) FEBS Lett 88:279−282
11. Heinemann U, Saenger W (1982) Nature 299:27−31
12. Inagaki F, Kawano Y, Shimada J, Takahashi K, Miyazawa T (1981) J Biochem (Tokyo) 89:1185−1195
13. Inagaki F, Shimada I, Miyazawa T (1985) Biochemistry 24:1013−1020
14. Kimura S, Matsuo H, Narita K (1979) J Biochem (Tokyo) 85:301−310
15. Kumar A, Wagner G, Ernst RR, Wüthrich K (1981) J Am Chem Soc 103:3654−3658
16. Kyogoku Y, Watanabe M, Kainosho M, Oshima T (1982) J Biochem (Tokyo) 91:675−679
17. Macura S, Wüthrich K, Ernst RR (1982) J Magn Reson 47:351−357
18. Marion D, Wüthrich K (1983) Biochem Biophys Res Commun 113:967−974
19. Maurer W (1972) Dissertation, Münster
20. Menke G (1984) Dissertation, Frankfurt

21. Nagai H, Kawata Y, Hayashi F, Sakiyama F, Kyogoku Y (1985) FEBS Lett 189:167–170
22. Nagayama K, Kumar A, Wüthrich K, Ernst RR (1980) J Magn Reson 40:321–334
23. Rüterjans H, Pongs O (1971) Eur J Biochem 18:313–318
24. Sugio S, Amisaki T, Ohishi H, Tomita K, Heinemann U, Saenger W (1985) FEBS Lett 181:129–132
25. Takahashi K, Moore S (1982) In: Boyer PD (ed) The Enzymes, vol. XV: Nucleic Acids, Part B (3rd Edn.). Academic Press, New York, pp 435–468
26. van Gunsteren WF, Kaptein R, Zuiderweg ERP (1984) In: Olson WK (ed) Proc. NATO/ CEGAM Workshop on Nucleic Acid Conformation and Dynamics, Orsay, pp 79–92
27. Wagner G (1983) J Magn Reson 55:151–156
28. Wüthrich K (1986) NMR of Proteins and Nucleic Acids. John Wiley & Sons, New York

Bioactive Sequences in Milk Proteins

E. Schlimme, H. Meisel and H. Frister

Institute for Chemistry and Physics, Federal Dairy Research Centre, Kiel, FRG

Nutritive value of proteins

Kaufmann has already summarized some important historical data on the development of knowledge about the "nutritive value" of proteins [16]. The documentation of this knowledge began 200 years ago with the first description of nitrogen, an element which scientists initially regarded as "azote," meaning lifeless, until it was recognized that, on the contrary, nitrogen is essential to life.

Assessment of the "biological value" of proteins based on the definition of Thomas, 1909, and Mitchell, 1924, considered the amino acid composition as well as nitrogen balances. It is current practice to include in the evaluation of food proteins their content of digestible protein as well as the availability of the absorbed essential and non-essential amino acids. Additionally it has to be taken into account that both the digestibility and the absorbability are markedly influenced by antagonistic and synergistic interactions with other dietary components.

The discovery in 1979 of the opioid activity of peptides derived from partial enzymatic digestion of milk proteins [1–3, 15, 17] followed the first isolation of endogenous opioid peptides called enkephalins in 1975. These findings introduced a new criterion in evaluating the "nutritive value" of food proteins. Peptides which are "hidden" in an inactive state within the protein sequence may be released by digestive processes in vivo and may act as potential physiological modulators of metabolism during the gastrointestinal passage of the diet.

Although animal as well as vegetable proteins contain potentially bioactive sequences, the following overview refers only to milk proteins because these are currently the main source of biologically active peptides. Furthermore, milk proteins constitute up to 30% of man's average dietary protein intake.

Strategies for seeking bioactive sequences

Chiba and Yoshikawa [9] described two strategies for identifying new bioactive peptide sequences in milk proteins:
1) searching proteins for amino acid sequences similar to those known to be bioactive, e.g., opioid peptides, followed by synthesis of these peptides and evaluation of their bioactive properties;
2) isolation and characterization of bioactive peptides from in vitro digests of proteins.

143

A third strategy of investigation was employed in the Dairy Centre in Kiel [12, 14, 19–21], i.e.

3) isolation and characterization of bioactive peptides from the gastrointestinal chyme of Göttingen minipigs fed with animal or vegetable proteins.

Bioactive sequences in casein

Most of the known bioactive peptides are derived from caseins. Casein fragments have been shown to behave like opioid agonists [1–3, 5–9, 11, 15, 18, 19], to modulate the gastrointestinal motility and to stimulate secretion processes [4, 10, 23–26].

β-Casein fragments

The heptapeptide Tyr-Pro-Phe-Pro-Gly-Pro-Ile isolated from a peptone digest of bovine β-casein was the first opioid peptide described [1–3]. This fragment and other C-terminally shortened peptide sequences derived from it were named β-casomorphins (Fig. 1). Since they originate from exogenous food sources and behave like morphine they are also termed exorphins or formons (from food hormones).

Table 1 lists a series of peptides which correspond to the 60–70th residues of bovine β-casein. These compounds were isolated from in-vitro digests [1, 8, 15] or from in-vivo duodenal chyme samples [19], or they were synthesized [17]. With the exception of the tripeptide Tyr-Pro-Phe, all of these β-casomorphins, including β-casomorphin-11 (the only one which could be isolated in vivo from duodenal chyme of minipigs [19]), exert naloxone inhibitable opioid activities. Morphiceptin, the amide of β-casomorphin-4, is the most active opioid peptide known so far [6]. Lack of the N-terminal Tyr, as in des-Tyr-β-casomorphin-7, results in a total absence of bioactivity [6].

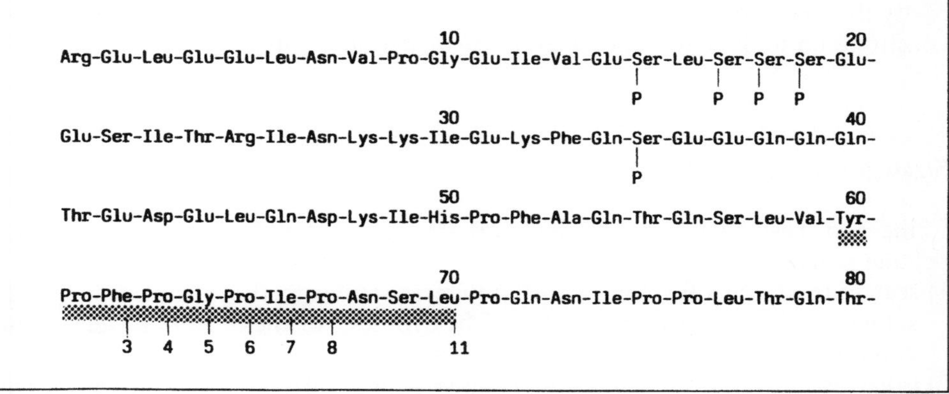

Fig. 1. Bioactive β-casein fragments: N-terminal region of bovine β-casein A²

144

Table 1. Bioactive β-casein fragments

Peptide	Residues	Bioactivity	Isolation		Syn-the-sized	References
			in vitro	in vivo		
β-casomorphin-3	60–62	inactive	–	–	+	[3, 6, 17]
β-casomorphin-4	60–63	opioid agonist immunoreactive	–	(+)[d]	+	[2, 3, 6, 17, 26]
β-casomorphin-4 amide (morphiceptin)	60–63	opioid agonist	+[a]	–	+	[3, 6, 8, 17]
β-casomorphin-5	60–64	opioid agonist	+[a]	–	+	[1–3, 6, 15, 17]
β-casomorphin-6	60–65	opioid agonist immunoreactive	–	(+)[d]	+	[17, 26]
β-casomorphin-7	60–66	opioid agonist	+[b, c]	–	–	[1, 2, 6, 15]
β-casomorphin-8	60–67	?	+[a]	–	–	[8]
β-casomorphin-11	60–70	opioid agonist	–	+[e]	–	[19]

[a] enzymatic digest; [b] casein hydrolysate; [c] peptone digest; [d] intestine contents after bovine milk ingestion in adult humans [26]; [e] jejunal chyme of minipigs

Similar peptides have been found in human milk proteins; the appropriate "hidden" sequences correspond to the 41–44th, the 51–58th, and to the 59–63rd residues of human β-casein [13, 24]. Among other bioactive fragments, the tetrapeptide amides valmuceptin (51–54th residues) and β-casorphin (41–44th residues) show opioid activities similar to morphiceptin, whereas the hexapeptide Tyr-Pro-Ile-Pro-Glu-Val (54–59th residues) for example – which has been purified from in-vitro digests – acts as an immunostimulant [24]. These findings suggest that anti-infectious immunostimulants are naturally released from human β-casein in the infant's gut [24].

α_s-Casein fragments

Figure 2 shows the bioactive peptides derived from bovine α_s-casein. The fragment which corresponds to the 90–96th residues of α_{s1}-casein could be isolated from in-vitro enzymatic digest and was named α-casein exorphin [18] because of its opioid activity. Very recently [21], a caseinophosphopeptide released from α-casein could be purified from jejunal chyme of minipigs fed with casein diet. It was characterized as a nonapeptide corresponding to the 66–74th residues of α-casein [21]. The calcium binding properties of this oligopeptide probably enhance calcium uptake in the intestine.

$\varkappa$-Casein fragments

Figure 3 lists a number of bioactive peptides derived from bovine $\varkappa$-casein [9]. The fragments which correspond to the 33–39th residues [22] could be isolated not only from in-vitro peptic digests but were also chemically synthesized [9]. The tetra-, penta-, and hexapeptides, named casoxins, were modified by methoxylation during

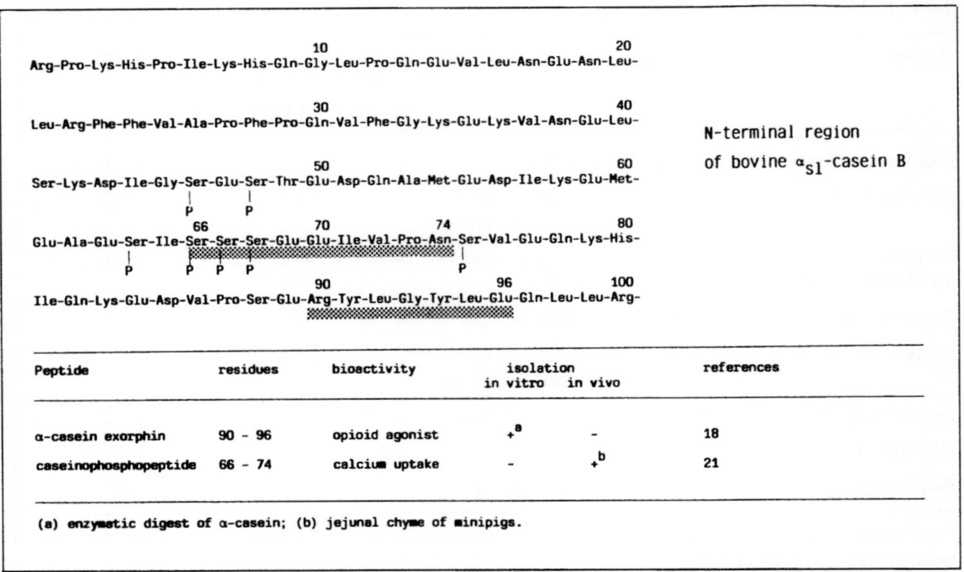

Peptide	residues	bioactivity	isolation in vitro	in vivo	references
α-casein exorphin	90 – 96	opioid agonist	+[a]	–	18
caseinophosphopeptide	66 – 74	calcium uptake	–	+[b]	21

(a) enzymatic digest of α-casein; (b) jejunal chyme of minipigs.

Fig. 2. Bioactive α_s-casein fragments: N-terminal region of bovine α_{s1}-casein B

Peptide	residues	bioactivity	isolation in vitro	in vivo	synthesized	references
casoxin-4	35 – 38	opioid antagonist	+[a]	–	+	9
casoxin-5	34 – 38	"	+[a]	–	+	9
casoxin-6	33 – 38	"	+[a]	–	+	9
hexapeptide[b]	33 – 38	less active	–	–	+	9

(a) enzymatic digest of κ-casein with pepsin; (b) non-methoxylated casoxin-6.

Fig. 3. Bioactive $\varkappa$-casein fragment: N-terminal region of bovine $\varkappa$-casein B

the isolation procedure. The methyl esters are much more active as opioid antagonists than the corresponding non-methoxylated natural peptides [9].

Bioactive sequences in whey proteins

Figures 4 and 5 show the sequences within the primary structures of bovine and human whey proteins [5, 11] which have potential opioid activity. The tetrapeptide

146

```
                                    10                                    20
Glu-Gln-Leu-Thr-Lys-Cys-Glu-Val-Phe-Arg-Glu-Leu-Lys-Asp-Leu-Lys-Gly-Tyr-Gly-Gly-
                   |
             Cys 120
                                    30                                    40
Val-Ser-Leu-Pro-Glu-Trp-Val-Cys-Thr-Thr-Phe-His-Thr-Ser-Gly-Tyr-Asp-Thr-Glu-Ala-
                       |
               Cys 111
                                 50          53                           60
Ile-Val-Glu-Asn-Asn-Gln-Ser-Thr-Asp-Tyr-Gly-Leu-Phe-Gln-Ile-Asn-Asn-Lys-Ile-Trp-
```

N-terminal region of bovine α-lactalbumin B

Peptide	residues	bioactivity	isolation		synthesized	references
			in vitro	in vivo		
α-lactorphin[a]	50 - 53	opioid agonist	-	-	+	9

(a) human and bovine α-lactalbumin

Fig. 4. Bioactive α-lactalbumin fragment: N-terminal region of bovine α-lactalbumin B

```
  102        105                     110                                   120
-Lys-Tyr-Leu-Leu-Phe-Cys-Met-Glu-Asn-Ser-Ala-Glu-Pro-Glu-Gln-Ser-Leu-Ala-Cys-Gln-

                                     130                                   140
Cys-Leu-Val-Arg-Thr-Pro-Glu-Val-Asp-Asp-Glu-Ala-Leu-Glu-Lys-Phe-Asp-Lys-Ala-Leu-

                                     150                                   160
Lys-Ala-Leu-Pro-Met-His-Ile-Arg-Leu-Ser-Phe-Asn-Pro-Thr-Gln-Leu-Glu-Glu-Gln-Cys-

  162
His-Ile
```

C-terminal region of bovine β-lactoglobulin B

Peptide	residues	bioactivity	isolation		synthesized	references
			in vitro	in vivo		
β-lactorphin[b]	102 -105	opioid agonist	-	-	+	9

(b) bovine β-Lactoglobulin.

Fig. 5. Bioactive β-lactoglobulin fragment: C-terminal region of bovine β-lactoglobulin B

amides α-lactorphin and β-lactorphin correspond either to the 50–53rd residues of α-lactalbumin or to the 102–105th residues of β-lactoglobulin. Both lactorphins were chemically synthesized and their opioid activity was established [9].

Conclusion

The discovery of biologically active peptides derived from food proteins has introduced a new criterion in defining the "nutritive value" of proteins. Although the real physiological role of these peptides as exogenous metabolic modulators during their gastrointestinal liberation is not yet well unterstood [9, 19, 26] and their activities are

147

contradictory to some extent, the following dietary applications are indicated for the future:
- supplementation of the diet with desirable synthetic bioactive peptides; and
- production of desirable bioactive peptides during food processing by use of genetically transformed microorganisms.

References

1. Brantl V, Teschemacher H, Bläsig J, Henschen A, Lottspeich F (1979) Novel opioid peptides derived from casein (β-casomorphins). I. Isolation from bovine casein peptone. Hoppe-Seyler's Z Physiol Chem 360:1211–1216
2. Brantl V, Teschemacher H, Bläsig J, Henschen A, Lottspeich F (1981) Opioid activities of β-casomorphin. Life Sci 28:1903–1909
3. Brantl V, Pfeiffer A, Herz A, Henschen A, Lottspeich F (1982) Antinociceptive potencies of β-casomorphin analogs as compared to their affinities towards μ and σ opiate receptor sites in brain and periphery. Peptides 3:793–797
4. Brantl V, Schmitzler G, Daniel H (1988) β-Casomorphine: Opioide aus Casein als Modulatoren gastrointestinaler Funktionen. 2. Effekte auf intestinale Sekretionsprozesse. Ernährungsumschau 35:160
5. Braunitzer G, Chen R, Schrank B, Stangl A (1973) Die Sequenzanalyse des β-Lactoglobulin. Hoppe-Seyler's Z Physiol Chem 354:867–878
6. Chang KJ, Killian A, Hazum E, Cuatrecasas P (1981) Morphiceptin: A potent and specific agonist for morphine (μ) receptors. Science 212:75–77
7. Chang KJ, Cuatrecasas JP, Wei ET, Chang JK (1982) Analgesic activity of intracerebroventricular administration of morphiceptin and β-casomorphins: Correlation with the morphine (μ) receptor binding affinity. Life Sci 30:1547–1551
8. Chang KJ, Su YF, Brent DA, Chang JK (1985) Isolation of a specific μ-opiate receptor peptide, morphiceptin, from an enzymatic digest of milk proteins. J Biol Chem 260:9706–9712
9. Chiba H, Yoshikawa M (1986) Biologically functional peptides from food proteins: New opioid peptides from milk proteins. In: Feeney RE, Whitaker JR (eds) Protein tailoring for food and medical uses. Marcel Dekker, New York and Basel, pp 123–153
10. Daniel H, Vohwinkel M, Brantl V (1988) β-Casomorphine: Opioide aus Casein als Modulatoren gastrointestinaler Funktionen. 1. Effekte auf die gastrointestinale Motilität. Ernährungsumschau 35:160
11. Findlay JBC, Brew K (1972) The complete amino acid sequence of human α-lactalbumin. Eur J Biochem 27:65–86
12. Frister H, Hagemeister H, Meisel H, Schlimme E (1987) Postprandiale intestinale Proteolyse. In: Forschungsbericht Bundesanstalt für Milchforschung Kiel (EWG Nr. 282/84-26.1). Auffindung von neuen Maßstäben zur Bewertung von Milcheiweiß im Vergleich zu pflanzlichem Eiweiß, S 2–29
13. Greenberg R, Groves ML, Dower HJ (1984) Human β-casein: Amino acid sequence and identification of phosphorylation sites. J Biol Chem 259:5132–5138
14. Hagemeister H, Pfeuffer M, Scholz-Ahrens K-E, Barth C (1987) Metabolische Effekte. In: Forschungsbericht Bundesanstalt für Milchforschung Kiel (EWG Nr. 282/84-26.1) Auffindung von neuen Maßstäben zur Bewertung von Milcheiweiß im Vergleich zu pflanzlichem Eiweiß, S 34–56
15. Henschen A, Lottspeich F, Brantl V, Teschemacher H (1979) Novel opioid peptides derived from casein (β-casomorphins). II. Structure of active components from bovine casein peptone. Hoppe-Seyler's Z Physiol Chem 360:1217–1224
16. Kaufmann W (1983) Role of milk proteins in human nutrition. Kieler Milchwirtschaftliche Forschungsberichte 35:241–245
17. Lottspeich F, Henschen A, Brantl V, Teschemacher H (1980) Novel opioid peptides derived from casein (β-casomorphin) III. Synthetic peptides corresponding to components from bovine casein peptone. Hoppe-Seyler's Z Physiol Chem 361:1835–1839

18. Loukas S, Varoucha D, Zioudrou C, Streaty RA, Klee WA (1983) Opioid activities and structures of α-casein-derived exorphins. Biochemistry 22:4567–4573
19. Meisel H (1986) Chemical characterization and opioid activity of an exorphin isolated from in vivo digests of casein. FEBS Lett 196:223–227
20. Meisel H, Hagemeister H (1987) Postprandiale Proteolyse von Casein und Sojaprotein. Milchwissenschaft 42:153–158
21. Meisel H, Frister H (1988) Chemical characterization of a phospho-peptide isolated from in vivo digests of casein. In: Barth C, Schlimme E (eds) Milk proteins: Structural, functional, technological and clinical aspects. Steinkopff, Darmstadt, Springer, New York, pp150–154
22. Mercier JC, Brignon G, Ribadeau-Dumas B (1973) Structure primaire de la caséine B bovine. Séquence complète. Eur J Biochem 35:222–235
23. Morley JE, Levine AS, Yamada T, Gebhard RK, Prigge WF, Shafer RB, Goetz FC, Silvis SE (1983) Effect of exorphins on gastrointestinal function, hormonal release, and appetite. Gastroenterology 84:1517–1523
24. Parker F, Migliore-Samour D, Floc'h F, Zerial A, Werner GH, Jollès J, Casaretto M, Zahn H, Jollès P (1984) Immunostimulating hexapeptide from human casein: amino acid sequence, synthesis and biological properties. Eur J Biochem 145:677–682
25. Schusdziarra V, Schick R, de la Fuente A, Holland A, Brantl V, Pfeiffer EF (1983) Effect of β-casomorphins on somastatin release in dogs. Endocrinology 112:1948–1951
26. Svedberg J, de Haas J, Leimenstoll G, Paul F, Teschemacher H (1985) Demonstration of β-casomorphin immunoreactive materials in in vitro digests of bovine milk and in small intestine contents after bovine milk ingestion in adult humans. Peptides 6:825–830

Isolation and Chemical Characterization
of a Phosphopeptide from In Vivo Digests of Casein

H. Meisel and H. Frister

Institute for Chemistry and Physics, Federal Dairy Research Centre, Kiel, FRG

Introduction

It is now accepted that peptides, and not amino acids, are the main degradation products of protein digestion. Moreover, it is becoming increasingly evident that a consideration of protein value must take into account the relationship between protein structure and the amount and composition of peptides released during digestion in the gastrointestinal tract. In particular, interactions with other nutrients, transport functions, and biological activities of peptides liberated from dietary protein might be of significance to its nutritive value.

Bioactive peptides have been identified as digestion products of several food proteins [3, 9, 16]. All bioactive peptides are hidden in an inactive state inside the polypeptide chain of a larger protein. Milk protein is a rich source of biologically active peptides [15] such as casomorphins and caseinophosphopeptides, which might have significant nutritional implications. Recently, it has been shown that phosphopeptides are formed in the small intestinal lumen during the digestion of food containing casein [14].

The aim of the present study is to identify caseinophosphopeptides as in vivo digestion products of bovine casein.

Experimental

The primary advantage of pigs as animal models is their marked anatomical and physiological similarity to humans with respect to the digestive tract and nutritional requirements [5]. Digests of the distal jejunum were quantitatively collected in two Göttingen-strain minipigs fitted with T-shaped cannulas after feeding with 200 g of a diet containing 15% casein [10].

The reversed phase (C 18) method for analytical and preparative peptide separation with a gradient of acetonitrile in 0.1% trifluoroacetic acid is described elsewhere [11].

Phosphopeptides were enriched by affinity chromatography using ferric ions immobilized on a chelating iminodiacetate-agarose gel [1]. The ferric-ion-loaded gel served as a group-specific sorbent for phosphorylated amino acid side chains. Phosphopeptides were strongly retained by this sorbent at acidic pH, but were readily eluted by increasing the pH [11].

The quantitative amino acid analysis of total hydrolysates and the determination of phosphoserine (SerP) in partial hydrolysates was performed by an ion-exchange

150

chromatography technique using an automated analyzer. For details see Meisel and Hagemeister [10] and Meisel and Frister [11].

Purified peptides were subjected to N-terminal group analysis by the dansyl chloride method and to C-terminal analysis by digestion with carboxypeptidase Y [9, 11]. Furthermore, the treatment with carboxypeptidase Y allowed the sequential release of amino acid residues from the C-terminus.

α- and ε-amino groups were determined by a modified OPA method using N,N-dimethyl-2-mercaptoethylammonium chloride as the thiol component [6]. The quotients calculated from OPA-sensitive amino groups before and after HCl hydrolysis allow the determination of the chainlength of purified peptides, provided that the amino acid composition is known.

The phosphorus content of phosphopeptides was measured after wet ashing [7] which converts organic phosphorus to inorganic phosphate. Inorganic phosphate was quantified as the phosphomolybdate complex by an ultramicro method [2].

Isolation and characterization

Since casein contains some clusters of phosphorylated serine residues, digestion of this protein probably liberates various phosphopeptides. Figure 1 shows the typical peptide pattern obtained by reversed phase HPLC of the soluble part of intestinal

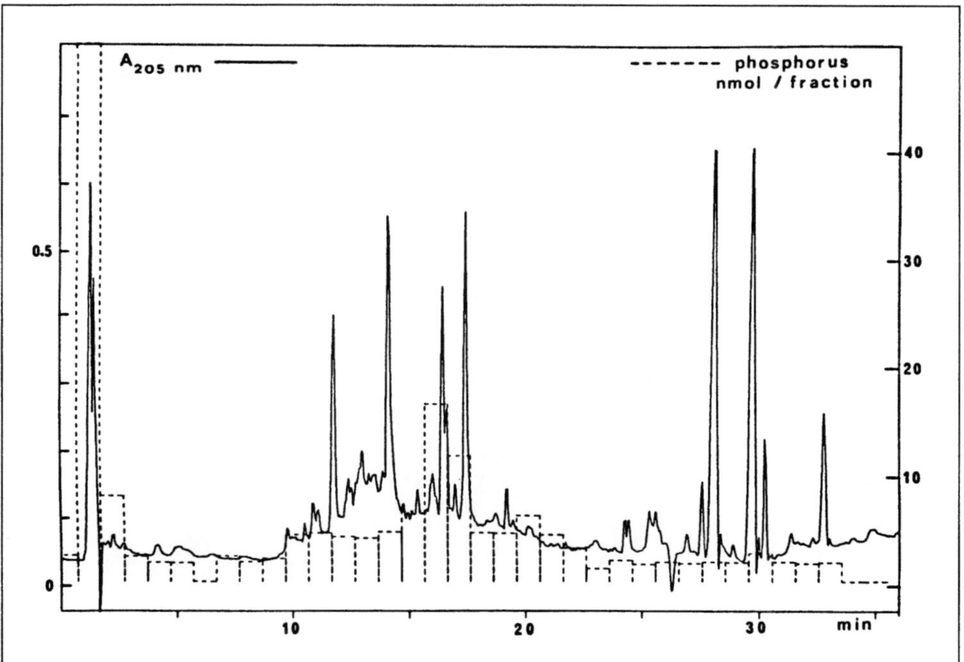

Fig. 1. HPLC (RP 18) elution pattern and phosphorus content in the peptide fraction of the soluble part in the jejunum after ingestion of a casein diet

151

chyme after ingestion of the diet containing casein. The inorganic phosphorus, as well as free amino acids, appeared in the first two fractions. Caseinophosphopeptides could be identified by the relatively high content of peptide-bound phosphorus in certain HPLC fractions. The main phosphopeptide fraction with a retention time between 16 and 17 min was isolated by several preparative HPLC runs. Qualitative amino acid analysis of the isolated caseinophosphopeptides indicated that phosphoserine was the only phosphorylated amino acid present.

The phosphopeptide-enriched fraction obtained after affinity chromatography was separated by preparative reversed phase HPLC (Fig. 2). The main phosphopeptide with a retention time of about 17 min was collected from several runs. After further rechromatography using a less steep gradient, the isolated phosphopeptide was pure enough for the determination of amino acid composition, endgroups, chain length, and phosphorus content. The results are summarized in Table 1.

The nearly equimolar concentrations of serine residues and organic phosphorus indicate that all of the serine is present in the phosphopeptide as phosphoserine. When the known amino acid sequence of the different caseins is examined for phosphorylated regions, it is clear that the analytical data are only consistent with a nonapeptide fragment corresponding to the sequence 66 to 74 in the primary structure of α_{s1}-casein (genetic variants A–E): SerP-SerP-SerP-Glu-Glu-Ile-Val-Pro-Asn.

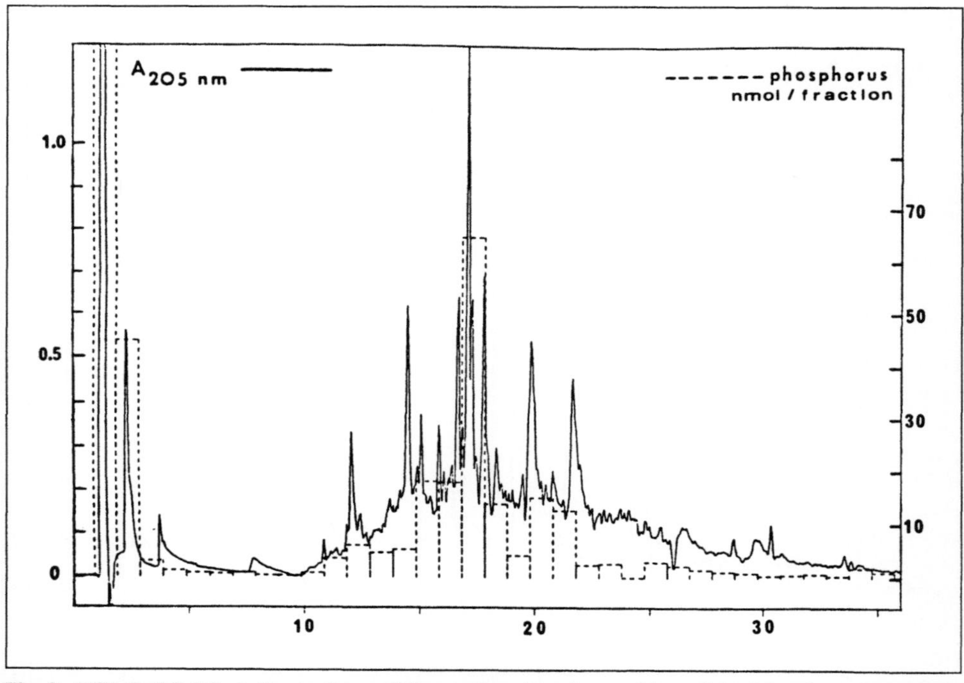

Fig. 2. HPLC (RP 18) elution pattern of the caseinophosphopeptide-enriched fraction obtained by (Fe^{3+})-affinity chromatography of jejunal chyme

Table 1. Amino acid composition, serine/phosphorus ratio, chain length, N- and C-termini and partial C-terminal sequence of the isolated caseinophosphopeptide

Component	Purified peptide (analytical data)	Theoretical peptide (deduced from analytical data)
Amino acids (molar ratio)		
Asp	1.05	1
Ser	2.73	3
Glu	1.98	2
Pro	1.00	1
Gly	0.30	
Ala	0.13	
Val	0.88	1
Ile	1.48	1
Ser/P (molar ratio)	1.10	1
Peptide chain length	9.30	9
N-terminus	Ser	Ser (P)
C-terminus	Asn	Asn
C-terminal partial sequence	-Ile-Val-Pro-Asn·OH	

Possible physiological effects of bioactive caseinopeptides

It is known that caseinophosphopeptides are highly resistent to further proteolytic breakdown by digestive enzymes. Furthermore, they exhibit a potent ability to form soluble complexes with calcium and prevent the precipitation of calcium phosphate [12]. Hence, caseinophosphopeptides are likely to enhance the rate of passive calcium absorption from the distal small intestine lumen by increasing the concentration of soluble calcium [14]. Caseinophosphopeptides can also form organophosphate salts with trace elements, such as iron, zinc, and copper, and may function as carriers for these particular minerals. Accordingly, they have already found interesting applications as dietary supplements and in pharmaceutical preparations [4].

Conclusions

In this study, the in vivo formation of a caseinophosphopeptide, has been proven for the first time and under physiological conditions. These findings therefore support the hypothesis that bioactive peptides derived from casein participate in the regulation of nutrient entry. Moreover, the molecular characteristics of peptides generated during proteolysis, as well as their influence on metabolic processes, offer new aspects of the evaluation of the nutritive value of milk proteins.

Acknowledgements

This work was supported by the European Community (VO No. 1150/86-14.5). The authors wish to thank Dipl.-Ing. J. Carstens for performing the HPLC separations and Dr. Tait for reading the manuscript. Sincere appreciation is expressed to Prof. E. Schlimme for helpful discussions during this work.

References

1. Andersson L, Porath J (1986) Isolation of phosphoproteins by immobilized metal (Fe^{3+}) affinity chromatography. Anal Biochem 154:250–254
2. Bartlett GR (1959) Phosphorus assay in column chromatography. J Biol Chem 234:466–468
3. Brantl V, Teschemacher H, Henschen A, Lottspeich F (1979) Novel opioid peptides derived from casein (β-casomorphins) I. Isolation from bovine casein peptone. Hoppe-Seyler's Z Physiol Chem 360:1211–1216
4. Brule G, Roger L, Fauquant J, Piot M (1982) Phosphopeptides from casein-based material. United States Patent 4, 358, 465. 9/1982
5. Dodds WJ (1982) The pig model for biomedical research. Fed Proc 41:247–256
6. Frister H, Meisel H, Schlimme E (1988) OPA method modified by use of N,N-dimethyl-2-mercaptoethylammonium chloride as thiol component. Fresenius Z Anal Chem 330:631–633
7. Genter PR, Haasemann A (1979) Phosphometrie – Eine Methode zur schnellen quantitativen Bestimmung von Phospholipiden nach dünnschichtchromatographischer Trennung. Fette, Seifen, Anstrichm 81:357–360
8. Henschen A, Lottspeich F, Brantl V, Teschemacher H (1979) Novel opioid peptides derived from casein (β-casomorphins) II. Structure of active components from bovine casein peptone. Hoppe-Seyler's Z Physiol Chem 360:1217–1224
9. Meisel H (1986) Chemical characterization and opioid activity of an exorphin isolated from *in vivo* digests of casein. FEBS Lett 196:223–227
10. Meisel H, Hagemeister H (1987) Postprandiale Proteolyse von Casein und Sojaprotein. Milchwissenschaft 42:153–158
11. Meisel H, Frister H (1988) Chemical characterization of a caseinophosphopeptide isolated from *in vivo* digests of a casein diet. Biol Chem Hoppe-Seyler, in press
12. Mellander O (1963) In: Wasserman RH (ed) The transfer of calcium and strontium across biological membranes. Academic Press, New York London, pp 265–276
13. Narita K, Matsuo H, Nakajima T (1975) In: Needleman SB (ed) Protein Sequence Determination. Springer, Berlin Heidelberg New York, pp 30–103
14. Sato R, Noguchi T, Naito H (1986) Casein phosphopeptide (CPP) enhances calcium absorption from the ligated segment of rat small intestine. J Nutr Sci Vitaminol 32:67–76
15. Schlimme E, Meisel H, Frister H (1988) Bioactive sequences in milk proteins. This volume, pp 143–149
16. Zioudrou C, Streaty RA, Klee WA (1979) Opioid peptides derived from food proteins. J Biol Chem 254:2446–2449

Expression of an α_{s1}-Casein cDNA-Clone in a Cell-free and Procaryote Model System

F. Niepold[1], P. Dovč[2] and O. J. Rottmann[3]

[1] Institute of Biochemistry, Biologische Bundesanstalt, Messeweg 11, D-3300 Braunschweig, FRG
[2] University E. Kardelj Ljubljana, Biotechnical faculty, Zootechnical department, 61 230 Domžale, Yugoslavia
[3] Institute of Animal Science, Technische Universität München, D-5080 Freising-Weihenstephan, FRG

Introduction

The protein fraction of milk consists of a few major lactoprotein species synthesized in the lactating mammary gland. In bovine milk caseins represent 80% of the total lactoproteins. Linkage studies [1, 2] and in situ hybridization [3, 5] revealed linkage between α_{s1}-Cn-α_{s2}-Cn-β-Cn-$\varkappa$-Cn genes, forming a gene cluster on a single chromosome, controlled by steroid and peptide hormones. Large amounts of casein-specific mRNAs in the lactating mammary gland make possible the successful application of recombinant DNA techniques to establish cDNA libraries [9]. Isolation of genomic sequences, coding for caseins, sequencing of cDNA and genomic DNA clones, and RFLP studies offer much information about the fine structure and possible mechanisms of regulation of that gene complex. In the following experiments we studied expression of the α_{s1}-casein-coding cDNA clone using two in vitro expression systems.

Material and methods

The cDNA sequence, coding for bovine pre-α_{s1}-casein B was isolated from the plasmid pBα_{s1}C184 [7]. For the in vitro translation experiment the plasmid pSPcas was constructed. The 1106 bp-BglII-PstI α_{s1}-Cn cDNA fragment was inserted into BamHI-PstI-BglII-PstI α_{s1}-Cn cDNA fragment was inserted into BamHI-PstI double restricted polylinked sequence of the vector pSP65 (Promega Biotec). After ligation with DNA T4 ligase (BRL) the derived plasmid pSPcas was propagated in *E. coli* HB101 and after linearization, either with PstI or BglI, endonuclease served as a template for in vitro transcription. In vitro transcription and capping of the transcripts were carried out using SP6 RNA polymerase and $m^7G(5')ppp(5')Gm$ cap structure [4]. The RNA transcripts were translated in a rabbit reticulocyte lysate in vitro translation system (BRL). The ^{35}S-methionine labeled translation products were run on SDS PAGE followed by autoradiography.

For the expression experiment in procaryotic cells the UV sensitive *E. coli* mutant SS3228c (maxi cell) was used. The vector pCAS was constructed by insertion of the 668 bp casein coding BglII-EcoRI sequence into the BamHI-EcoRI double restricted plasmid pUC9 [8]. After UV irradiation and cycloserin-D treatment the pCAS-transformed maxi cells were starved for sulphate. After addition of ^{35}S methionine and IPTG, *E. coli* was incubated for 1 h and subsequently lyzed using lysozyme- and ultra sonic-treatment. Cell-lysate was immunoprecipitated and run on SDS PAGE. Radioactive-labeled protein products were identified by autoradiography.

Results and discussion

The plasmid pSPcas containing the cDNA sequences coding for bovine pre-α_{s1}-casein B was set under transcriptional control of the strong bacteriophage SP6 promoter (Fig. 1a) in order to have an efficient in vitro transcription. RNA transcripts were capped and in vitro translated using rabbit reticulocyte lysate in vitro translation system. Specific in vitro translation products showed in SDS PAGE slightly higher molecular weight (MW) compared to the chemically purified α_{s1}-casein, due to the additional leader peptide sequence. The site of linearization of the DNA template had no influence on the translation products. As previously shown [6] the poly-A tail is not essential for in vitro translation using reticulocyte system.

Successful integration of the cDNA fragment coding for α_{s1}-casein into the lac Z gene-region of the plasmid pUC9 (Fig. 1b) was tested using the x-gal test. Protein products of transformed maxi cells precipitate with anti-casein antibodies (Fig. 2). The reason for the unexpected high MW of the protein showing the casein-like antigenic properties might be a "read through", due to reading frame change on the 3'-end of the insert.

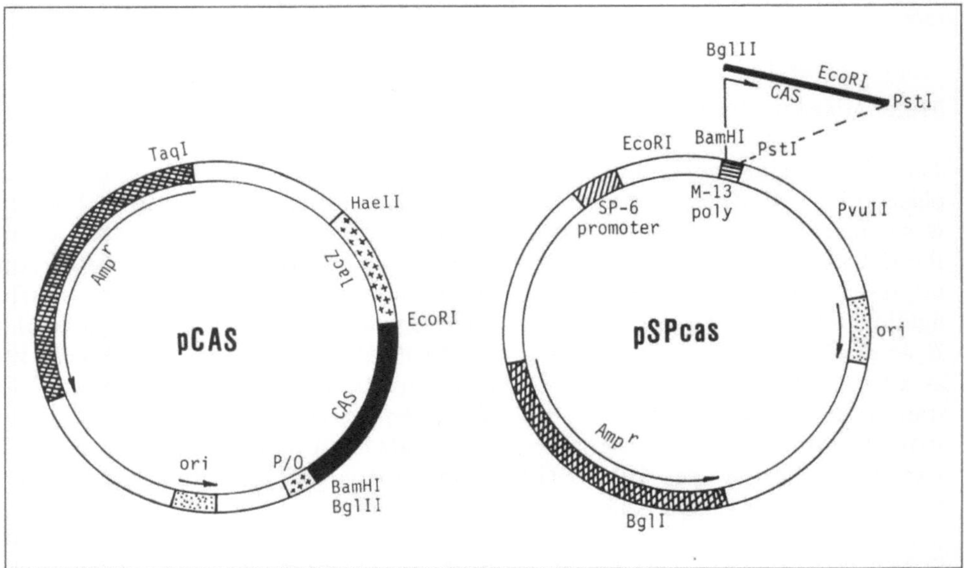

Fig. 1. Construction of the expression vectors: a) pSPcas; b) pCAS

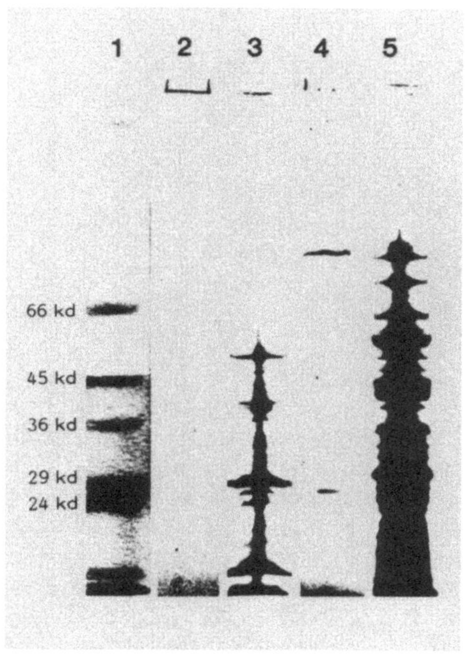

Fig. 2. SDS PAGE autoradiogram from immuno-precipitated bacterial protein products: 1) molecular weight marker; 2) immuno-precipitation of the pUC9 transformed maxi cells; 3) pUC9 transformed maxi cells; 4) immuno-precipitation of pCAS transformed maxi cells showing a specific immunoprecipitation line; 5) pCAS transformed maxi cells

References

1. Grosclaude F, Mercier J-C, Ribadeau-Dumas B (1973) Genetic aspects of cattle casein research. Neth Milk Dairy J 27:328–340
2. Grosclaude F, Jourdier P, Mahe MF (1979) A genetic and biochemical analysis of a polymorphism of bovine α_{s2}-casein. J Dairy Res 46:211–213
3. Gupta P, Rosen JM, Eustachio PD, Ruddle FH (1982) Localization of the casein gene family to a single mouse chromosome. J Cell Biol 93:199–204
4. Melton AD, Krieg PA, Rebagliati MR, Maniatis T, Zinn K, Green MR (1984) Efficient in vitro synthesis of biologically active RNA and RNA hybridization probes from plasmids containing a bacteriophage SP6 promoter. Nucl Acid Res 12(18):7035–7056
5. Mercier J-C, Gaye P, Soulier S, Hue-Delahaie D, Vilotte J-I (1985) Construction and Identification of recombinant plasmids carrying cDNAs coding for ovine α_{s1}, α_{s2}- β-, $\varkappa$-casein and β-lactoglobulin. Nucleotide sequence of s1-casein cDNA. Biochemie 67:959–971
6. Noma Y, Sideras P, Naito T, Bergstedt-Lindquist S, Azuma C, Severinson E, Zanabe T, Kinashi T, Matsuda F, Yaoita Y, Honjo T (1986) Cloning of cDNA encoding the murine IgG1 induction factor by a novel strategy using SP6 promoter. Nature 319:640–646
7. Stewart AF, Willis IM, Mackinlay AG (1984) Nucleotide sequences of bovine α_{s1}- and $\varkappa$-casein cDNAs. Nucl Acid Res 12(9):3895–3907
8. Viera J, Messing J (1982) The pUC plasmids, an M13mp7-derived system for insertion mutagenesis and sequencing with synthetic universal primers. Gene 19:259–268
9. Willis IM, Stewart AF, Caputo A, Thompson AR, Mackinlay AG (1984) Construction and identification by partial nucleotide sequence analysis of bovine casein and β-lactoglobulin cDNA clones. DNA 1:375–386

Heat Stability of Milk with Increased Whey Protein Content

P. Jelen and J. Patocka

Department of Food Science, University of Alberta, Edmonton, Canada

Introduction

To minimize hard-clot formation in the human stomach upon ingestion of regular cow milk, a product with increased whey protein content has been proposed [1]. Investigations of heat stability of this modified milk showed differences due to the use of ultrafiltration (UF) produced whey protein retentates (UFR) from cottage cheese or rennet-type wheys [2]. Coagulation of milk with cottage cheese whey UFR was observed, but the cause was not determined. To elucidate this problem, investigations of the heat-induced coagulation in blends of skim milk and cottage cheese whey UFR were carried out in comparison to similar blends with UFR from unheated and pasteurized (72 °C for 15 s) rennet wheys and from a sweet whey obtained by ultracentrifugation (UC) of skim milk.

Methodology

The same approach as in [2] was followed for preparation of the modified milks. Cottage cheese whey from an industrial processor, laboratory-prepared rennet whey, and the UC whey obtained according to [3], were ultrafiltered on an Amicon 8400 (Amicon Corp., Lexington, Massachusetts, USA) UF unit with a 10 000 D membrane. Some of the cottage cheese whey was decalcified before the UF by an ion exchange treatment described in [4]. Heating was in test tubes submerged in a boiling water-bath for 30 min. Heat stability was taken as time needed for appearance of visually distinctive coagulation. Sodium dodecyl sulphate polyacrylamide gel electrophoresis (SDS-PAGE) before and after heating was carried out as in our previous work [4]. Selected samples embedded in agar gel were subjected to transmission electron miocroscopy observations using established methodology [5].

Results and discussion

Heat stability data for various skim milk-cottage cheese whey protein retentate mixtures (Table 1) indicate that when the whey protein-casein ratio increased from 1:4, as in milk, to about 3:4 (i.e., mixtures of about 70% milk and 30% UFR), the heat stability was impaired. At these higher whey protein concentrations the heat induced β-Lg and $\varkappa$-casein interaction does not have the usual stabilizing effect against the β-Lg precipitation. On the contrary, the higher calcium ion concentration

158

Table 1. Heat stability of milk mixtures with cottage cheese whey UF retentates; heating time 30 min max at 93 °C (come-up time excluded)

Skim milk/whey UF retentate (% v/v)	Heating time to coagulation (min)	Protein content[a] (% w/w)
0/100	0	0.8
10/ 90	0	0.8
20/ 80	2	0.9
40/ 60	4	0.9
50/ 50	8	1.0
70/ 30	> 30	3.3
90/ 10	> 30	3.2
100/ 0	> 30	3.4

[a] By Kjeldahl of centrifugation supernatants after heating (average protein content in mixtures before heating was 3.2, range 3.1–3.4)

in UFR seems to induce casein co-precipitation and destabilization of the whole protein system, as shown by the protein content (Table 1) and the SDS-PAGE of supernatants from the coagulated samples and the heat-stable mixtures (Fig. 1 a, b).

The Ca content of the cottage cheese whey UFR (0.11%) decreased drastically (to 0.03%) after heating the pH 6.5 adjusted retentate and coagulum removal by centrifugation. In similar experiments with decalcified acid whey UFR (Ca content 0.01%), a partial precipitation was seen after 10 min of heating at 93 °C. Unheated or pasteurized rennet whey (Ca = 0.04%) or UC sweet whey (Ca 0.05%) UFR appeared to be more heat-resistant and no heat-induced coagulation was observed in the conditions of the test. All combinations of milk and UFR from these wheys were heat stable, with the exception of mixtures with unpasteurized rennet whey UFR which behaved similarly to those made with regular cottage cheese whey UFR. However, in this case the coagulation may have been due to the residual rennet activity in the UF concentrates.

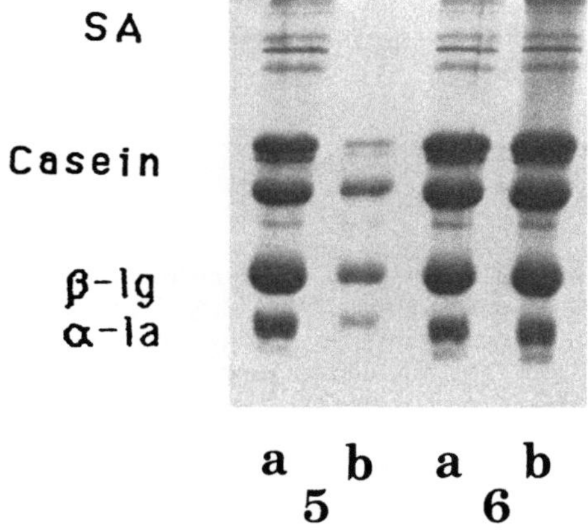

Fig. 1. Examples of SDS-PAGE patterns of heat-unstable and heat-stable milk with varying whey protein/casein content before (a) and after (b) heating at 93 °C for 30 min. Skim milk/cottage cheese whey UF retentate ratios: (5) – 50:50; (6) – 70:30. Below 50:50 all patterns were similar to (5) with progressively fainter casein bands as the casein concentration decreased; above 50:50, all patterns were the same as in (6)

Electron microscopy observations confirmed that heating the rennet whey UFR alone produced large, well defined particles but no aggregates typical of coagulated protein (Fig. 2a). The visual appearance of this retentate changed upon heating to resemble regular skim milk. The 10:90 mixture of the unpasteurized UFR with skim milk coagulated upon heating, similar to the cottage cheese whey UFR mixtures; correspondingly, large aggregated clusters typical of coagulated protein can be seen (Fig. 2b). Microstructure of the heat-stable 70:30 mixture (Fig. 2c) was similar to regular milk after heating, with clearly distinguishable casein micelles and no whey protein aggregation.

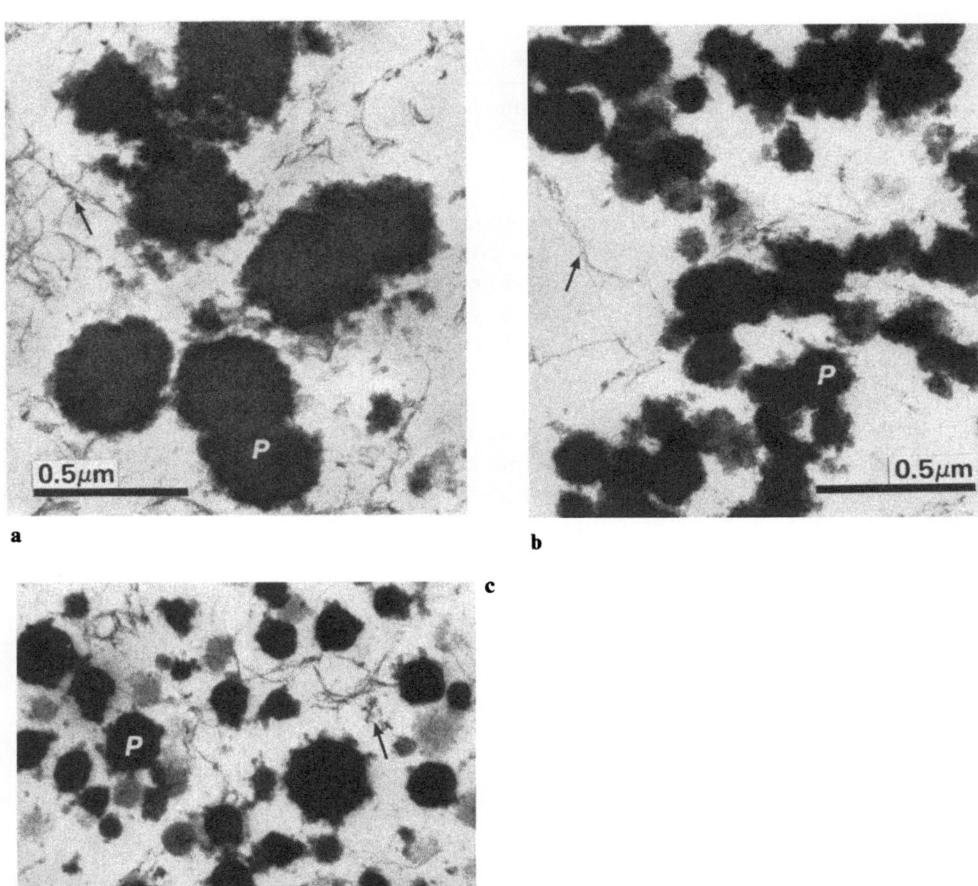

Fig. 2. Transmission electron micrographs of the unpasteurized rennet whey UFR/skim milk preparations after heating for 30 min at 93 °C. The arrow indicates agar gel used for embedding of the samples; P=milk protein. (a)=rennet whey UFR with no skim milk; (b)=10/90 skim milk/UFR; (c)=70/30 skim milk/UFR

Conclusions

The heat instability of milk with increased whey protein content appears to be due to calcium-modulated co-precipitation of β-Lg with casein. After calcium removal from acid-type whey, or in UC, or pasteurized rennet-type wheys with genuinely low Ca-content, the skim milk/UFR mixtures were heat stable.

Acknowledgement:

The electron microscopy work was carried out at the Electron Microscopy Centre, Agriculture Canada, Ottawa, by Dr. M. Kalab; his contribution is acknowledged with sincere thanks.

References

1. Buchheim W, Peters K-H, Kaufmann W (1986) Milchwissenschaft 41:139–141
2. Jelen P, Buchheim W, Peters K-H (1987) Milchwissenschaft 42:418–421
3. Morr CV (1973) J Dairy Sci 56:544–552
4. Patocka J, Drathen M, Jelen P (1987) Milchwissenschaft 42:700–705
5. Kalab M, Emmons DB, Sargant AG (1976) Milchwissenschaft 31:402–408

Ultrastructural Characterization of High Molecular Weight Milkfat Globule Membrane Glycoproteins

W. Buchheim

Institute for Chemistry and Physics, Federal Dairy Research Centre, Kiel (FRG)

Introduction

During the secretion mechanism of intracellular milkfat droplets from the lactating cell into the glandular lumen an envelopment by the apical plasma membrane takes place which results in the well known high emulsion stability of the fat phase in fresh milk. This initial milkfat globule membrane (MFGM) is mainly composed of various types of polar lipids, (glyco-)proteins, and enzymes; its structural organization exhibits three characteristic zones, i.e., an (inner) proteinaceous coat, a unit membrane, and a glycocalyx [1, 5, 8]. MFGM glycoproteins show a considerable variation among mammalian species. Mucin-like glycoproteins with an estimated molecular weight of ca. 500 kD and a carbohydrate content of 50–80% (w/w) – mainly galactose, fucose, glucos- and galactosamine, and sialic acids – have been isolated from human MFGM, but have not been found e.g., in bovine MFGM [6, 7]. Only in the presence of these mucin-like glycoproteins has an unusually well developed glycocalyx, i.e., a filamentous exterior coat, been demonstrated by transmission electron microscopy applying the freeze-etching technique [1–3, 8]. Because these characteristic proteinaceous constituents of some milks have so far received little attention as to their possible biological significance some methodical aspects and some recent findings will be summarized.

Materials and methods

Fat globules from freshly drawn milk are transferred to distilled water or dilute buffer solutions, e.g., by means of repeated centrifugal creaming and redispersion (final concentration of fat phase ca. 20–50% (v/v)). Cross-linking reactions, e.g., by glutaraldehyde, may be applied for stabilizing membrane structure.

Prior to electron microscopy preparation these globule dispersions may be further treated, e.g., with heating, with proteolytic or glycolytic enzymes, or with specific markers like cationized ferritin or lectin-ferritin complexes. Small droplets of fat globule dispersion are rapidly deep-frozen by direct immersion into coolants, e.g., melting Freon 22 (–160 °C). Deep-frozen specimens are further handled in a freeze-fracture apparatus, i.e., cleaved (e.g., at –105 °C) and subsequently freeze-etched (e.g., for 2–5 min) in order to sublime ice to a depth of roughly 1 µm. The freeze-etched surface of the specimen is replicated by unidirectional or rotary shadowing with platinum/carbon plus pure carbon. Cleaned replicas are studied in a transmission electron microscope.

162

Results and discussion

Milkfat globules from species like humans, horses, or pigs which show high molecular weight glycoproteins in electrophoresis exhibit a pronounced filamentous glycocalyx in contrast to others without such glycoproteins like globules from cow, goat or rabbit milk (Fig. 1).

The filaments originate from the membrane surface and extend up to $0.5-1$ µm into the aqueous phase. Apparently these filaments represent mucin molecules in an unfolded state, the degree of unfolding possibly depends on the ionic strength of the buffer solution. Pretreatment of milks with glutaraldehyde stabilizes the structure of the membrane but does not affect the appearance of the filaments.

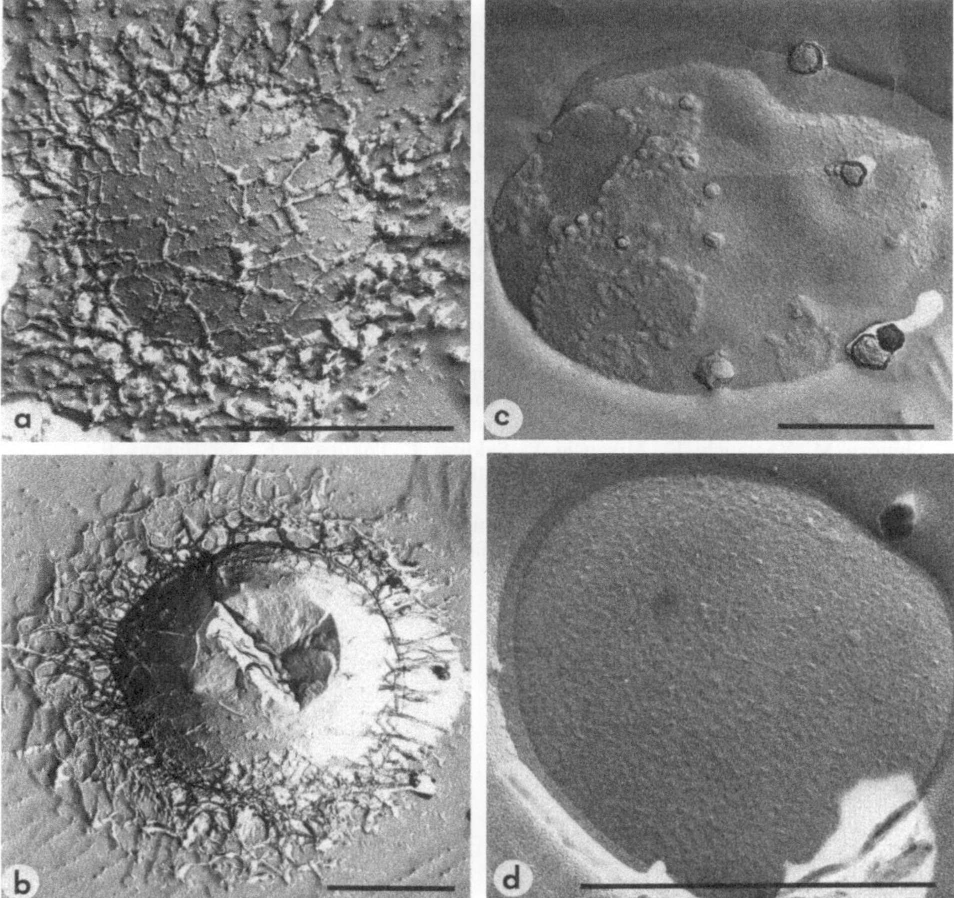

Fig. 1. Freeze-etch electron micrographs of milkfat globules from milks of humans (a), horses (b), cows (c), and goats (d), demonstrating the species-dependent occurrence of a filamentous glycocalyx, i.e., mucin-like glycoproteins. (Scale bars equal 1 µm)

The glycoprotein filaments are removed from the MFGM surface by heating or by certain proteolytic enzymes [1–3, 6]. Heavy binding of cationized ferritin demonstrates the negatively charged nature of the filaments and the membrane surface [2]; binding of various lectins demonstrates the presence of certain carbohydrates on the filaments [2, 8].

It is concluded that the direct visualization of these MFGM-bound mucin-like glycoproteins is supportive for biochemical and immunological characterizations. The species-dependent presence of such glycoproteins on the MFGM raises questions about their biological significance and possible function either in the mammary gland or also in the nutrition of the newborn [3, 6]. In the field of cancer research monoclonal antibodies raised against these mucins of human MFGM have received major attention because of their significance for the diagnosis and prognosis of breast cancer [4].

References

1. Buchheim W (1986) Membranes of milk fat globules. Ultrastructural, biochemical and technological aspects. Kieler Milchw Forsch Ber 38:227–246
2. Buchheim W, Welsch U, Patton S (1988) Electron microscopy and carbohydrate histochemistry of the human milk fat globule membrane. In: Hanson LA (ed) Biology of human milk. Nestlé Nutrition Workshop Series, Vol 15, Nestec Ltd, Vevey/Raven Press, New York, pp 27–44
3. Buchheim W, Welsch U, Huston GE, Patton S (1988) Glycoprotein filament removal from human milk fat globules by heat treatment. Pediatrics 8:357–365
4. Hilgers JHM, Hagemann PC (1986) A review on antigens of milk fat globule membranes and their value for diagnosis and prognosis of breast cancer and other carcinomas. Keio J Med 35:235–244
5. Keenan TW, Dylewski DP, Woodford TA, Ford RH (1983) Origin of milkfat globules and the nature of the milk fat globule membrane. In: Fox PF (ed) Developments in dairy chemistry – Vol 2. Applied Science Publishers, London, pp 83–118
6. Shimizu M, Yamauchi K (1982) Isolation and characterization of mucin-like glycoprotein in human milk fat globule membrane. J Biochem (Tokyo) 91:515–525
7. Shimizu M, Yamauchi K, Miyauchi Y, Sakurai T, Tokugawa K, Mc-Ilhenny RAJ (1986) High-M_r glycoprotein profiles in human milk serum and fat globule membrane. Biochem J 233:725–730
8. Welsch U, Buchheim W, Schumacher U, Schinko I, Patton S (1988) Structural, histochemical and biochemical observations on horse milk fat globule membranes and casein micelles. Histochemistry 88:357–365

Colostrum as a Source of Antibodies

G. L. de Crignis, H. H. Mank, H. Elbertzhagen, and H. Klostermeyer

Süddeutsche Versuchs- und Forschungsanstalt für Milchwirtschaft,
Institut für Chemie und Physik, TU München-Weihenstephan, FRG

Introduction

Vaccination of pregnant cattle for immunization of calves by pinocitary resorption of colostral antibodies vs pathogenous microorganisms of infectious breeding diseases has been applied for years in veterinary medicine [2].

The application of antibodies from the colostrum against defined protein antigens for therapeutic or analytic purposes by immunization of cows was seldom applied up to now; only Fey et al. [3] and Hammer et al. [4] have dealt with this subject more intensively and produced antibodies in bovine colostrum against ovalbumin, rabbit serum albumin, rabbit IgG, and horse ferritin.

Compared to the conventional production of antibodies the production of colostral antibodies has the following advantages:

1) No special animal housing is necessary;
2) The yield of antibodies is essentially higher and more uniform; about 80% of the immunoglobulins of bovine colostrum consist of IgG_1, whereas in the serum it is only 50%. The concentration of IgG_1 in colostrum varies between $35-75$ g/l; in serum the concentration is 11 g/l. The average quantity of colostrum per cow amounts to $4-5$ l, corresponding to a yield of IgG of $136-375$ g per animal per birth;
3) A veterinarian is only required for the immunization; colostrum is obtained simply by milking;
4) Invasive manipulations for the extraction of blood and physical and psychical stress of the cows are minimized.

We have applied chemically modified 7 S-soyprotein fraction (conglycinin), or chemically modified bovine β-lactoglobulin to cows in late gestation [1, 5].

Deriving β-lactoglobulin was necessary to make it more specific for the bovine immune system. A chemical modification of the soy protein was made for improved solubility.

Colostrum taken two and ten hours post partum was defatted, caseins were removed by renneting, and the colostral whey was tested for the following properties:

1) test of successful immunization by the qualitative and quantitative determination of specific antibodies;
2) cross reactivity of antibodies;
3) enrichment of antibodies;
4) suitability of antibodies for analytical purposes.

Methods

Immunogens:

a) conglycinin succinylated according to Hoagland [1, 5];
b) conglycinin oxidized with performic acid according to Klostermeyer et al. [1, 5];
c) β-lactoglobulin was converted to reduced S-carboxymethyl-β-lactoglobulin
   ($=$RCM) according to McKenzie et al. [1, 5] and thereafter it was succinylated
   ($=$S-RCM) according to Habeeb [1, 5]; a peptide produced by BrCN-cleavage
   from RCM, according to Otani [1, 5], was coupled with glutaraldehyde and
   human serumalbumin according to Avrameas [1, 5].

Immunization:

Every animal was immunized eight, five and two weeks ante partum (calculated from
the date of insemination). The total quantity of immunogens was 150 mg of S-RCM
and 220 mg of HSA-RCM per cow; 1 000 mg of the soy protein derivates were
applied. Freund's incomplete adjuvant was used as immunostimulant in order to
avoid a tuberculine-positive reaction of the animals.

Determination of antibodies:

a) double radial immunodiffusion according to Ouchterlony;
b) immunoelectrophoresis according to Grabar-Williams;
c) counter currency electrophoresis according to Gocke and Howe;
d) single radial immunodiffusion according to Mancini et al. [1, 5].

Enrichment of colostral IgG:

The following methods were tested for their suitability:
a) isoelectric dialysis precipitation at low ionic strength;
b) ammonium sulfate precipitation;
c) ethanol-precipitation;
d) polyethyleneglycol precipitation;
e) Congo Red precipitation.

Results

1) Neither in the colostrums nor in the precipitates taken from the colostrum could
   precipitating antibodies against S-RCM and HSA-RCM be detected. Probable
   reasons: the derivated allogenic β-lactoglobulins are not behaving sufficiently like
   a xenogenic antibody and therefore, they are only immunogenically weak in the
   bovine immune system. Incomplete Freund's adjuvant cannot counterbalance

this deficiency; therefore, the quantities of the immunogenes used were not high enough.

2) Colostrums of animals which were immunized by the soy protein derivates contained antibodies, but the concentrations were so low that for further investigations enriched precipitation preparations were used.

3) Of the five methods tested, precipitation by ammonium sulphate (33% saturation) and precipitation with 25% ethanol (pH 6.3, −5 °C) showed the best results with regard to enrichment of antibodies.

4) Colostral antibodies show cross-reaction with either native soy protein or with native protein extracts from pollens of various species of leguminosae; if the latter were succinylated or oxidized by performic acid, no cross-reaction could be detected.

5) Because of the different protein content the detection limit depends on the kind of soy product in use. We found a detection limit of 0.07−0.13 mg/ml which agrees well with values cited in the literature for serum antibodies. A Tris/Glycine buffered urea solution (pH 8.6) was favored as the best-suited extraction medium. Textured and structured soy proteins could only be detected qualitatively.

6) Recovery is greatly influenced by addition of polysaccharides and proteins from other sources. Determination of soy protein in food stuff is therefore complicated and only accomplished by the use of corresponding reference products.

References

1. de Crignis GL (1985) Immunologischer Nachweis von Sojaproteinen. Diss Techn Univ München
2. Fey H (1972) Colibacillus in calves. Hans Huber Publishers, Bern
3. Fey H, Marti F (1972) Antiprotein-Immunisierung bei der hochträchtigen Kuh. Schweiz Arch Tierheilk 114:33−48
4. Hammer DK, Kickhöfen B, Henning G (1968) Molecular classes and properties of antibodies in cattle serum and colostrum synthesized during the primary and secondary response to protein antigens. Eur J Biochem 6:443−454
5. Mank HH (1987) Versuche zur Gewinnung präzipitierender Antikörper unter kritischen Randbedingungen. Diss Techn Univ München

Expression of the Bovine α_{s1} – Casein cDNA in CHO Cells

P. Dovč[1], H. Elbertzhagen[2], F. Niepold[3], and O.J. Rottmann[4]

[1] University E. Kardelj, Ljubljana, Biotechnical faculty, Zootechnical Department, Domžale, Yugoslavia
[2] Institute of Dairy Science, Technische Universität München, Freising-Weihenstephan, FRG
[3] Institute of Biochemistry, Biologische Bundesanstalt, Braunschweig, FRG
[4] Institute of Animal Science, Technische Universität München, Freising-Weihenstephan, FRG

Introduction

The major effort in molecular biology during the last two decades was to explain the structure and function of the eucaryotic genome. The investigators were able to establish a variety of recombinant DNA techniques which have a great potential for the application in animal breeding. cDNA- and genomic DNA-libraries and gene transfer techniques are the most important tools to diagnose and manipulate the genome. The first indication for possible application of gene transfer was successful expression of the MMT-growth hormone fusion gene in transgenic mice [7], but the following experiments in rabbits, pigs, sheep, and cattle [2, 4, 1] reduced the great expectations. The recent data about controlled, tissue specific expression of the transferred milk protein gene for ovine β-lactoglobulin [8] in mice suggest that manipulation of milk composition and changing of technological properties of milk proteins could be of interest for practical animal breeding [3].

In our experiments we tried to express the cDNA, coding for bovine α_{s1}-casein in eucaryotic cells under control of H-2K promoter from murine histocompatibility complex.

Material and methods

For the expression experiments in eucaryotic Chinese hamster ovary (CHO) cells the vector pH-2Kcas was constructed. The 2kb promoter region (kindly provided by D. Morello), from murine major histocompatibility complex gene region was isolated from the plasmid pSB1/H-2K [6] as a HindIII-BamHI fragment. The 1 106 bp BglII-PstI cDNA sequence, coding for bovine pre α_{s1}-casein B, originally cloned into the plasmid pB α_{s1}C184 [9], was ligated in an equimolar ratio with the H-2K promoter region using DNA T4 ligase. The H-2K/cas fusion gene was inserted into HindIII/PstI double restricted vector pSP65 (Promega Biotec), giving the plasmid pH-2Kcas.

Co-transfection plasmid pMMTneoK, coding for neomycin resistance in eucaryotic cells was derived from plasmid pdBVP-MMTneo342-12 [5] after deletion of the

168

BPV (Bovine papilloma virus) sequence. CHO cells were transfected with 5 µg pH-2Kcas, 1 µg pMMTneoK and 10 µg calf thymus DNA/500 µl precipitate, using the Ca-phosphate precipitation method. Cells were selected for G418 resistance and resistant colonies were isolated. The chromosomal DNA from resistant cells was prepared and digested with EcoRI endonuclease; 5 µg of digested chromosomal DNA were electrophoresed on a 0.8% agarose gel and transferred to nitrocellulose filter. The nick-translated plasmid $pB\alpha_{s1}C184$ ($> 1.4 \cdot 10^8$ cpm specific activity) served as the probe.

The supernatant of the transfected CHO cells was collected and concentrated by ultrafiltration (Amicon) and lyophilization. 0.8 mg of the lyophilized probe was dissolved in 100 µl 7 M urea and 5 µl of this solution was used for immunelectrophoresis. The electrophoresis was run on 1.6% agarose gel on the Gel-Bound-Foil (FMC Corporation, Marine Colloids Division). First-direction was running 2 h at 300 V and 10 °C, the second direction for 6 h at 170 V. To the second run gel 2.4% rabbit anti-bovine casein serum (Behring-Werke AG) was added; then the gel was dried and stained with silver dye.

Results and discussion

Agarose gel electrophoresis of DNA restriction fragments served as control for the construction of the vector pH-2Kcas (Fig. 1). The Southern blot analysis of the chromosomal DNA of one transfected cell clone showed hybridization signal in the range of 9.0 and 5,7 kb, respectively, displaying the integration of the transferred DNA. By using immunelectrophoresis we could show specific precipitation curves, indicating casein-specific antigenic determinants in the supernatant of transfected cells. This could be expected because H-2 antigens are expressed in almost all differentiated cells. Consequently the recent gene transfer experiments showed regulation of expression of transferred genes under control of the H-2K promoter, similar to expression of the endogenic H-2K gene complex [6]. The occurrence of more than one precipitation curve can be explained with the different phosphorylation and

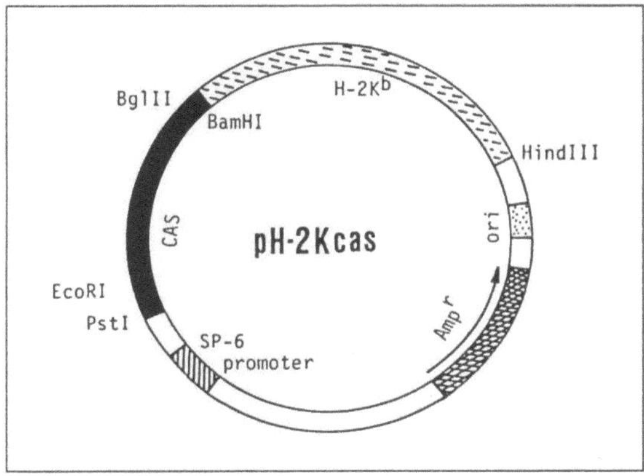

Fig. 1. Construction of pH-2Kcas

169

glycosylation rate, as a post translational modification, due to CHO-cell protein-kinases, and unspecific for bovine casein. This experiment showed the possibility for expression of milk protein genes in unspecialized tissues.

References

1. Brem G (1987) Possibilities of growth manipulation by gene transfer techniques. 38th Annual Meeting of the EAAP, Lissabon, Portugal, 28. 9.–1. 10. 1987
2. Hammer RE, Pursel VG, Rexroad CE Jr, Wall RJ, Bolt DJ, Ebert KM, Palmiter RD, Brinster RL (1985) Production of transgenic rabbits, sheep and pigs by microinjection. Nature 315:680–683
3. Kang Y, Jimenez-Flores R, Richardson T (1986) Casein genes and genetic engineering of the caseins. In: Evans JW, Hollaender A (eds) Genetic engineering of animals. An agricultural perspective. Basic life sciences, Plenum Press, New York, 37:95–111
4. Kraemer DC, Minhas B, Loskutoff N (1986) Strategien für einen Einsatz des Gentransfers in der Rinderproduktion. Züchtungskunde 58:32–37
5. Law MF, Byrne JC, Howley PM (1983) A stable bovine papillomavirus hybrid plasmid that expresses a dominant selective trait. Mol Cell Biol 3:2110–2115
6. Morello D, Moore G, Salmon AM, Yaniv M, Babinet C (1986) Studies on the expression of an H-2K/human growth hormone fusion gene in giant transgenic mice. EMBO J 5:1877–1883
7. Palmiter RD, Brinster RL, Hammer RE, Trumbauer ME, Rosenfeld MG, Birnberg NC, Evans RM (1982) Dramatic growth of mice that develop from eggs microinjected with metallothionein-growth hormone fusion genes. Nature 300:611–615
8. Simons JP, McClenaghan M, Clark AJ (1987) Alteration of the quality of milk by expression of sheep β-lactoglobulin in transgenic mice. Nature 328:530–532
9. Stewart AF, Willis IM, MacKinlay AG (1984) Nucleotide sequences of bovine α_{s1}- and $\varkappa$-casein cDNAs. Nucl Acids Res 12:3895–3907

Screening Methods for Genetic Variants of Milk Proteins

I. Krause, J. Buchberger, G. Weiß, and H. Klostermeyer

Institut für Chemie und Physik, Süddeutsche Versuchs- und Forschungsanstalt für Milchwirtschaft, Freising-Weihenstephan, FRG

Introduction

Genetic variants of milk proteins are interesting because: 1) There are correlations between the genetic variants and protein content, protein composition, fat content, functional properties (e.g., cheese making ability, heat stability), and the fitness of the cow; and, 2) breeding for yield of milk, fat and protein may have altered gene frequencies; studying breeds which have barely undergone selection, or breeds with a small population, is therefore very important. Furthermore, electrophoretic methods provide fast and reliable phenotyping of great numbers of milk samples because of the high resolution obtained, especially by isoelectric focusing.

Methods

1) Cellulose acetate gel electrophoresis of whey proteins in veronal buffer; pH 8.9, 200 Vh;
2) Isoelectric focusing in immobilized pH-gradients (IPG-IEF) in $260 \times 125 \times 0.5$ mm polyacrylamide gels; pH 5.2–5.7; 1800 Vh;
3) Carrier ampholyte isoelectric focusing (CA-IEF) in $265 \times 125 \times 0.25$ mm polyacrylamide gels; pH 2.5–8.0, 8 M urea, 5 000 Vh.

Results

When phenotyping milk samples of "Murnau-Werdenfelser" breed for β-lactoglobulin variants, a new variant provisionally called β-lactoglobulin W was detected (Fig. 1) [5]. In a recent study of several breeds we applied IPG-IEF (Fig. 2) and

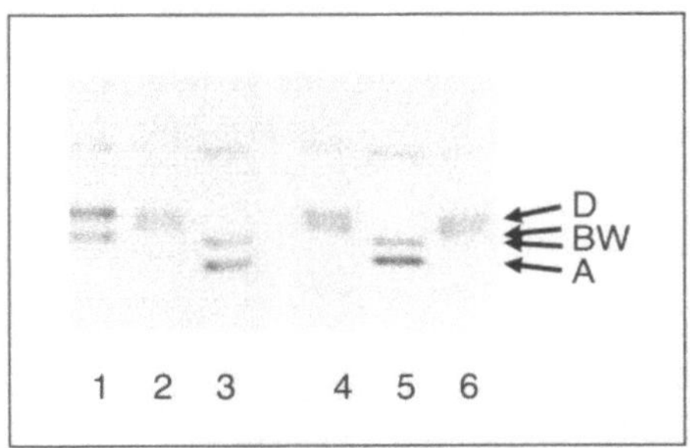

Fig. 1. Cellulose acetate gel electrophoresis at pH 8.9 of bovine β-Lg variants from the milk of individual "Murnau-Werdenfelser" cows.
(1) β-Lg BD; (2) β-Lg WD; (3) β-Lg AB; (4) β-lg WD; (5) β-Lg AB; (6) β-Lg WD. Capital letters and arrows denote the migration position

171

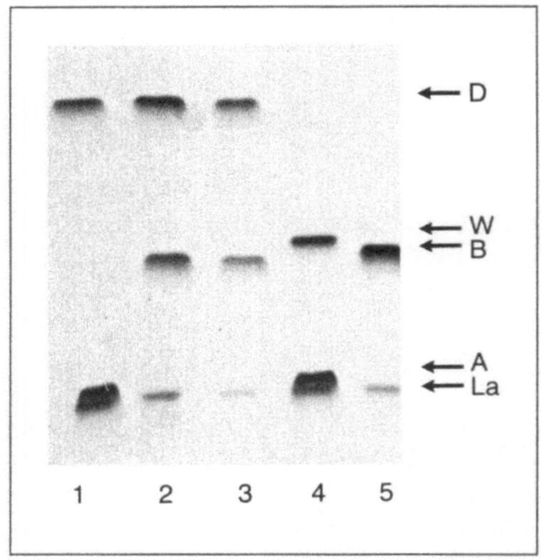

Fig. 2. IPG-IEF at pH 5.2–5.7 (resolution = 0.025 pH/cm). Total whey proteins of individual "Murnau-Werdenfelser" cows.
(1) β-Lg AD; (2), (3) β-Lg BD; (4) β-Lg AW; (5) β-Lg BB. La = α-Lactalbumin B

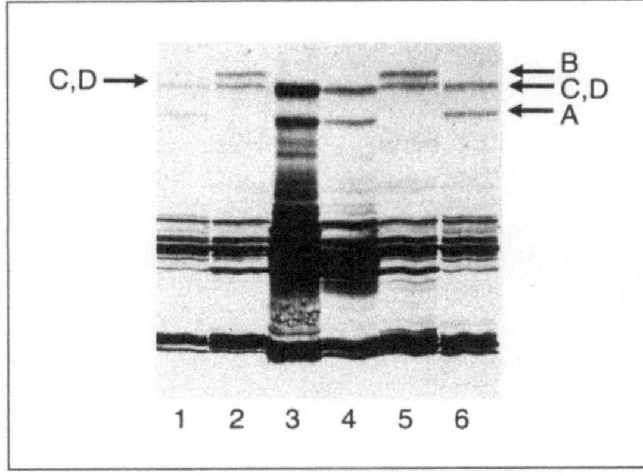

Fig. 3. CA-IEF at pH 2.5–8.0. Caseins from different breeds with rare kappa-casein C or D variant. Samples with kappa Cn:
(1), (6) AC (Angler), (2), (5) BC (Angler × Simmental) (3) AC (a gift from Prof. C. Corradini, Univ. Piacenza, Italy), (4) AD (Simmental breed, a gift from Dr. G. Erhardt, Univ. Giessen, FRG)

confirmed the new variant [2]. It was found in 19 Murnau-Werdenfelser cows, in two Jersey cows, and in two Simmental × Red Holstein crossbred cows.

Upon CA-IEF of caseins of different breeds a new k-variant was observed (Fig. 3) and it proved to be identical with k C found by Di Stasio and Merlin [1], as well as by Mariani [3], and identical with k D described by Seibert et al. [4].

We observed the C variant in one purebred Angler cow, in one crossbred cow (Angler × Simmental), in one crossbred cow (Angler × Brown Swiss), and in one purebred Brown Swiss cow.

Conclusions

The results indicate that breeds which have scarcely undergone selection (e.g., Murnau-Werdenfelser), or which are not widely distributed (e.g., Jersey, Angler)

172

show greater genetic variants than more common breeds. With the high resolution obtainable by the new electrophoretic methods there is a chance to discover new rare variants even in those breeds that have been studied extensively.

References

1. Di Stasio L, Merlin P (1979) Polimorfismi biochimici del latte nella razza bovina Grigio Alpina. Riv Zootecnia Veter 2:64–67
2. Krause I, Buchberger I, Weiß G, Pflügler M, Klostermeyer H (1988) Isoelectric focusing in immobilized pH gradients with carrier ampholytes added for high-resolution phenotyping of bovine β-lactoglobulins: characterization of a new genetic variant. Electrophoresis (in press)
3. Mariani P (1983) Sulla presenza di una terza k-caseina nel latte di vacche di razza Bruna. Sci tecnica lattiero-casearia 34:174–181
4. Seibert B, Erhardt G, Senft B (1987) Detection of a new k-casein variant in cow's milk. Anim Genet 18:269–272
5. Weiß G, Buchberger J, Klostermeyer H (1980) Ein neues β-Lactoglobulin des Rindes. Meeting of the German Society of Dairy Science, Kiel, FRG, March 26–28, poster presentation

Rheological Properties of Heat-Induced Whey Protein Gels

M. Paulsson and P. Dejmek

Department of Food Technology, University of Lund, Lund, Sweden

Introduction

Whey proteins have several excellent functional properties and one of their most important functions is to improve consistency of foods by forming thermally induced gels. β-Lactoglobulin, which is the major whey protein, has good gelation properties and the aim of this study was to follow the formation of heat-induced gels of an industrially fractionated β-lactoglobulin isolate at different pH and assess their properties using dynamic rheometry.

Materials

Industrially fractionated high purity β-lactoglobulin isolate was obtained from INRA (Rennes, France). In the INRA process, whey is freed of phospholipids through aggregation with Ca and heat and microfiltration. α-Lactalbumin fraction is heat-aggregated at $55\,^{\circ}\mathrm{C}$, pH 3.8 and centrifuged. The supernatant is diafiltered and freeze-dried [1].

A β-lactoglobulin preparation (L-6879) from Sigma Chemical Co. was used for comparison. NaCl, HCl, and NaOH were all of analytical grade. Newly prepared protein solutions were used in all experiments.

Methods

The protein was dissolved in 2% NaCl-solution to a concentration of 3.5% (w/w) and pH was adjusted to appropriate pH value with 1 M HCl or 1 M NaOH.

Gelation of the protein solution was induced by heating in situ in the rheometer cup (DIN 53 019) of Bohlin VOR Rheometer (Bohlin Rheology, Lund, Sweden). The sample in the rheometer is contained between a Couette-type cup (the outer cylinder) and a bob system (the inner cylinder, diameter 14 mm). The bob is suspended in a torsion bar. When the cup is oscillating, the moment couple of the bob is sensed through the bob's angular deflection and the signal is fed to the computer for conditioning, corrections, and calculations of moduli via fast Fourier transform. For the relaxation measurement a given angular deflection is applied to the cup and the decay of the resulting couple is measured.

The temperature profile of the protein solution in the cup is shown in Fig. 1.

174

The first oscillation test: gel formation is followed during the final stage of temperature gradient and at 90 °C by oscillation measurement at 1 Hz, nominal strain 0.02.

The second oscillation test: gel dynamic moduli are measured at frequencies 0.01 Hz − 1 Hz, nominal strain 0.016 after cooling to 40 °C and an equilibrium period of 9 min.

The relaxation test: gel relaxation is measured after a ramp increase of strain to 0.007 in 0.1 s.

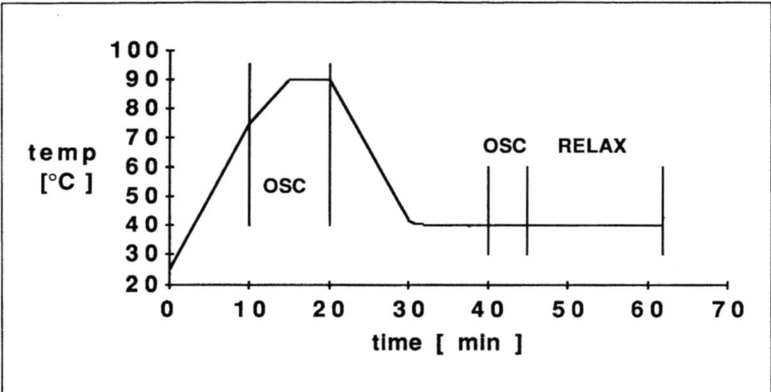

Fig. 1. The temperature profile of the β-lactoglobulin solution during gel formation and measurement

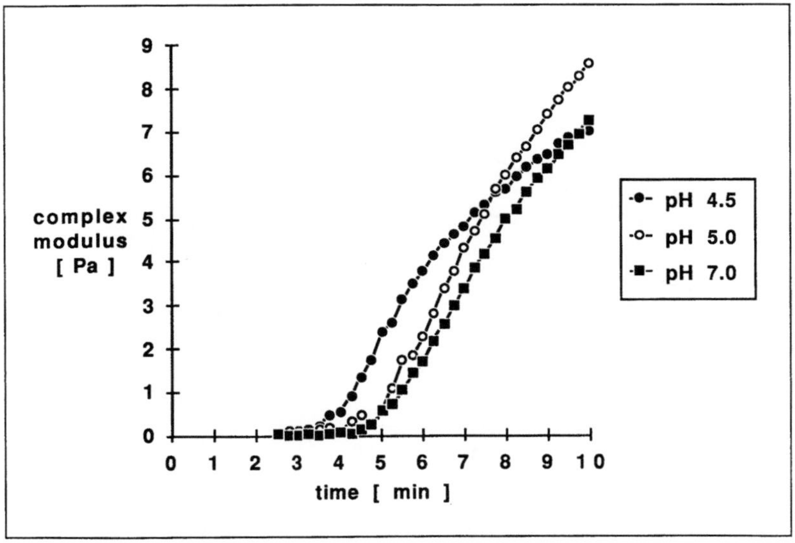

Fig. 2. Complex modulus, G*, at 1 Hz as a function of time (first oscillation test)

175

Results

The effect of pH on gel formation of β-lactoglobulin is shown in Fig. 2. The temperature of the onset of gelation is largely independent on pH (within 2 °C), despite the known large variation of β-lactoglobulin denaturation temperatures (17 °C) in the pH range 4.5–7 [2].

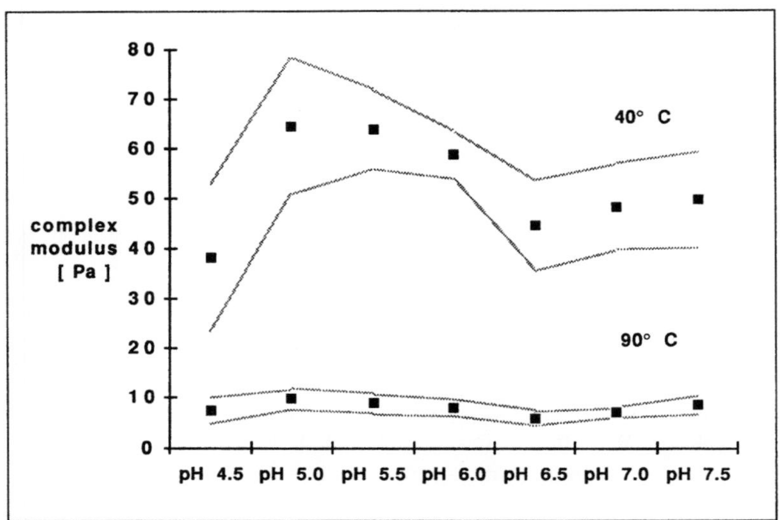

Fig. 3. Complex modulus, G*, at 1 Hz as a function of pH at 90° and 40 °C. Average value $\pm$ standard deviation

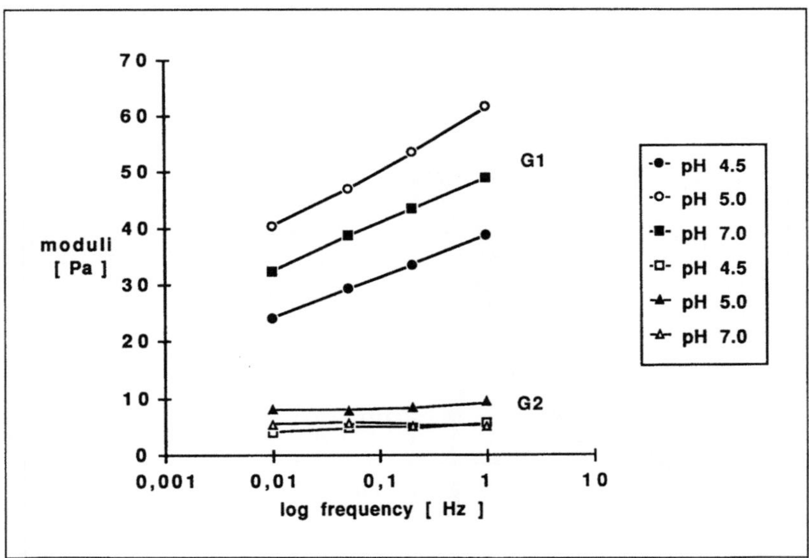

Fig. 4. Elastic modulus, G1, and viscous modulus, G2, as a function of frequency at 40 °C

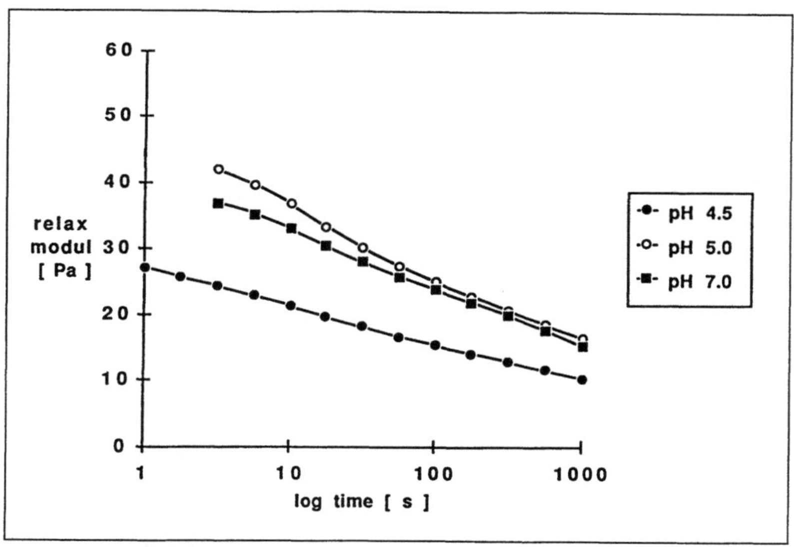

Fig. 5. Relaxation modulus, G, as a function of time at 40 °C

The highest gel strengths are reached at pH 5.0 – 6.0 and significantly lower values are found both at more acidic (pH 4.5) and more neutral pH (pH 6.5). Gel strength increases again at alkali pH values. The pH dependence of gel strength is the same at both 90 °C and after cooling to 40 °C (Fig. 3). β-Lactoglobulin from Sigma starts geling at a somewhat lower temperature (2 °C) and gives a somewhat stronger gel (15–20%) than the β-lactoglobulin from INRA.

Both elastic and viscous modulus of the gel at different pH are measured in the frequency range 0.01 – 1.0 Hz and Fig. 4 shows that the elastic component of moduli dominates at all pH values, i.e., the gels of β-lactoglobulin are elastic.

Relaxation behavior of the gels at different pH values at times up to 1 000 s (Fig. 5) indicates a reduction of the elastic component of the β-lactoglobulin gels linear with log time, in accordance with the oscillation experiments, and shows no evidence of residual permanent elasticity.

References

1. Maubois J-L, Pierre A, Fauquant J, Piot M (1986) IDF Conference of Whey Protein, Chicago, USA
2. Paulsson M, Hegg P-O, Castberg HB (1985) Thermochim Acta 95:435–440

General Discussion:
Milk Proteins – Structural and Genetic Aspects

Conclusions by the chairman, H. Klostermeyer

Süddeutsche Versuchs- und Forschungsanstalt für Milchwirtschaft.
Institut für Chemie und Physik, TU München-Weihenstephan, FRG

Milk seems to be the best known food with respect to its biosynthesis, composition, and chemical structure of the individual components. However, there is still further need for fundamental research.

Especially, knowledge about the influence of the physical properties on the biological behavior, on the chemical reactivity, as well as on the technological suitability, is rather limited. This might change during the next years. At least the industry is already highly interested in the physical properties of their products, since properties like mouthfeeling and eatability are of higher importance for selling a product than, for instance its nutritive value.

At the moment, data on the relationships between chemical structure, physical properties, and nutritive value are few. Does a protein with a certain amino acid composition have the same biological value if it is placed in solution, in a membrane, at the surface of a fat globule, within a foam, or in a textured matter?

Thanks to the fundamental work of Prof. Ribadeau-Dumas and his colleagues we know many details about the primary structure of milk proteins, and now many details about the tertiary structure of the whey proteins are also known. Clearly, with little changes in primary structure, nature manages important biological problems, however, our understanding of such phenomenons is still rather limited.

New methods and possibilities for genetic techniques and biotechnology may help us to get further information in the near future, and protein engineering may offer a way for studying structure effects and structure-reactivity relationships, especially in the milk protein system. The udder seems to be a suitable, if not a good system to express artificial genes by mammalian cells.

Our knowledge of the structure of proteins and its parts (domains) in solutions, up to now, comes mainly from studies with crystallized materials or from research with one-dimensional techniques like spectroscopy. Prof. Rüterjans demonstrates how it is possible to find out the real three-dimensional structure of a protein in solution, namely by knowledge of the primary structure (one dimension) and combining this with two-dimensional NMR-spectroscopies. This new possibility will again stimulate research in our field.

Comparing this year's Kiel-Milk-Protein-Symposium with that of 1983, we observed a certain shift of the subjects; in 1988 more papers were based on the known chemical molecular structure of the subjects. In the future, physical chemistry may play an important role in a meeting like ours. However, one might presume that in 1988, the right persons discussed the right problems at the right moment.

Protein Modification: Effects on Functional Properties and Digestibility

J. E. Kinsella

Institute of Food Science, Cornell University, Ithaca, New York, USA

Introduction

The modern food industry is rapidly changing from a commodity-handling industry toward a market-driven consumer products industry. Today's informed consumer is increasingly demanding food products that are compatible with a busy, healthy lifestyle that include convenience, that are balanced in calories and nutrients, that are safe (less saturated fatty acids and cholesterol), more wholesome, of consistent high quality, appropriately portioned, and attractively packaged. Quality attributes, flavor, odor, color, taste, texture, mouthfeel are expected. More fabricated food products will increasingly be manufactured placing a premium on ingredients with versatile but consistent functional properties and compatible with automated fabrication. In addition to nutrition, these developments have dramatized the need for a range of functional ingredients, in particular low-calorie and structure-forming macromolecules like polysaccharides and proteins [27].

Proteins represent a most important class of functional ingredients because they possess a range of dynamic functional properties (Table 1); they show versatility during processing, they can form networks and structures, and they provide essential amino acids, i.e., they fulfill functional and nutritional requirements. In addition, they interact with other components and improve quality attributes [29].

Many sources of ingredient functional proteins are used in foods and dairy proteins are a major source. The bulk of these are used in products that are relatively tolerant of variability in the ingredient proteins but with refinements in product formulae and with automated formulation the food industry is becoming more

Table 1. Functional properties of proteins in foods

General property	Functional criteria
Organoleptic	Color, flavor, odor
Kinesthetic	Texture, mouthfeel, smoothness, grittiness, turbidity
Hydration	Solubility, wettability, water sorption, swelling, thickening, gelling, syneresis
Rheological	Viscosity, gelation
Surface	Emulsification, foaming, film formation
Binding	Lipid, flavor-binding
Structural	Elasticity, cohesiveness, chewiness, adhesiveness, network formation, aggregation, dough formation, texturizability, extrudability
Enzymatic	Coagulation (rennet), tenderization (papain), mellowing (proteinases)

demanding not only for compositional but also for functional specifications of ingredient proteins. This has dramatized the need for standard methods to determine and quantitatively describe the functional properties of proteins for applications and also help elucidate structure function relationships [25, 32].

The dairy industry will increasingly depend on the food industry as a market for functional ingredients, hence it must pay careful attention to meeting the varied and exacting needs of the food industry and to the methods of preparation (separation, dehydration, fractionation) of functional ingredients. Research is warranted to show the unique functional attributes of individual dairy proteins, protein blends, or modified dairy proteins [31]. In this context much more information concerning the relationships between the physical properties of milk proteins (which are reasonably well understood) and functional behavior is needed. In addition, the particular physical properties that meet the functional requirements in particular food applications must be elucidated so that rational decisions can be made in selecting the best proteins for specific application or determining what modifications are required for improving a particular function.

Different applications require quite different functional properties and many products require an array of properties and in some cases depend on changes in properties during actual processing or preparation, e.g., in foam formation and stabilization some molecular unfolding and subsequent protein:protein interaction must occur [25]. In reformed meats adhesion and binding are important and adhesive proteins (unlike myofibrillar proteins) that can function without added salt are needed. In comminuted meat products solubility, viscosity, emulsifying, water-holding, and gelation are required to ensure good texture, shape retention, cutting characteristics, and smooth mouthfeel. Few single proteins possess the appropriate range to perform all these functional properties and usually a mix of proteins is needed. In emulsions or foams the protein(s) should possess good interfacial activity and form a strong cohesive, elastic film. To achieve this a protein must perform a sequence of functions which few proteins can achieve satisfactorily. Casein and caseinates being amphipathic are distinctly surface active, readily form films and satisfactory emulsions and foams, but in some products these films lack sufficient interactions and critical viscoelastic properties for stability. Proteins which retain more tertiary structure, that possess molecular flexibility and a balance of hydrophilic and hydrophobic interactions form stronger films, e.g., β-lactoglobulin and bovine serum albumin (BSA) [22, 51]. However, the best films are obtained with mixtures of proteins, e.g., egg white or caseins and whey proteins blended with basic proteins [52].

Modification and properties

Knowledge of the physicochemical characteristics required for particular uses are only partly understood. In gelation limited unfolding of polypeptides to facilitate interactions and a balance of attractive and repulsive forces are desirable [43, 24]. In foams and emulsions amphiphillic flexible molecules which readily orient at an oil-water interface with maximal protein:protein interaction is required to form a strong film [24, 27]. However, while the features required may be known, the current

180

state of knowledge is inadequate in terms of specifying the best protein structures to use for particular uses. Much of the available information has been obtained by empirical practices and currently the rational design of functional protein is not possible. However, a large amount of general information which is indicative of particular molecular structures that perform best in certain applications is available, as exemplified by gelatin gels; egg-white proteins for whipping, caseins for cheese curds, and gluten for viscoelastic properties in leavened bread [32]. The structure, conformation, size, surface charge, hydrophobic:hydrophilic balance, molecular flexibility, etc., affect functional properties, hence, modification of these can be exploited to alter the functional behavior of proteins. Because of the increasing, though incomplete, knowledge of structure:function relationships the modification of proteins by physical, thermal, chemical, enzymatic, and genetic procedures or by blending of different proteins (and other macromolecules) is being gradually developed to improve the nutritional and functional behavior of proteins [27, 5, 6, 44].

Physical: Fractionation and separation is widely used in the dairy industry to prepare amphipathic caseins for use as heat-stable surfactants in coffee-whiteners. Ultrafiltration is used to prepare various protein fractions. Heat treatments are used to modify or improve functional behavior of products. e.g., "low-," "medium-," "high-fat" milk powders [24]. Low-heat powders contain more soluble whey proteins and less β-lactoglobulin/$\varkappa$-casein complexes and are useful in fortifying yogurts, ricotta cheese, and in rennet casein manufacture. High-heat powders are required to avoid loaf depression in bakery products [26].

Heat-induced interaction between β-Lg and the $\varkappa$-casein on casein micelles improves the gel structure of yogurts and reduces syneresis [17]. Heat combined with pH adjustment can be exploited to separate β-Lg or α-lactalbumin from whey [48]. Heat treatment of whey protein or β-Lg is used to prepare networks with a range of rheological and textural properties from soft curds to hard gels [43, 12].

Charge: Adjustment of pH by altering net charge on proteins can be exploited to alter functional behavior of milk proteins and to prepare acid casein [42]. The heat-stable surface active properties of caseins cannot be exploited as such, however, the caseinates are universally used in foods where solubility, heat stability, water holding, and surface activity are required [42, 24, 39].

The pH greatly affects the functional behavior of whey proteins [24, 40, 41]. Thus, whey proteins and β-Lg form stronger, clearer heat-induced gels above pH 7.0 [43]. The emulsifying properties of β-Lg improve as the pH is increased, i.e., surface activity, film thickness, and film strength is greatly enhanced at pH 9.0 compared to pH 5.0 reflecting greater molecular interaction and better film formation [57].

Acid stability of β-Lactoglobulin: Resistance to Proteolysis: β-lactoglobulin (β-Lg) avidly binds retinol (Ka 10^8 M^{-1}) and has been proposed as a specific carrier of retinol from maternal milk to the neonate [49, 47]. The β-Lg molecule possesses a barrel structure with a hydrophobic pocket or calyx which forms a high affinity binding site. Because the binding affinity of a ligand is sensitive to the conformation of the host protein the delivery of the retinol to the small intestine of the neonate would require that β-Lg remain stable at the low pH values (pH < 2) encountered in the stomach. To assess this, the conformational stability of β-Lg at progressively decreasing pH values was examined by [20]. The UV-difference spectra of β-Lg revealed a progressive increase as the pH was reduced from 7.5 to 1.5. This was consistent with the transfer of some chromophores (tryptophan, tyrosine/phenylalanine) into the more apolar interior with tighter folding and decreased molecular flexibility as observed by Shimizu et al. [57].

The conformational stability was determined by studying the equilibrium thermal unfolding of β-Lg at various pH values. The T_m (i.e., temperature at which 50% of the protein is unfolded) progressively increased with decreasing pH (Table 2). The pH stability curves indicated a sharp reversible transition at specific temperatures and the associated thermodynamic parameters were calculated from the apparent equilibrium constants (Kapp). As the pH was adjusted down from 7.0 to 1.5 the temperature of maximum stability (T_{max}) increased and the magnitude of the apparent free energy (ΔG°_{app}) at T_{max} increased. The relative change in the free energy of unfolding became more positive with decreasing pH reflecting enhanced stability by >12 kJ/mol and a T_m increase of $18.5\,°C$ (Table 3). At pH 1.5, β-Lg possesses about 18 positive charges yet this does not destabilize the protein. Conceivably the protonation of certain carboxylic acid groups may contribute to extra hydrogen bonding and stabilize the protein at pH 2.0 [21]. This stable conformation may effect the peptic digestibility of β-Lg hence this was examined.

Digestibility. The structure and conformation of proteins affect their digestibility and hence their nutritional value [37]. Caseins with limited secondary structure are facilely digested compared to whey proteins. Other proteins in milk which perform specific biological functions during milk synthesis (e.g., α-lactalbumin) or in the protection of the neonate (lactoferrin, lysozyme, immunoglobulins) may be resistant to proteolytic digestion because of their structures [61, 56, 58, 18]. In this regard β-Lg (which is absent from human milk), because of its compact structure at pH 2.0, may be resistant to pepsin. This may present a problem in humans because of its allergenicity. In animals β-Lg is resistant to gastric digestion and apparently remains intact after passing through the stomach [38, 69]. Jakobsson et al. [15] identified

Table 2. Effect of pH transition temperature (T_m) of β-lactoglobulin

pH	T_m (°C)	ΔT_m
7.5	64.8	0
7.0	69.2	$+ 4.4$
6.5	71.9	$+ 7.1$
3.0	77.2	$+12.4$
2.5	78.8	$+14.0$
2.0	82.5	$+17.7$
1.5	83.3	$+18.5$

Table 3. The increased enthalpy (ΔH°_{app} Kcal/mole) and entropy (ΔS°_{app} entropy units) of thermal unfolding of β-lactoglobulin at decreasing pH values

pH	H°_{app}	ΔS°_{app}
7.5	47.80	141.22
7.0	51.74	151.34
6.5	53.53	155.23
3.0	57.70	164.73
2.5	59.65	169.12
2.0	63.68	178.95
1.5	65.98	185.07

[a] These values were calculated with those at pH 7.5 as references (at $70\,°C$)

182

immunoreactive bovine β-Lg in human milk and correlated its presence to the development of colic in breast-fed babies.

In addition, β-Lg may function as a retinol-carrier protein to provide vitamin A via specific receptors in the small intestine of the neonate [47]. This would require that the protein retain its structure (and binding capacity) [7] and resist pepsin under the acidic conditions of the stomach. Therefore, the structural and conformational behavior of native, heat modified, and reduced β-Lg under acidic conditions were correlated with its susceptibility to pepsin and α-chymotrypsin hydrolysis [55].

Native β-Lg is quite resistant to pepsin hydrolysis at pH 2.0 and heating at 50°, 60°, and 70 °C did not significantly alter its resistance to pepsin. Heating at 80 °C increased the hydrolysis to some extent, while heating at 90 °C resulted in rapid hydrolysis of β-Lg (Fig. 1). In contrast, cleavage of the two S–S bonds via sulfitolysis markedly increased the digestibility of β-Lg.

The effects of heating at pH 6.8 on the conformation of β-Lg was assessed by fluorescence emission spectroscopy at pH 2.0. Structural changes were negligible following heating at 50°, 60°, and 70 °C but fluorescence quenching dramatically decreased above 80 °C reflecting exposure of previously buried tryptophan groups. Cleavage of S–S bonds caused marked quenching of fluorescence and a red-shift in the absorption maxima with a concomitant increase in emission λ_{max} reflecting extensive conformational changes. Calculated emission transition energy (Vf) and surface polarity (Z) values for extrinsic fluorescence of ANS conjugates showed a decrease upon heating and S–S bond cleavage at pH 2.0 indicating exposure of hydrophobic groups to the aqueous polar medium; this was associated with improved accessibility to pepsin and enhanced digestion [55].

The chymotryptic digestibility of native, heat-treated, and S–S cleaved β-Lg increased with time in all cases, especially following heating at 80° and 90 °C and cleavage of the S–S bonds (Fig. 1).

Protein digestibility is influenced by the conformation of protein and the accessibility of susceptible bonds to the enzyme. Pepsin has a specificity for tryptophan, tyrosine, phenylalanine, leucine, and isoleucine residues [4]. The resistance of native β-Lg to peptic digestibility indicates that these groups are not accessible to enzyme because of the tightly folded state of β-Lg which internalizes most of the pepsin specific bonds in native β-Lg at pH 1.5. Heating β-Lg at temperatures above 70 °C and cleavage of S–S bonds by causing unfolding and exposure of hydrophobic residues greatly increased proteolysis by pepsin. Both heat treatment above 70 °C and disulfide cleavage caused marked conformational changes in β-Lg. This reflects dissociation and unfolding of β-Lg to expose hydrophobic residues. This was conducive to more extensive proteolysis by both pepsin and chymotrypsin which shows a specificity for the phenylalanine, tyrosine, and tryptophan residues that were exposed following heat treatment and reduction. These data indicate that the disruption of native structure especially by S–S cleavage, exposes susceptible peptide bonds and significantly decreases the resistance of β-Lg to pepsin. Thus, processes such as heating (>90 °C) or cleavage of S–S bonds may be used to improve the digestibility of β-Lg and possibly reduce its allergenicity to sensitive babies.

These data that show the resistance of β-Lg to pepsin proteolysis are consistent with a post secretory biological function that requires molecular integrity in the small intestine. Thus β-Lg may act as carrier for the insoluble retinol in bovine milk.

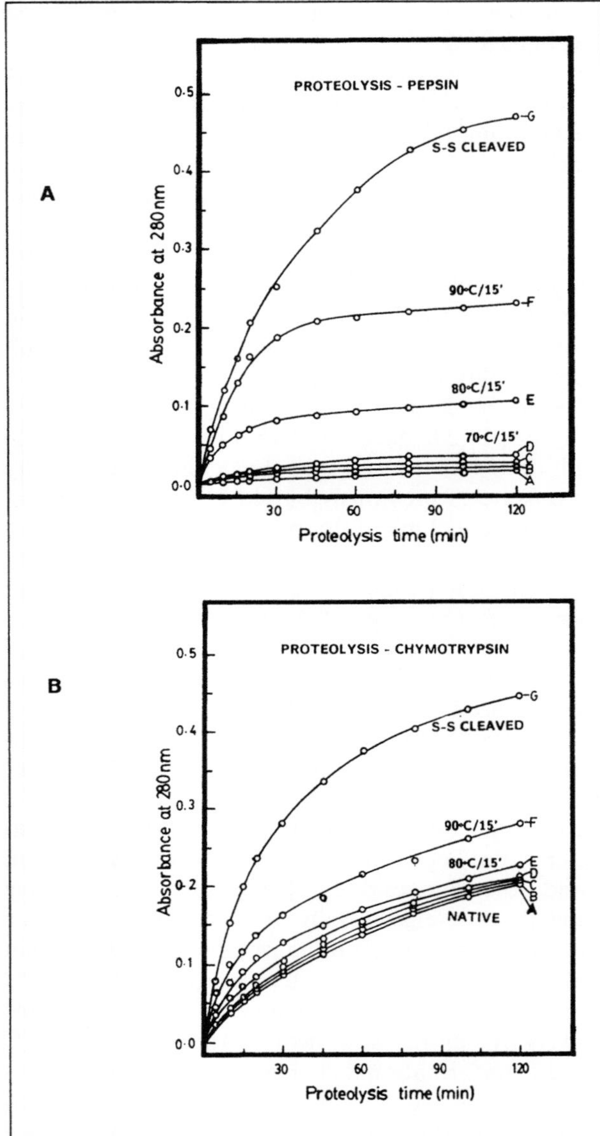

Fig. 1.

However, since milk fat contains β-carotene and Vitamin A the importance of this proposed carrier function is questionable, though it is possible that β-Lg may facilitate the esterification of retinol in neonatal tissue as observed with other retinol binding proteins [46].

pH and Thiol Reactivity of β-Lg: The free thiol group in β-Lg is partly responsible for the thermal instability of β-Lg via irreversible aggregation [9] which is a problem

184

in preparing functional proteins from whey [24]. In addition, the interaction of β-Lg with micellar $\varkappa$-casein may destabilize the casein micelles in UHT milks and impair chymosin action in milks reconstituted from heated milk powders.

The thiol group on residue 121 is located on the surface β-barrel sheet of the β-Lg monomer that is normally occluded by the association in the dimer [2]. The thiol group of β-Lg is not very reactive at pH 2.0. However, as the pH is increased, especially above pH 7.0, the thiol group becomes more reactive especially upon heating [9]. Therefore agents which stabilize the structure of β-Lg may reduce the reactivity of the thiol groups of β-Lg. The reactivity of the thiol group of β-Lg under different solution conditions of pH, heat, and anions were determined using dithio-nitrobenzene titration [21].

The reactivity of the thiol group was negligible at 25 °C at pH 6.8, however increasing the pH caused an increase in thiol exposure and above pH 8.4 the reactivity increased markedly, i.e., 0.9 moles of thiol/mole of β-Lg reacted within 30 min. The pKa of the thiol group occurred at pH 9.3 which is higher than in other proteins reflecting the fact that it is normally buried in the hydrophobic region of the dimer. Dissociation of the β-Lg dimer using increasing concentrations of chaotropic agents (e.g., guanine hydrochloride or urea above 1.5 M), increased the reactivity of the thiol group, i.e., more than 75% of the thiol groups reacted in 3 min reflecting dissociation of the β-Lg dimer.

Heating β-Lg (pH 6.8) from 50–80 °C progressively increased thiol exposure at pseudo first order rates which increased with temperature. The dissociation had positive ΔH^+, ΔS, and ΔG values at all temperatures (Table 4) [19].

The addition of various structuring anions decreased the rate of thiol exposure (Table 5), probably by enhancing the stability of dimeric β-Lg and retarding disso-ciation. Citrate had the most stabilizing effect as reflected in the relative change in activation free energy Δ (ΔG dimer) [where $(\Delta G) = -RT \cdot \ln \cdot ks/ko$]. These salts stabilized the β-Lg in the order citrate > phosphate > sulfate > chloride as deter-mined by circular dichroic-spectroscopy. The mechanism apparently involves the structuring of water and the unfavorable thermodynamic changes (negative entropy) required to expose non-polar groups to the water and to create hydration cavities [13]. Thus, by enhancing hydrophobic interactions within the β-Lg dimer and con-comitantly strengthening hydrogen bonding and van der Waals interactions these

Table 4. Effect of temperature on the thermodynamic activation parameters for the dissociation of β-lactoglobulin dimer, i.e., accessibility of thiol group at pH 6.85

Temperature (°C)	Rate K (sec^{-1})	ΔH^+_+ (cal/mol)	ΔS^+_+ (cal/mol/K)	ΔG^+_+ (cal/mol)
50	5.58×10^{-4}	31 200	22.9	23 800
55	1.10×10^{-3}	31 190	22.7	23 740
60	2.29×10^{-3}	31 180	22.7	23 620
65	$5,60 \times 10^{-3}$	31 170	23.0	23 390
70	1.08×10^{-2}	31 160	22.9	23 290
75	1.93×10^{-2}	31 150	22.7	23 240
80	3.42×10^{-2}	31 140	22.7	23 180

Table 5. Effect of various anions on the rate constant (k) and relative change in free energy of activation, $\Delta(\Delta G_+^+$ dimer) for the dissociation of the β-lactoglobulin dimer and exposure of thiol group at 70 °C

Salt (0.2 M)	k (sec^{-1})	$\Delta(\Delta_+^+$ dimer) (cal/mol)
None	1.20×10^{-2}	0
Chloride	9.80×10^{-3}	140
Tartrate	7.61×10^{-3}	312
Sulfate	4.80×10^{-3}	620
Phosphate	4.07×10^{-3}	730
Citrate	3.33×10^{-3}	875

anions increase the thermal stability of β-Lg and decrease the reactivity of the thiol groups. These observations may be further exploited to improve the heat stability and reduce the reactivity of β-Lg in milk.

Chemical modification: The principal functional groups of proteins that can be modified are the free amino and carboxylic groups; however, in addition the hydroxyl, phenolic, sulfhydryl, and disulfide groups may occasionally be modified [33]. The most common methods that have been used for modification of milk proteins have included acylation or alkylation of free amino groups; esterification and amidation of carboxyl groups; glycosylation of amino and carboxylic groups; phosphorylation of free hydroxyl (mostly serine) and amino groups; reduction of disulfide bonds; cross-linking using thiol/disulfide interchange, and also transglutaminase (γ-glutamyl carboxyl to ε lysine condensation) and proteolytic modification [23, 44, 39, 5, 6].

Acylation: Acyl anhydrides facilely modify the ε-NH$_2$ group of lysine to form acylated proteins with increased negative charge depending on the group added [33, 8, 25]. Thus, succinylation of caseins and whey proteins enhances their solubility, hydration and rheological properties and in the case of whey proteins and β-Lg limited succinylation improved surface activity and heat stability. Acylation of palmitic acid to the free amino groups of caseins via palmitoyl hydroxysuccinimide ester improved the surface active properties [10]. The incorporation of methionine via the carboxyanhydride reduced the solubility of casein [1]. Acylation generally reduces the nutritional value of proteins by reducing the availability of lysine, however at the levels used in foods this should not be significant.

Phosphorylation of the serine hydroxyl and ε-NH$_2$ lysine of casein and β-Lg respectively has been achieved using phosphorus oxychloride [35, 67]. Phosphorylation altered structure, increased solubility, water binding, viscosity, emulsifying activity, and it enhanced heat stability. It improved network formation via calcium cross-linking which induced gelation and improved the stability of emulsions [68]. The effects depend upon the extent of phosphorylation. This is a harsh method and may be impractical, however, enzymatic phosphorylation may become of practical use with the production of enzymes and deserves further study.

Glycosylation of caseins via reductive alkylation improved their solubility at acidic pH values, significantly enhanced viscosity, and slightly improved foaming properties [6]. Maltose and β-cyclodextrin were coupled to the lysine residues of β-Lg using the cyclic carbonates [65, 62, 66]. Glucoseamine was linked to carboxyl groups via carbodiimide. These yielded a heterogeneous population of β-Lg products with altered conformation, charge and solubility characteristics. The viscosity and water-holding properties were enhanced and although surface activity was not significantly altered, some of these glycosylated β-Lg had useful emulsion and foaming properties. Increasing glycosylation enhanced in vitro digestibility reflecting extensive unfolding of these proteins [63, 64]. Glycosylation of β-Lg with gluconic and melibionic acid caused improved solubility over the entire pH range and enhanced heat stability by approximately 5 °C [34]. Generally, the introduction of charged functional groups or bulky substituents into proteins causes extensive changes in the conformation of these proteins and alters physicochemical properties, e.g., viscosity, surface hydrophobicity, solubility and, network forming behavior.

The free carboxylic groups of glutamic and aspartic acid can be readily esterified and/or amidated to alter the isoionic pH of β-Lg and increase it from 5.2 to above pH 9.0. These esterified proteins improved surface activity and emulsion stability, perhaps by reducing charge and enhancing hydrophobicity [36]. These cationic proteins readily interact with other food proteins and improve foaming properties [52].

Surface activity: The capacity of proteins to form interfacial films and stabilize foams is an important functional attribute in many applications and is particularly important in the fabrication of low-calorie desserts, whipped toppings, aerated pastries, etc. Some proteins that are amphipathic and possess dynamic flexible structures are very surface active and can form effective foams, e.g., caseins, but generally these foams are unstable, particularly to heat [25]. The criteria required in a "foaming" protein are solubility, rapid adsorption of part of the molecule at the interface with concomitant partial unfolding, spreading in the interface, and extensive protein:protein association in the air, aqueous and air:water interface to form a continuous cohesive film. Ideally, these films should possess mechanical strength and viscoelastic properties to enable them to encapsulate air bubbles, resist rupture and gas leakage, and effectively retain water in the foam capillaries. Few individual proteins can perform all the dynamic functions required for formation of a stable foam – usually a mixture of proteins, e.g., egg white perform best [50].

The caseins have good whipping and overrun characteristics but the films formed tend to be weak and hence the stability of these foams is unreliable. Whey proteins and particularly β-lactoglobulin have excellent foaming properties that are superior to egg white in overrun [50]. They form thicker stronger films reflecting extensive intermolecular interactions. However, these have limited heat stability for many applications. Poole et al. [54] showed that the addition of basic proteins, clupeine, and lysozyme improved the foaming behavior of several proteins including β-Lg. Proteins with pI values greater than pH 9.0 progressively improve foaming behavior of other proteins.

Modification of milk proteins to improve their film forming properties has not been very successful [23], however, recently it has been observed that modification of β-Lg via amidation or esterification to increase its isoelectric point to pH > 9.0

makes it a good complementary foaming agent, i.e., when mixed with β-Lg at a ratio of 1 : 10 at pH 7.0 it markedly enhanced the foaming behavior (overrun and stability) of β-Lg [52]; clupeine and lysozyme also enhances foaming and whey protein isolates and improves their heat stability [50]. The glycine and arginine esters of β-Lg also markedly enhanced the foaming properties of food proteins and this was related to their net positive charge which facilitated maximum interactions with other proteins [52]. This observation provides a new approach for improving the functional properties of milk proteins. These observations are consistent with the suggestion that thicker interfacial films form more stable foams [25, 51].

Disulfide Bonds: Disulfide bonds maintain the structure and stability of many proteins including k-casein, α_{s2}-casein, β-lactoglobulin, α-lactalbumin, serum albumin, and immunoglobulins of milk. While the disulfide bonds in these proteins may be important in maintaining biological activity, they may actually impair functional applications. Thus controlled cleavage of the disulfide (S–S) bonds of these proteins may be an approach for improving their functional properties.

The cleavage of S–S bonds by reduction is of limited use in foods because of rapid reoxidation unless the thiol groups are protected, e.g., with iodoacetamide. However, most of these SH protective reagents are not allowed in foods, therefore alternative cleavage methods are needed, e.g., sulfitolysis.

A facile method for the controlled cleavage of S–S bonds involves treating a protein solution containing an S–S concentration of 40–400 µM in phosphate buffer at pH 7.0 with sodium sulfite (0.1 M), ammoniacal cupric sulfate 0.1 M, and bubbling oxygen through at 45 °C. The progress can be monitored by quantifying S–S bonds [20]. In this reaction the free SH groups are sulfonated by oxygen in a cupric ion mediated oxidation/reduction cycle. The rate of the reaction can be controlled by the copper concentration with rates of 1.2×10^{-3} and 1.6×10^{-2} min^{-1} obtained at 40 and 400 µM. The rates of cleavage varies with protein but can be accomplished without denaturants [20].

Sulfite reduction of whey proteins, serum albumin, and β-Lg greatly increased their viscosity and at low levels of sulfitolysis their surface active properties can be improved. Emulsions were made using bovine serum albumin with approximately 50% and 100% of the S–S bonds reduced by sulfitolysis. This caused extensive unfolding of the protein as reflected in the specific viscosity which increased from 0.05 to 0.12 and 0.3 at 50% and 80% sulfitolysis, respectively. The isoelectric point was decreased by the additional SO_3^- groups and the overall conformation of the protein changed with an increase in conformational flexibility as indicated from fluorescence and CD spectra [22]. The cleavage of disulfide bonds improved the emulsifying behavior of BSA as reflected in the decrease in emulsion droplet size and increased surface area per gram of protein (Table 6). These data suggest that the increased flexibility and unfolding of the protein following sulfitolysis improved the capacity of BSA to form interfacial films for these emulsions.

Enzymatic modification: Because chemical modification lacks specificity, control, and causes concerns about safety and nutritional value, enzymatic modification should be systematically explored because of specificity, mild conditions, safety and control. Furthermore, with advances in biotechnology a wide range of enzymes

Table 6. Effects of sulfitolysis on emulsifying behavior of bovine serum albumin

S–S cleaved %	Oil fraction					
	Average droplet size D(μ)			Surface area (m^2/g protein)		
	0.2	0.4	0.6	0.2	0.4	0.6
O Native	2.1	4.6	10.1	260	225	160
50%	2.0	3.7	9.2	240	300	175
100%	1.9	3.0	6.9	270	380	250

should become available for hydrolyzing, acylating, reduction (disulfide), oxidation (thiol), cross-linking, and glycosylation of proteins.

Limited proteolysis improves the foaming properties of proteins, especially aggregated denatured proteins [60]. Partial hydrolysis of α_s-casein with pepsin, plasmin, or chymosin released 23 amino acid N-terminal residue that had a high surface activity, facilitated emulsion formation but failed to stabilize emulsions.

Transglutaminase catalyzes a calcium dependent acyl transfer in which the carboxyamide group of glutaminyl residues of protein (donor) is linked to a nucleophilic group (acceptor) such as the ε-NH$_2$ of lysine. This enzyme can be used to incorporate amino acids/peptides, e.g., methionine or lysine esters, into proteins [14] and to cross-link proteins and increase viscosity, heat stability and induce proteins to form cold-set gels [59, 45]. This enzyme may be used to introduce a wide range of groups, e.g., glycosyl residues, peptides, etc. into proteins to alter functional properties.

There is a need for enzymes such as disulfide reductase and sulfhydryl oxidases to manipulate intra- and intermolecular crosslinking and improve functional behavior of milk protein in foods.

Genetic modification: An understanding of structure:functional relationships is desirable in selecting or modifying proteins for food applications and in the rational design of proteins via recombinant DNA technology. Unfortunately, with the exception of β-lactoglobulin (and to a lesser extent α-lactalbumin and caseins) little is known of the native molecular structures and the particular conformation which are important in specific functions. However, for many food applications general structure function relationships should be adequate in directing rational modification. Thus, in the case of β-Lg, deletion of cysteine 121 or insertion of a pairing cysteine to form a new disulfide link should improve the thermal stability of β-Lg and reduce its interaction with $\varkappa$-casein. The recent successful cloning of the β-Lg gene will facilitate site-directed mutagenesis [16]. Modification of this protein to study structure function relationships and eventually to design proteins that can be produced in lactating mammary of transgenic animals may now be feasible. Other research is focusing on isolating and cloning genes for caseins with a view to altering their properties via site directed mutagenesis [5]. In this context, the mammary tissue and milk may gradually become the vehicle for an array of additional food and therapeutic products via recombinant DNA and gene transfer, e.g., tailored proteins, enzymes, immune globulins, lactoferrin, vitamins, etc. Mammary tissue could become a major synthetic source of many genetically engineered mammalian products.

Acknowledgement

This work received the support of the National Dairy Board.

References

1. Chobert J-M, Bertrand-Harb C, Nicolas M-G, Baertner HF, Puigserver AJ (1987) J Agric Food Chem 35:638
2. Creamer LK, Parry DAD, Malcolm GN (1983) Arch Biochem Biophys 227:98–105
3. Feeney RE (1977) Food Proteins: Improvement through chemical and enzymatic modifications. In: Feeney RE, Whitaker JR (eds) Am Chem Soc, Wash DC, p 3
4. Fersht A (1977) Enzyme structure and mechanisms. Freeman, San Francisco
5. Flores RJ, Richardson T (1987) Food Biotech. Book
6. Fox, Mulvihill (1982) J Dairy Res 49:679
7. Fugate RD, Soon PS (1980) Biochem Biophys Acta 625:28
8. Haque Z, Kito M (1984) J Agric Food Chem 32:1392
9. Haque Z, Kinsella JE (1988) J Dairy Res 55:67–80
10. Haque Z, Kito M (1983) J Agric Food Chem 31:1225
11. Haque Z, Kito M (1983) J Agric Food Chem 31:1231
12. Harwalkar VR, Kalab M (1985) Milchwissenschaft 40:665
13. Horvath W, Melander C (1977) Arch Biochem Biophys 183:200
14. Ikura K, Goto M, Yoshikawa M, Sasaki RL, Chiba H (1984) J Agric Biol Chem 48:2347
15. Jakobsson I, Lindberg T, Benediktsson B, Hansson BG (1985) Acta Pediatr Scand 74:341
16. Jamieson AC, Vandeyar MA, Kang YC, Kinsella JE, Batt CA (1987) Gene 61:85–90
17. Kalab M, Emmons DB, Sargent AG (1976) Milchwissenschaft 31:402
18. Kato et al. (1985) J Agric Food Chem 33:931
19. Kella N, Kang Y, Kinsella JE (1988) J Protein Chem (in press)
20. Kella NKD, Kinsella JE (1985) J Biochem Biophys Methods 11:251
21. Kella NKD, Kinsella JE (1988) Biochem J (in press)
22. Kinsella JE, Phillips P (1988) Structure function of proteins in films and foams. Proceeds. AOCS Conf. Am. Oil Chem. Soc., Champaign, IL
23. Kinsella JE, Whitehead DM (1987) Film, foaming and emulsifying properties of some food proteins: Effects of modification. In: Protein Films. ACS Symp. Proceed Vol. 343, p 629–646
24. Kinsella JE (1984) Proceed. Int. Food Congr. Boole Press, Dublin, p 226
25. Kinsella JE (1981) Food Chem 7:273–288
26. Kinsella JE (1971) Adv Food Res 19:147
27. Kinsella JE, Whitehead DM (1987) Modification of milk proteins to improve functional properties and applications. In: Reidel D (ed) Milk, the vital force, p 791–804, Boston, MA
28. Kinsella JE (1987) Food Tech 41:62
29. Kinsella JE (1976) Crit Rev Food Sci Nutr 7:219
30. Kinsella JE (1984) IDF Symp. Proceed Danish Conf. on Functional Properties of Milk. Helsinger, Denmark, p 12–30
31. Kinsella JE (1984) CRC Crit Rev Food Sci Nutr 21(3):197–262
32. Kinsella JE (1982) In: Fox P, Codon J (eds) Food Proteins. Appl Sci, London, p 51
33. Kinsella JE, Shetty J (1979) ACS Symp. 92, Wash DC, p 3–50
34. Kitabatake N, Cuq JL, Cheftel JC (1985) J Agric Food Chem 33:125
35. Matheis G, Penner MH, Feeney RE, Whitaker JR (1983) J Agric Food Chem 31:379
36. Mattarella NL, Richardson T (1983) J Agric Food Chem 31:972

37. Mihalyi E (1978) Application of proteolytic enzymes to protein structure studies, CRC: West Palm Beach, FL. Vol. 1, p 153
38. Miranda G, Pelissier JP (1983) J Dairy Res 504:27
39. Modler HW (1985) J Dairy Sci 68:2195
40. Morr C (1982b) ACS Symp Ser 147. Washington, DC, p 201
41. Morr C (1982a) In: Fox PF (ed) Developments in Dairy Chemistry – 1. Appl Sci Press, London, PP 375–399
42. Mueller L (1982) In: Fox P (ed) Food Proteins. Appl Sci Press, London
43. Mulvihill DM, Kinsella JE (1987) Food Tech 41(a):102
44. Mulvihill DM, Fox PF (1987) Bulletin, IDF. No 209, p 3–11
45. Nio et al. (1985) Agric Biol Chem 49:2283
46. Ong DE, Kakkad B, McDonald PN (1987) J Biol Chem 262:2729
47. Papiz MZ, Sawyer L, Eliopoulos EE, North ACT, Findley JBC, Sivaprasada Rao R, Jones TA, Newcomer ME, Kraulis PJ (1986) Nature 324:383–385
48. Pearce RJ (1983) J Dairy Tech (Aust) 38:144
49. Pervaiz S, Brew K (1985) Science 228:335
50. Phillips L, Kinsella JE (1988) J Food Sci 52(4):1047–1049
51. Phillips M (1981) Food Technol 35:50
52. Poole S, Fry JC (1987) In: Developments in food protein, vol. 5, p 257
53. Poole S, West S, Fry CT (1987) Food Hydrocolloids 1:227
54. Poole S, West S, Walter S (1984) J Sci Food Agr 35:701
55. Reddy M, Kella NKD, Kinsella JE (1988) J Agric Food Chem
56. Rham O, Isliker H (1977) Int Arch Allergy Appl Immun 55:61
57. Shimizu M, Saito M, Yamauchi K (1985) Agric Biol Chem 49(1):189
58. Stone SS, Phillipps M, Kemeny LJ (1979) Am J Vet Res 40:607
59. Tanimoto, Kinsella JE (1988) J Agric Food Chem 36
60. To B, Helbig NB, Nakai S, Ma CY (1985) Can Inst Food Sci Technol J 18:150
61. Underdown BJ, Dorrington KJ (1974) J Immunol 112:949
62. Waniska R, Kinsella JE (1984) Intl J Protein Peptide Res 23:467
63. Waniska RD, Kinsella JE (1987) Colloid Interfac Sci 117:251
64. Waniska R, Kinsella JE (1989) In: Kinsella JE, Soucie W (eds) Protein structure and function. Am Oil Chem Soc, Champaign, IL
65. Waniska R, Kinsella JE (1981) J Agr Food Chem 29:826–831
66. Waniska R, Kinsella JE (1984) J Agr Food Chem 32:1042
67. Woo SL, Creamer LK, Richardson T (1982) J Agric Food Chem 30:65
68. Woo SL, Richardson T (1983) J Dairy Sci 66:984
69. Yvon M, Van Hille I, Pelissier JP, Guilloteau P, Toullec R (1984) Reprod Nutr Develop 24:835

Structural Changes in Milk Proteins

H. E. Swaisgood

Department of Food Science, North Carolina State University, Raleigh, North Carolina, USA

Introduction

With the growing body of knowledge of protein structure has come the realization that proteins are not rigid, inflexible structures. Rather their structures are dynamic and are characterized by marginal stability and conformational equilibria. Their structures represent a minimization of the sum of the free energies resulting from noncovalent interactions between amino acid residues within a polypeptide chain, between residues in contacting polypeptide chains, and between residues and solvent molecules. Because of their marginal stability, protein structures are sensitive to environmental change and alteration in their structure may be reversible or irreversible. Neglecting postsynthetic covalent change, proteins will refold to their native conformation from a random coil state under conditions which mimic the physiological environment. This result is a consequence of the thermodynamic nature of their structural stabilization. Recent evidence suggests that folding pathways as well as the native structures are determined by the primary structure and the consequent thermodynamics of noncovalent interactions between residues and residues plus solvent [1].

Often however, structural changes are irreversible. Obviously this would be true in instances where postsynthetic covalent change had occurred, such as limited proteolysis. Frequently, however, the structure of a protein is kinetically trapped in a non-native structure because the refolding conditions do not adequately mimic the physiological state. For example, refolding in solutions allows interchain interaction of reactive residues, whereas in vivo, physical isolation of nascent folding polypeptide chains would favor intramolecular interactions [2]. Also, in the case of thermally denatured proteins, the denatured state is often separated from the native state by a high activation energy due to multiple intermolecular interactions. Thus, if covalent changes have not occurred, a transition from the "irreversibly" thermally denatured state to the native structure can be achieved by first eliciting a random coil structure in a chaotropic solvent [3].

As a consequence of the dynamic nature and marginal stability of protein structure, subtle changes in the primary structure or environmental conditions can significantly alter their structure and their resulting functionality and nutritional quality. This paper will review some of the structural characteristics of milk proteins, the relationship of structure to proteolytic susceptibility, and some of the structural changes that may occur during processing and storage. Finally, the effect of these structural changes on the protein's digestibility will be discussed.

192

General aspects of protein structure

In recent years, our concept of protein structure has dramatically expanded, largely as a result of the rapid increase in the numbers of proteins whose three-dimensional structures have been solved by x-ray crystallography. Of the major milk proteins only the three-dimensional structures of α-lactalbumin and β-lactoglobulin have been solved at a high resolution [4–6]. Nevertheless, the principles derived from analysis of many known structures provide insight and allow a more probing discussion of protein structures based on other characteristics. For example, although it is still not possible to derive the three-dimensional structure from the primary structure, certain predictions of secondary structure can be made with reasonable accuracy [7]. It should be recalled that the primary structures of all the major milk proteins are known [8]. Furthermore, recent combination of x-ray crystallography with site-directed mutagenesis has permitted unequivocal analysis of the contribution of specific noncovalent interactions to protein structure and stability [9]. The derived principles and relative contributions of various noncovalent interactions provide a growing comprehension of structure and acquisition of structure in proteins.

Hydrophobic interactions result from an increase in the surface free energy and structure of water adjacent to nonpolar groups and thus the thermodynamic necessity to minimize the nonpolar surface area in contact with water. These interactions appear to be a major factor in the initiation of protein folding and are responsible for the globular nature of protein structure [1]. However, other than required complimentarity of surface topology, hydrophobic interactions are not specific or directional. Thus, biospecifically interacting hydrophobic surfaces often include hydrogen bonds and/or ion-pairs which provide the specificity and correct orientation. Hydrogen bonds, on the other hand, are very directional as the free energy is greatly dependent upon the bond angle and, hence, are responsible for formation and maintenance of regular backbone structure (secondary structures such as α-helix, β-sheet, and β-turn structures). An isolated backbone hydrogen bond, however, would not be stable in competition with potential H-bonds of similar energy with water. Consequently the requirement for a minimum number of H-bonds to simultaneously form results in cooperation with structure formation, i.e., the "velcro" effect. Of course, interactions between side chain residues protruding from the backbone are also important for stabilization of the secondary structure, consequently certain sequences favor α-helix, β-sheet, or β-turn structures and some sequences do not permit formation of an H-bonded secondary structure.

Folding most likely proceeds from initial formation of hydrophobic globules resulting from short range interactions, through acquisition of secondary structures, and finally interaction of these structures yields supersecondary structures and, ultimately, the final tertiary structure. The last step in the stabilization of native structure for secreted proteins containing disulfide bonds is oxidation. Disulfide bond formation pathways have been elucidated for several proteins [10, 11]. Formation of non-native disulfide bonds has been observed in the preferred folding pathway which must subsequently rearrange through sulfhydryl-disulfide interchange to give the native structure. Physiologically, formation of disulfide bonds in proteins is most likely catalyzed by protein : disulfide isomerase [12] and/or sulfhydryl oxidase

[13]. The latter enzyme will catalyze de novo synthesis of disulfide. Therefore, its participation, either directly or indirectly, seems highly probable through provision of oxidizing equivalents.

The tertiary structure of a protein consists of one or more domains which may fold independently and may have different stability. Noncovalent interactions (hydrophobic interactions, H-bonding, ion-pair formation) and complimentarity of surface topology will lead to association of supersecondary structures and domains yielding a folded native protein. In some cases, additional association of folded protein subunits occurs to yield quaternary structure and also superstructure. A fundamental question of biology addresses how the biospecificity of these interactions is achieved since similar interactions participate in all levels of structure. It appears that the primary structure is not sufficient in all cases to specify all levels of structure since some proteins cannot be refolded from random coils in solution [14]. Moreover, intracellular synthesis and storage of protein in recombinant microorganisms commonly yields inactive proteins [15]. The answer may lie in the physical [2] and/or temporal [16] isolation of nascent folding polypeptide chains. Recently it has been proposed that pauses occur at crucial times during biosynthesis which would allow certain domains to fold before synthesis of adjoining domains [15]. Previously, we have proposed that polypeptide chains are substantially folded prior to release from ribosomes or membranes in the case of secreted proteins and thus are physically separated while they are folding [2]. Consequently, temporal isolation would prevent non-native intrachain interactions while physical isolation would prevent non-native interchain interactions.

It is important to recognize that conditions for folding to yield native structures are not obtained when foods are restored to ambient temperature, neutral pH, or physiological salt concentrations. Hence, the protein structures obtained are often quite different from those in the original biological material.

Structure of caseins

In comparison with typical globular proteins, such as the whey proteins, the structures of the caseins are quite unique. Perhaps the most unusual feature is the amphiphilicity of their primary structures [8]. This characteristic is enhanced, particularly in the α_{s1}-, α_{s2}-, and β-caseins, by phosphorylation of seryl or threonyl residues in the polar domains. For example, at pH 6.6 the C-terminal 53-residue sequence of $\varkappa$-casein has a net charge of -10 to -17, the sequence $41 \rightarrow 80$ in α_{s1}-casein has a net charge of -21, the C-terminal 47-residue sequence of α_{s2}-casein has net charge of $+9.5$ while the N-terminal 68-residue sequence has a net charge of -21, and the N-terminal 21-residue sequence of β-casein would have a net charge of -12 [8]. Those regions of 10-residue segments with a net charge frequency of 0.5 or greater are illustrated in Fig. 1. Such sequences are unique to the caseins. The remaining sequences, particularly for α_{s1}-, β-, and $\varkappa$-casein, are rather hydrophobic and contain a low frequency of charged residues that are about equally divided between cations and anions.

As a result, the tertiary structures of caseins are most likely composed of two domains: a very hydrophobic globular domain and a highly charged, polar domain

194

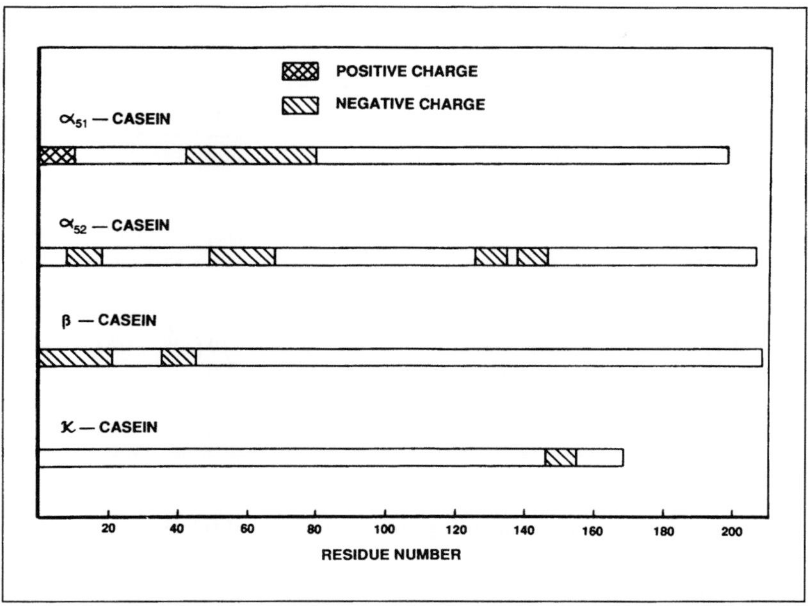

Fig. 1. Amino acid sequences in bovine milk proteins exhibit a net charge density of 0.5 or greater at the pH of milk

[8]. The caseins, vis-a-vis typical globular proteins, contain less secondary structure; nevertheless, some is predicted from the primary structure and by analysis of CD spectra (Table 1). Most of the secondary structure is likely to be present in the hydrophobic domains. The hydrophobic domain of $\varkappa$-casein in particular contains appreciable secondary structure allowing for the possibility of supersecondary structure such as a $\beta\alpha\beta$ unit [18]. Some secondary structure is also predicted in the polar domains; however, the predictions depend upon the probability assignments for phosphoserine [19] which are not well established at the present time. Although the polar domains contain fewer residues than the hydrophobic domains, the polypeptide chain in these regions probably exhibits much greater flexibility and thus would occupy a greater share of the molecular volume. Consequently, these proteins would not be expected to behave hydrodynamically as a typical globular protein. This result has been well established [8].

Because of their amphiphilic structure and the large size of their hydrophobic domains, the casein monomers cannot sufficiently remove their hydrophobic surfaces from contact with water. Consequently, their hydrophobic surfaces interact to form polymers in the case of isolated individual caseins or to form submicelles as these proteins exist in their natural state [25, 26]. Submicelles appear to have a variable composition of each of the caseins and are roughly spherical (10 nm diameter) with the polar domains of the various caseins on the surface [25, 27, 28]. Their composition is presently controversial [25–29]; however, the average composition is a 3:0.8:3:1 ratio of $\alpha_{s1}:\alpha_{s2}:\beta:\varkappa$-caseins [25] and it appears they can be separated into several fractions including those rich in $\varkappa$-casein and those rich in α_s and/or β-casein [27].

195

Table 1. Secondary structures of milk proteins

Protein	Secondary structure				
	Sequence prediction [a]			CD spectra [a]	
	α-helix %	β-sheet %	β-turn %	α-helix %	β-sheet %
$\varkappa$-casein	23 [17] 16 [18]	31 [17] 17 [18]	24 [17]	14 [17]	31 [17]
α_{s1}-casein	22 [19]	7.5 [19]		13–20 [19]	17 [19]
β-casein	12 [19] 10 [20]	11 [19] 13 [20]		12–20 [19] 1–10 [20]	0 [19] 13–16 [20]
β-lactoglobulin	10–50 [21] 7 [6][b]	20–30 [21] 51 [6][b]	17–24 [21] 11 [6][b]	10 [24]	43 [24]
α-lactalbumin	35 [22][c]	18 [22][c]	18 [22][c]	26 [23]	14 [23]

[a] The number in parentheses is the reference
[b] Calculated by the author from the crystallographic structure [6]
[c] Calculated from energy minimization and homology to the structure of lysozyme

Various models have been proposed for micelle structure. The submicelle models of Slattery and Evard [30] and Schmidt [25] are particularly attractive since these models account for most of the observations. The major distinction between the two models is that the Slattery model places emphasis on the hydrophobic interactions linking the submicelles together, whereas Schmidt's model emphasizes the interaction between submicelles through calcium phosphate. In both models, submicelles rich in $\varkappa$-casein are on the surface of the micelle in agreement with various experimental observations [31, 32].

As noted above, hydrophobic interactions are likely to be the major stabilizing force in submicelle structure. However, anionic clusters of phosphoryl groups are a unique feature of casein structure and I suggest that these clusters form a recognition interface on the surface of submicelles that selectively bind certain divalent metal cations or their salts, especially Ca^{2+} and $Ca_9 (PO_4)_6$. Submicelles rich in $\varkappa$-casein, which does not have an anionic cluster, would have fewer binding sites capable of interaction through calcium phosphate with other submicelles. Therefore, these micelles would form an unreactive surface which would limit growth of the micelle. Binding of Ca^{2+} to the anionic clusters may discharge these sites, lower the polarity, and allow less directional hydrophobic interactions to occur between submicelles. Thus as Ca^{2+} is added to milk, more dense and less hydrated aggregates of submicelles are formed, eventually leading to precipitates of the most hydrophobic submicelles. Furthermore, considering that some anionic clusters in native micelles may be occupied by Ca^{2+}, submicelles that are rich in the more hydrophobic β-casein, and which thus contain fewer anionic cluster linkage sites may interact with other submicelles hydrophobically. Such submicelles would be released at low temperatures due to weakening of hydrophobic interactions. However, their release would not disrupt the overall micelle because these interactions are less directional than anionic cluster interactions and do not form the matrix structure of the micelle.

Finally, it should be noted that the structure of caseins and the resulting micelle structure has great nutritional significance. Due to their more flexible structure the individual caseins are more susceptible to proteolysis (see below) than typical globular proteins and the open structure of micelles is easily penetrated by exo- and endopeptidases. Furthermore, the curd formed by crosslinking of micelles, from which the stabilizing surface polar domain of $\varkappa$-casein has been removed, remains in the stomach for long periods thus allowing slow release of nutrient [33]. Experimental evidence also suggests that curd firmness is a function of the concentration of anionic clusters (phosphoseryl residues) in the micelle [33] as would be anticipated from the model presented above. It also appears that the bioavailability of divalent metal ions is related to these anionic clusters [33].

Proteolysis and protein structure

It is well established that peptide bonds susceptible to hydrolysis by exo- or endopeptidases are those bonds which are exposed and flexible [34]. This requires that the polypeptide chain is not stabilized in some secondary or tertiary structure; therefore, the rate of hydrolysis is proportional to the fraction of the substrate protein molecules in an unfolded conformation about that particular bond [35]. Quantitatively this relationship may be stated as

$$v = k f_d [S]$$

where v is the velocity of proteolysis, f_d is the fraction of the molecules with the peptide bond in an unfolded flexible conformation, $[S]$ is the substrate protein

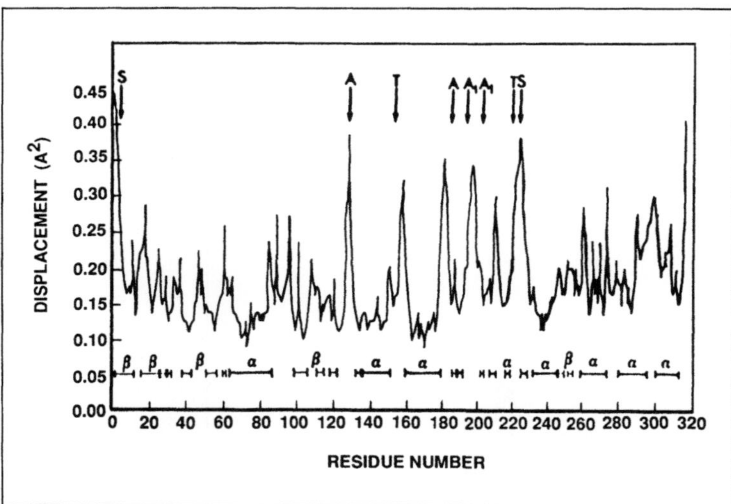

Fig. 2. Average main-chain temperature factors along the polypeptide chain of thermolysin. Arrows indicate sites of limited proteolysis or autolysis: S, sites for cleavage by subtilisin; A, cleavage by autolysis in 1.5 m*M* CaCl₂ and 1 m*M* or 10 mM EDTA; T, cleavage by thermal autolysis. (Taken from [36] with permission)

concentration, and k is a pseudo first-order rate constant. Consequently, peptidases can be used to probe the structure and stability of proteins.

The correlation between sites of limited proteolysis in thermolysin and chain flexibility, as measured by crystallographic temperature factors which are indicative of mean atomic displacements, is illustrated in Fig. 2. Only the regions having the greatest chain flexibility were susceptible, thus indicating the importance of the ability of the polypeptide chain to adapt to the conformation of the enzyme's active site.

Since the rate of peptide bond hydrolysis is proportional to f_d, proteolysis can be used to monitor structural transitions in proteins. There have been numerous examples of this application [35–37]; for example, the thermal transition of lysozyme at low pH measured by the rate of pepsinolysis [37]. As illustrated by data in Fig. 3, there is excellent agreement between the proteolytic method and classical spectral methods for monitoring this structural transition.

More recently, immobilized peptidases have been used to characterize structural transitions [35, 38]. Use of immobilized peptidases to monitor a transition has the following advantages: a) autolysis is prevented; b) the structural stability of the enzyme can be increased; c) the extent of the reaction is easily controlled; and d) the reaction mixture is not contaminated with the peptidase or its autolysis products [35]. Data in Fig. 4 show the thermal transition of ribonuclease as measured for the C-terminal region by hydrolysis with the immobilized exopeptidase carboxypeptidase A and for the overall structure as measured spectrally by exposure of tyrosyl residues [38]. Such experiments were not possible with soluble carboxypeptidase

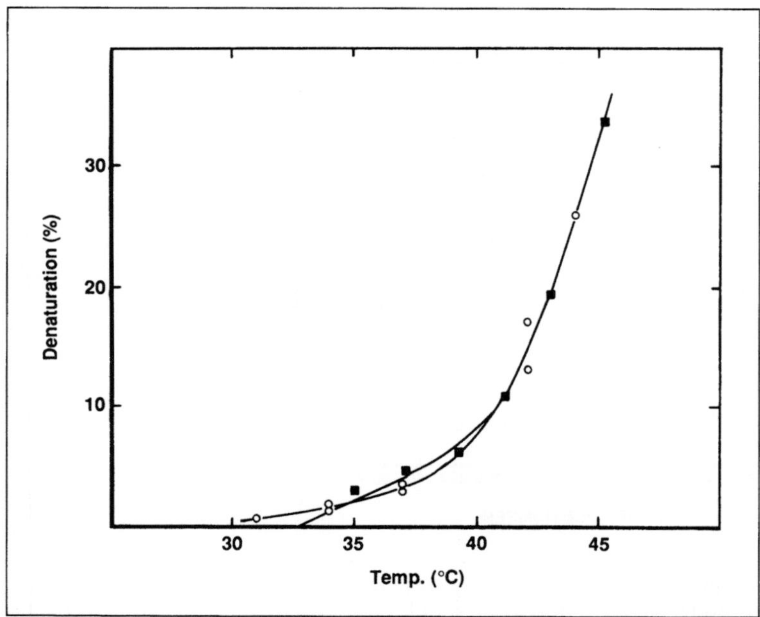

Fig. 3. Thermal transition curve of lysozyme at pH 2.0. ■; values obtained by classical methods, monitored by optical rotation; o, values obtained by pepsinolysis. Taken from [37] with permission

198

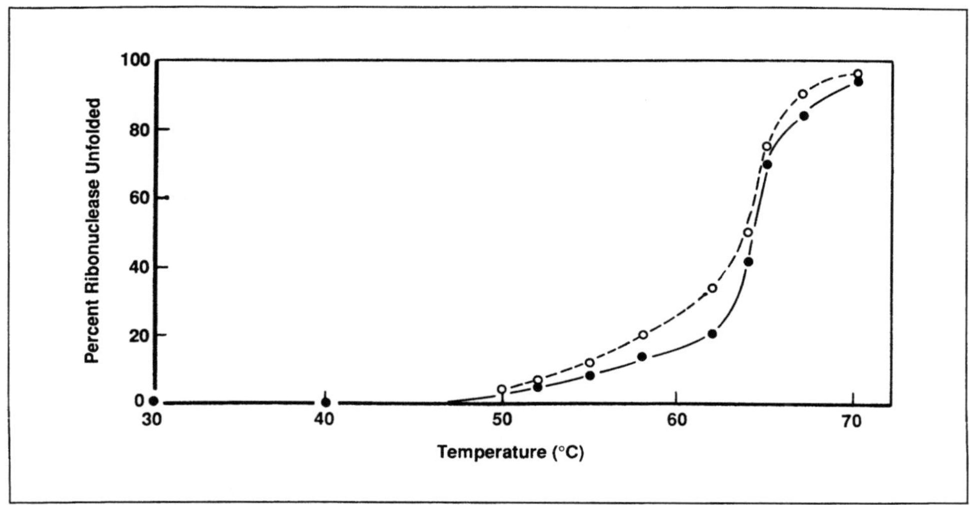

Fig. 4. Thermal unfolding of ribonuclease as measured by ultraviolet difference spectroscopy and optical rotation (o) and by hydrolysis with immobilized carboxypeptidase A (•). (Taken from [38] with permission)

since the enzyme does not remain active in the temperature region of the transition. In agreement with other independent studies, some exposure of tyrosyl residues in other parts of the ribonuclease molecule is indicated prior to unfolding of the C-terminal segment. Also, we have used immobilized *S. griseus* Pronase to monitor structural transitions of proteins in chaotropic solvents [39]. Results for the structural stabilities of whole casein, β-lactoglobulin, and lysozyme in urea are shown in Fig. 5. Obviously, the caseins are the least stable but, as expected from the discussion above, some structure is indicated in native casein.

It is apparent that some denatured proteins are more rapidly hydrolyzed by peptidases than their native counterparts and thus should be more digestible. Most proteins are less stable under acid conditions such as that in the stomach so the evolution of such conditions for initial digestion of proteins is not surprising. The types of denaturation or structural change which lead to a loss in digestibility will be addressed below.

Changes in protein structure resulting from processing and storage

The changes in protein structure that occur during a processing operation may be reversible or irreversible. Of course only irreversible changes will be of any consequence to subsequent properties. Nevertheless, it should be kept in mind that a reversible change may become irreversible when other components are added to the system. Irreversible changes vary from rearrangement of noncovalent interactions to changes in the protein's covalent structure. Both changes may be of nutritional significance; the first by altering the protein's digestibility and the second by changing both the digestibility and the bioavailability of certain amino acids.

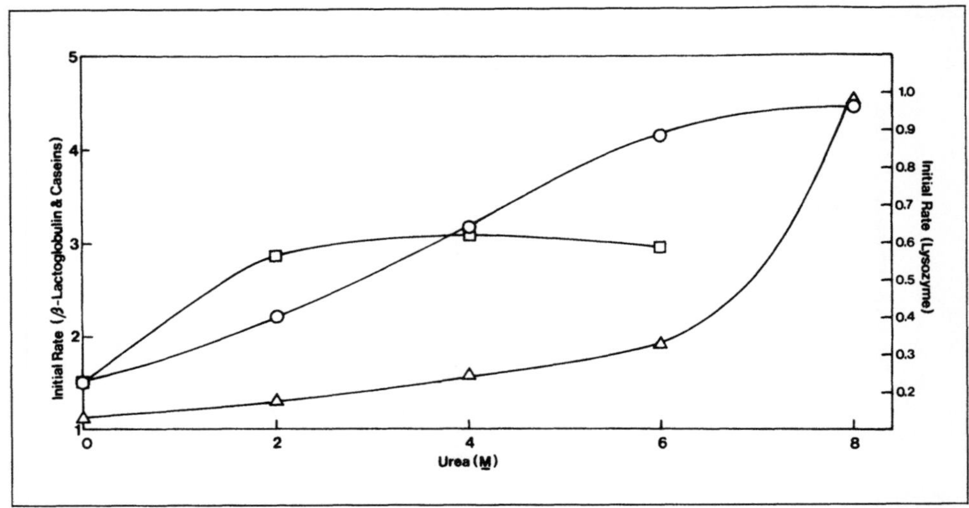

Fig. 5. Initial rates as a function of urea concentrations for hydrolysis of proteins by immobilized *S. griseus* pronase. The observed rates were corrected for small reversible losses of enzyme activity at the higher urea concentrations. The initial rates are given in µmol α-amino groups released per minute; (□), whole casein; (○), β-lactoglobulin; (▵), lysozyme. (Taken from [35] with permission)

Physical changes primarily resulting from rearrangement of noncovalent interaction

Irreversible structural change often occurs when unfolding is induced by a change in pH or temperature because of subsequent interaction of the unfolded chains either under the altered conditions or when the perturbing condition is relaxed to the original. Thermal denaturation of milk proteins is commonly observed since most processing of dairy foods involves some type of heat treatment. Whey proteins are the most susceptible of the milk proteins to thermal denaturation. Their heat stabilities, in decreasing order are α-lactalbumin > β-lactoglobulin > serum albumin > immunoglobulins (40–42). For example, a heat treatment of milk for 30 min at 70 °C caused irreversible denaturation of roughly 90% of the immunoglobulins, 50% of the serum albumin, 30% of the β-lactoglobulins and only 5% of the α-lactalbumin [40]. In a recent comparison of various types of processing treatment [43] the percent whey protein denaturation measured by several methods was 62–68% for a UHT indirect method (145 °C; 3 s), 51–54% for a UHT direct method (142 °C; 3 s), 30–36% for an HTST method (80 °C; 30 s), and 20–30% for a batch pasteurization (63 °C; 30 min).

Environmental factors including pH, Ca^{2+} concentration, and the concentration of sugars effect the thermal denaturation of whey proteins. Furthermore, in whey the influence of these factors will be different for thermal precipitation, resulting from unfolding followed by association, versus thermal unfolding alone. In whey at neutral pH or above, sulfhydryl-disulfide interchange will occur between the sulfhydryl group in β-lactoglobulin, exposed by unfolding, and disulfides in α-lactalbumin and serum albumin [44, 45] leading to formation of large complexes. Noncovalent interactions between the unfolded chains are also very strong under these conditions and

200

precipitates will form in whey and whey protein concentrates [41, 46]. However, proteins in acid whey below pH 3.7 are very stable to precipitation [46]. Nevertheless, thermal unfolding occurs at lower temperatures at acid pH; therefore, stability to precipitation probably results from increased electrostatic repulsion and prevention of sulfhydryl-disulfide interchange at these pHs [47]. Certainly α-lactalbumin will be less stable at low pH since it will exist as the apo-form [8] which is a less stable, expanded structure. However, its thermal transition would be partially reversible, with only those molecules that interact with other unfolded proteins being trapped in non-native states, thus the extent of its irreversible denaturation should be dependent on concentration in WPC.

Addition of sugars, e.g., sucrose [45], lactose [45, 46] and various monosaccharides [45, 46] to whey or WPC stabilizes the proteins to thermal precipitation. Sugars stabilize protein structure by inducing preferential hydration of either the native structure or expanded structures, thus, contradictory results on the stabilization of native structure have been obtained [45, 46]. However, it is clear that intermolecular interactions are prevented by added sugars.

It is often stated that the caseins are extremely thermostable, but this refers only to their stability to thermoprecipitation. Actually, micelle structure is quite sensitive to changes in environmental conditions which often lead to apparently irreversible structural alterations. For example, the dissociation of β-casein from micelles at cold temperature is well known and explains the steady increase in γ-caseins that occurs during refrigerated storage of raw milk since dissociated monomers of β-casein are more susceptible to proteolysis by the milk proteinases [48]. Thermal processing of milk leads to interaction of denatured whey proteins with the surface of the micelle [29, 31]. This interaction is primarily between unfolded β-lactoglobulin and $\varkappa$-casein and most likely results from sulfhydryl-disulfide interchange [49]. A surface interaction is implied by the observed interaction between immobilized β-lactoglobulin and $\varkappa$-casein [49] and the predominance of $\varkappa$-casein on the micelle surface as shown by a reversible immobilization technique [32] and by immunohistochemical localization [31]. As a result of this interaction, micelles become more resistant to thermally-induced and proteolytically-induced coagulation and are less susceptible to storage gelation [50].

Thermal processing, particularly UHT treatment, also causes a change in the micelle size distribution [29, 50–52] and a decrease in casein solubility [31]. An increase in the size and numbers of large micelles is observed as well as an increase in the number of small particles (submicelles) that are not sedimented by high speed centrifugation (100 000 × g). The increase in size is probably due to surface interaction with denatured β-lactoglobulin, whereas the small casein particles may result from dissociation of submicelles due to a change in their interaction with calcium phosphate.

The nutritional significance of the above changes in protein structure lies in the affects on digestibility. Denatured proteins with more unfolded, expanded structures should be more susceptible to both exo- and endopeptidases and hence more digestible. In fact, both in vitro and in vivo data have indicated that the ranking of digestibilities of milks are UHT > HTST > Raw [31]. Furthermore, in a feeding study of human infants, more rapid weight gain was realized with UHT as compared to pasteurized milk [53].

Covalent modification of protein structure

More severe processing and storage conditions can cause substantial change in protein structure which will result in both reduced digestibility and bioavailability. The most common structural changes resulting from thermal processing and/or exposure to alkaline pH are racemization of amino acid residues, crosslinking of protein chains, and modification of residues (particularly lysine), by Maillard-type reactions with carbohydrates [54–57]. The potential consequences of these covalent modifications on digestibility, absorption, and utilization are illustrated in Fig. 6.

Both racemization and crosslinking are initiated by abstraction of a proton from the α-carbon which, through a carbanion intermediate, results in racemization or, through subsequent β-elimination, yields the dehydroalanylprotein [54]. Consequently, the rates of these reactions vary greatly with pH where much faster rates are observed under alkaline conditions. The rate of racemization is related to σ^*, a parameter which reflects the electron inductive effects of the amino acid side chain

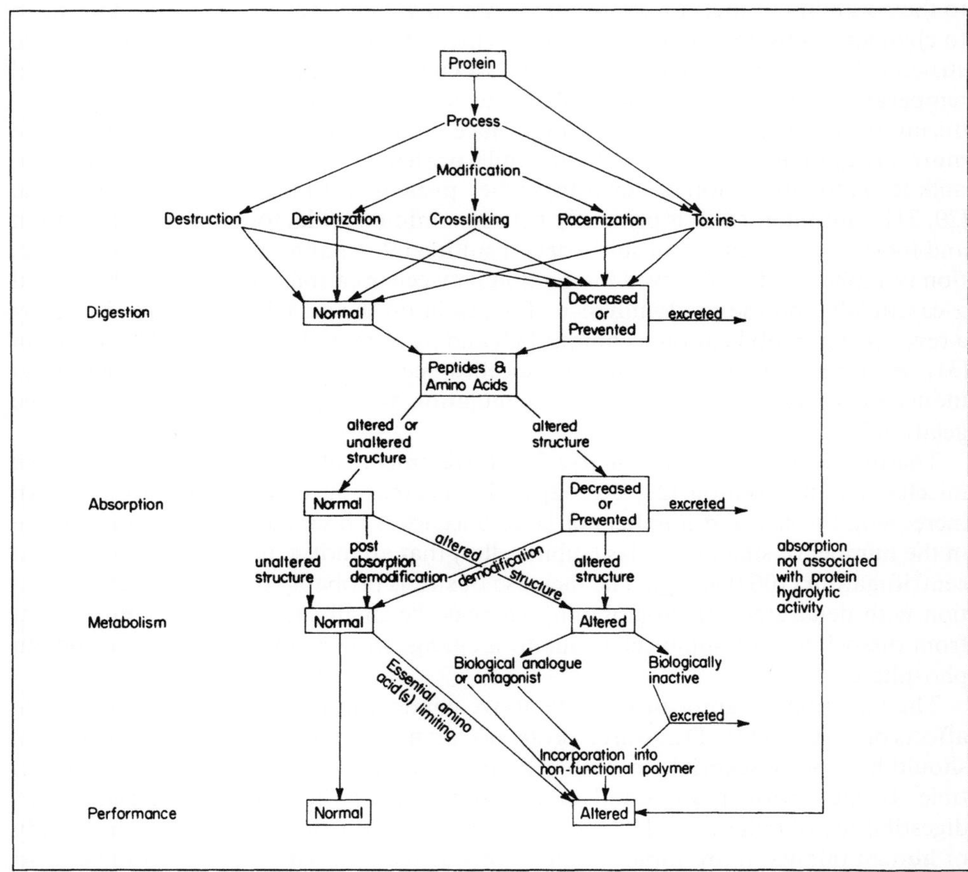

Fig. 6. Possible consequences of metabolism of modified proteins. Taken from [61] with permission

[58, 59]. Plots of $\log r_i / \log r_{Ala}$, where r_i is the rate for amino acid i and r_{Ala} that for *Ala*, versus σ^* are linear (corr. coeff. 0.61–0.97 [58]). However, the linear free energy relationship observed for racemization of various amino acid residues within a protein are not observed when comparisons are made between proteins [58, 59]. For example, a lower rate of racemization is observed for residues in α-lactalbumin as compared to soy protein [60]. The structure of the protein, directed by the amino acid sequence, may thus dictate a particular residue's susceptibility to racemization [58]. Consequently, a unique order for decreasing susceptibility to racemization cannot be given, but for milk proteins the order is roughly Cys, Ser, Asp > Thr, Met, Phe > Glu, Tyr, Lys > Ala > Val, Leu, Ile, Pro [58, 59].

Dehydroalanylprotein may be formed by β-elimination of the disulfide in cystinyl residues, the phosphoryl group in phosphoseryl residues, or the OH group in seryl residues. Reaction of cystinyl residues are probably most important in the whey proteins, whereas β-elimination of phosphoseryl residues are likely to be the major source of dehydroalanyl residues in the caseins. Calcium ion apparently enhances the rate of β-elimination of phosphoseryl residues [55]. Dehydroalanyl residues are quite reactive and will undergo vinyl-type additions, particularly with the ε-amino group of lysine to yield the lysinoalanyl (LAL) crosslink and with the sulfhydryl group of cysteine to give the lanthionyl crosslink [61, 62]. Both of these residues have been measured in various foods. For example, the LAL content of evaporated milk varies from about 200 to 900 ppm [55]. Also, a 6% loss of total half-cystine was noted in UHT processed milk (150 °C; 1.5 s), some of which may have been converted to dehydroalanine [63].

Although formation of LAL and racemization share a similar initiation mechanism, the extent of these two reactions in a protein appears to be inversely related. For example, α-lactalbumin and serum albumin are very susceptible to LAL formation, whereas the racemization rates in these proteins are low in comparison to soy and wheat protein isolates where the inverse relationship was observed [60]. Casein also appears to racemize more rapidly than α-lactalbumin [54]. It has been suggested that proteins with less stable structure may racemize more rapidly since their backbones would be more accessible, whereas some remaining structure may promote LAL formation [60]. Of the milk proteins, β-lactoglobulin seems to be the least susceptible to either crosslinking or racemization [60].

Nutritionally, the carbonyl-amine reactions are probably the most significant of the covalent changes that occur in dairy foods, particularly the Maillard reaction of carbohydrates with the ε-amino group of lysine [57]. The high content of lactose and lysyl residues in milk and milk proteins and the importance of milk proteins as a source of this essential amino acid make dairy foods particularly vulnerable to nutritional damage by Maillard reactions. Potentially, any carbonyl group, such as aldehydes derived from lipid oxidation, could react with lysine; however, lactose and lactulose are most reactive because they are present at much higher concentrations. Furthermore, since these reactions proceed at measurable rates under common storage conditions, they are of increased importance due to lengthening shelf-lives accompanying growth in aseptic technology.

During thermal processing some of the lactose is converted to lactulose [64] and, in fact, the amount of conversion has been used to indicate the amount of heat treatment received by milk products [64]. Typical values in various dairy foods are:

UHT milk, 10–51 mg/100 ml; in-bottle sterilized milk, 87–137 mg/100 ml; and spray dried powder, 17 mg/100 ml. Values of lactulose in infant food ranged from 86 to 187 mg/100 ml depending on the length of storage. Its presence in infant food is considered beneficial due to its stimulating effect on the growth of *Bifidobacterium bifidum* [63].

Both lactuloselysine and fructoselysine have been identified in processed milk products [65]. Although their concentrations are at or below the level of detection in pasteurized and UHT milk, significant quantities can develop during storage of UHT milk [66]. For example, a six-month storage of UHT milk reduced the available lysine by 5% at 4 °C and by 14% at 37 °C [66]. Typically, available lysine losses are 1–2% for pasteurization, 2–4% for UHT processing, 6–10% for in-bottle sterilization, and 20% for evaporation [67]. In a study of UHT skim milk (143 °C; 10 s), nearly 1.5% of the lactose was incorporated into the casein micelles, with the greatest amount attached to $\varkappa$-casein [68]. The significance of this modification of the micelle surface on its stability is not known.

The predominate factors affecting the extent of Maillard reaction in milk powders are the temperature and moisture content. A maximal rate of reaction is observed between 7% and 15% moisture [57]. Loss of available lysine can reach 45% in powders stored for four years at 20 °C.

Changes in protein digestibility

Changes in protein structure which cause unfolding or expansion leading to increased flexibility will be accompanied by increased digestibility. As previously noted, this effect has been observed in vitro as well as in vivo. Covalent changes in structure, on the other hand, may decrease the ability of exo- and endopeptidases to hydrolyze peptide bonds involving the modified residues. Moreover, such modification may affect not only the hydrolysis of peptide bonds with the modified residue, but also may prevent or decrease the hydrolysis rate of bonds involving adjacent residues. In the author's view, this possibility has not received sufficient attention. Thus, the common observation that the biological value of a processed protein may be reduced to a greater extent than that accounted for by loss of available lysine, for the case of Maillard reactions, in fact may be caused by an overall reduction in digestibility [57].

Various methods have been developed for in vitro assessment of digestibility of food and food proteins using soluble enzymes. These include the pepsin digest residue (PDR) method [69], the pepsin pancreatin digest (PPD) index [70]; the pepsin pancreatin digest dialysate (PPDD) index [71], the pH-drop rate assay using trypsin, chymotrypsin, intestinal peptidases, and microbial protease [72], and a pH stat rate method using similar enzymes [73]. We have developed an extent assay which uses the digestive enzymes as immobilized forms for the reasons previously listed [60]. This method, known as the immobilized digestive enzyme assay (IDEA) attempts to mimic the digestive process by measuring the extent of peptide bond hydrolysis. Some proteins, for example egg proteins, may exhibit rather slow hydrolysis rates but the extent of hydrolysis eventually reaches a value consistant with their digestibilities measured in vivo.

A number of studies have indicated that digestibilities determined with the IDEA system agree quite well with those measured by in vivo methods [74, 75]. For example, the comparison with in vivo digestibilities determined by two independent laboratories in a cooperative study of a number of foods and food proteins is given in Table 2. Correlation with these in vivo determinations is characterized by $r \sim 0.8$. When the IDEA digestibilities and in vivo measurements were made in the same laboratory a slightly better correlation was obtained using a number of proteins that had received alkali or heat treatment [75].

The digestibility of a protein exhibits a very good correlation with the extent of Maillard reaction or the degree of crosslinking and racemization of residues. For example, the correlation between loss of digestibility and loss of available lysine due to Maillard reaction with fructose is shown in Fig. 7 for a number of proteins, including β-lactoglobulin and α-lactalbumin [60]. It should be noted that the correlation does not appear to intersect the axis at zero, suggesting that formation of small amounts of fructoselysine ($<20\%$) does not detectably lower digestibility. At low levels of modification perhaps only trypsin activity would be affected.

Since racemization and crosslinking of residues occur simultaneously upon exposure of proteins to alkaline conditions, the correlation with loss in digestibility was tested by assuming their contributions to be additive. Hence, a measure of the sum

Table 2. Comparison of digestibilities of protein in various food and food components determined with the IDEA system with that determined in vivo (taken from data given by [74])

Food products	Digestibilities		In vivo [c]
	IDEA		
	D^a (in vitro)	Predicted [b] in vivo	
Animal:			
ANRC Casein	0.471 ± 0.049	92.5	96.8; 99
Nonfat dry milk	0.338 ± 0.036	89.3	92.6; 95
NFDM (Heated)	0.105 ± 0.012	83.7	90.1; 90
Sausage	0.446 ± 0.012	91.9	93.9; 94
Animal-vegetable mixture:			
Macaroni & cheese	0.352 ± 0.090	89.7	95.5; 94
Vegetable:			
Soy isolate	0.453	92.1	92.0; 98
Rolled oats	0.357 ± 0.041	89.8	93.8; 91
Chick peas	0.386 ± 0.043	90.5	88.0; 89
Pea protein concentrate	0.448	92.0	93.5; 92
Wheat cereal	0.361 ± 0.023	89.9	91.4; 91
Pinto beans	0.027 ± 0.035	81.9	72.7; 79
Rice-wheat gluten	0.509 ± 0.048	93.5	92.7; 95

[a] Defined as the fraction of the total peptide bonds that was hydrolyzed. The standard deviations listed are for triplicate measurements
[b] Calculated from the in vitro values using the regression: $Y = 81.14 + 24.19\,D$ (see [75])
[c] In vivo values determined independently in two laboratories using the same sample (see [74])

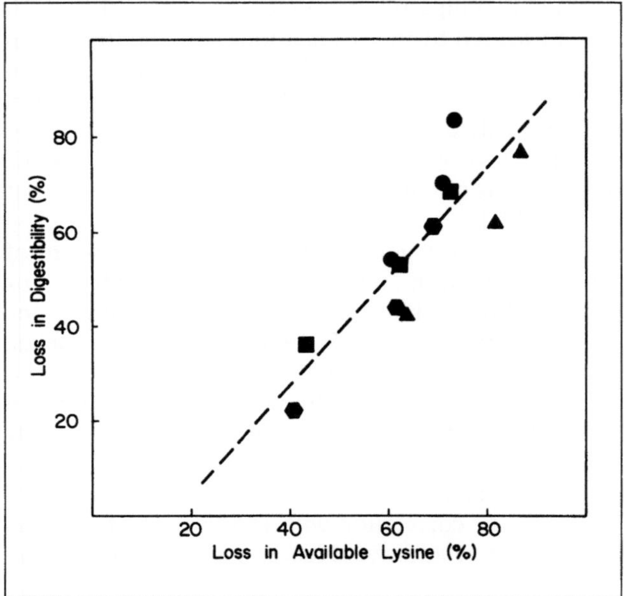

Fig. 7. Relationship between percent loss in digestibility as measured with the IDEA system and percent loss in available lysine: ●, soy protein isolate; ▲, bovine serum albumin; ■, β-lactoglobulin; ●, α-lactalbumin. The dashed line represents the linear regression given by $Y = 19.2 + 1.14\,X$, correlation coefficient 0.87. (Taken from [60] with permission)

of LAL formation and phenylalanine racemization is plotted versus the digestibility loss in Fig. 8. As for the case of Maillard reaction, the correlation with digestibility loss appears rather good. In this case the fit intersects very close to zero. This result should not be surprising since racemization and crosslinking will affect all enzyme activities, including those with extensive substrate-binding requirements. It has been shown that synthetic epimeric tripeptides containing D-amino acids as the central residue are not hydrolyzed by the enzymes used in the IDEA method nor are they utilized in rat feeding trials [76, 77]. The minimum size of peptides containing crosslinked or racemized residues that are recalcitrant to digestion is not presently known.

If comparisons are made for the same protein system, so that the composition of essential amino acids is not a factor, then measurement of digestibilities for samples of protein which received different treatments may allow prediction of their in vivo nutritional quality. Thus, data in Table 3 indicate a direct relationship for each individual protein source between digestibility and NPU or PER [78]. Such a relationship could be useful for prediction of PER or NPU values for a dairy food that may have been subjected to various processing or storage treatments; for example, comparison of various milk powders or UHT milks, etc.

Finally, it should be recalled that Maillard reactions, and therefore digestibilities, are greatly dependent upon temperature and water activity during storage. In a recent study of a model casein-glucose powder, it was observed that increasing the

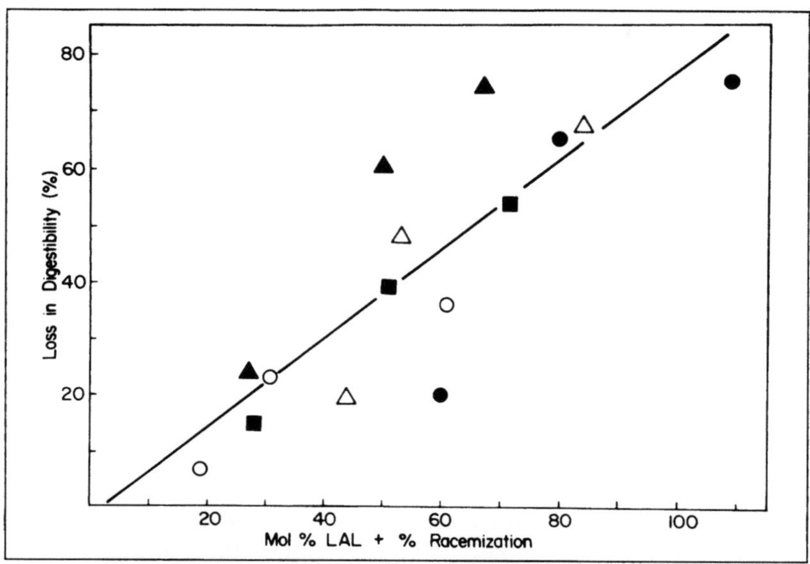

Fig. 8. Relationship between percent loss in digestibility as measured with the IDEA system and the sum of effects of LAL formation and racemization. The sum of these effects was calculated as the mol% of LAL plus percent racemization of Phe: o, β-lactoglobulin; •, α-lactalbumin; △, bovine serum albumin; ▲, soy protein isolate; ■, wheat protein isolate. The line represents the linear regression given by $Y = -2.2 + 0.789 \, X$, correlation coefficient 0.84. (Taken from [60] with permission)

Table 3. Comparison of predicted digestibilities determined with the IDEA method to NPU and PER values for native and alkali- or fructose-treated proteins (taken from data in [78])

Food protein	Digestibility[a] (predicted)	NPU	PER
Casein			
Native	94.5	87.3	3.35
Alkali treated[b]	91.7	82.2	2.89
Fructose treated[c]	89.9	81.6	2.69
Egg white			
Native	90.0	83.6	3.81
Alkali treated[b]	87.9	74.9	2.30
Fructose treated[b]	88.9	79.3	2.80
Soy protein			
Native	93.3	81.8	2.67
Alkali treated[b]	87.7	72.0	1.93
Fructose treated[c]	89.9	74.0	2.30
Whey protein			
Native	94.9	88.0	3.21
Alkali treated[b]	87.6	82.0	1.70
Fructose treated[c]	89.5	86.5	2.50

[a] Calculated using the regression: $Y = 81.14 + 24.19 \, D$ in vitro [75]

[b] Treated with $0.2 \, N$ NaOH $40\,°C$ for 4 hr

[c] Treated with $0.2 \, M$ fructose at pH 7 and $90\,°C$ for 4 hr

Table 4. Digestibilities of a casein-glucose model mixture stored for one month under various conditions (taken from [79]). The carbohydrate concentration was 3 mol glucose per mol of protein lysyl residues

Water activity	Temperature (°C)	Digestibility[a] (predicted)
0.2	30	94.0
0.5	30	89.2
0.2	60	89.9
0.5	60	85.7

[a] Digestibilities were measured with the IDEA method and predicted with the regression equation: $Y = 81.14 + 24.19\,D$ [75]

water activity from 0.2 to 0.5 decreased the digestibility by 50%, see Table 4 [79]. As expected, increasing the storage temperature also decreased the digestibility.

These considerations of the potential effects of processing and storage on digestibility emphasize the importance of avoiding extreme temperatures and pH during processing of dairy foods. If long storage periods are anticipated, the temperature, carbohydrate concentration and water activity (in the case of powders) should be considered in order to minimize nutritional damage. The nutritional quality of a dairy food may be considerably less than the expected value if care is not taken during processing and storage.

References

1. Fetrow JS, Zehfu MH, Rose GD (1988) Biotechnology 6:167–171
2. Janolino VG, Swaisgood HE, Horton HR (1985) J Appl Biochem 7:33–37
3. Klibanov AM, Mozhaev VV (1978) Biochem Biophys Res Commun 83:1012–1017
4. Smith SG, Lewis M, Aschaffenburg R, Fenna RE, Wilson IA, Sundaralingam M, Stuart DI, Phillips DC (1987) Biochem J 242:353–360
5. Stuart DI, Acharya KR, Walker NPC, Smith SG, Lewis M, Phillips DC (1986) Nature 324:84–87
6. Papiz MZ, Sawyer L, Eliopoulos EE, North ACT, Findlay JBC, Siraprasadarao R, Jones TA, Newcomer ME, Kraulis PJ (1986) Nature 324:383–385
7. Garnier J, Osguthorpe DJ, Robson B (1978) J Mol Biol 120:97–120
8. Swaisgood HE (1982) Chemistry of milk protein. In: Fox PF (ed) Developments in dairy chemistry-1. Applied Science Publishers, pp 1–59
9. Fersht AR (1987) Biochemistry 26:8031–8037
10. Creighton TE (1977) J Mol Biol 113:275–293
11. Konishi Y, Ooi T, Scheraga HA (1982) Biochemistry 21:4734–4740
12. Lambert N, Freedman RB (1983) Biochem J 213:225–234
13. Janolino VG, Swaisgood HE (1987) Arch Biochem Biophys 258:265–271
14. Janolino VG, Sliwkowski MX, Swaisgood HE, Horton HR (1978) Arch Biochem Biophys 191:269–277
15. Harris TJR (1983) In: Williamson R (ed) Genetic Engineering, vol. 4. Academic Press, London, pp 127–185
16. Purvis IJ, Bettany AJE, Santiago TC, Coggins JR, Duncan K, Eason R, Brown AJP (1987) J Mol Biol 193:413–417
17. Loucheux-Lefebvre MH, Aubert JP, Jolles P (1978) Biophys J 73:323–336

18. Raap J, Kerling KET, Vreeman HJ, Visser S (1983) Arch Biochem Biophys 221:117–124
19. Creamer LK, Richardson T, Parry DAD (1981) Arch Biochem Biophys 211:689–696
20. Andrews AL, Atkinson D, Evans MTA, Finer EG, Green JP, Phillips MC, Robertson RN (1979) Biopolymers 18:1105–1121
21. Deckmyn H, Preaux G (1978) Arch Int Physiol Biochim 86:938–939
22. Warme PK, Momany FA, Rumball SV, Tuttle RW, Scheraga HA (1974) Biochemistry 13:768–782
23. Robbins FM, Holmes LG (1970) Biochim Biophys Acta 221:234–240
24. Townend R, Kumosinski TF, Timasheff SN (1967) J Biol Chem 342:4538–4545
25. Schmidt DG (1982) Association of caseins and casein micelle structure. In: Fox PF (ed) Developments in dairy chemistry-1. Applied Science Publishers, pp 61–86
26. Slattery CW (1976) J Dairy Sci 59:1547–1556
27. Ono T, Takagi T (1986) J Dairy Res 53:547–555
28. Britten M, Boulet M, Paquin P (1986) J Dairy Res 53:537–584
29. Morr CV (1983) Physco-chemical basis for functionality of milk proteins. In: Kaufmann W (ed) Role of milk proteins in human nutrition. Th Mann KG Publisher, Gelsenkirchen-Buer, F. R. Germany, pp 333–344
30. Slattery CW, Evard R (1973) Biochim Biophys Acta 317:529–538
31. Farrell HM, Douglas FW (1983) Effects of ultra-high-temperature pasteurization on the functional and nutrition properties of milk proteins. In: Kaufmann W (ed) Role of Milk Proteins in Human Nutrition. Th Mann KG Publisher, Gelsenkirchen-Buer, F.R. Germany, pp 345–356
32. Heth AA, Swaisgood HE (1982) J Dair Sci 65:2047–2054
33. West DW (1986) J Dairy Res 53:333–352
34. Mihalyi E (1978) Application of Proteolytic Enzymes to Protein Structure Studies, vol 1. CRC Press, Boca Raton, FL, USA
35. Swaisgood HE, Catignani GL (1987) Methods Enzymol 135:596–604
36. Fontana A, Fassina G, Vita C, Dalzoppo D, Zamai M, Zambonin M (1986) Biochemistry 25:1847–1851
37. Matthyssens GE, Simons G, Kanarek L (1972) Eur J Biochem 26:449–454
38. Burgess AW, Weinstein LI, Gabel D, Scheraga HA (1975) Biochemistry 14:197–200
39. Church FC, Catignani GL, Swaisgood HE (1982) Enzyme Microb Technol 4:317–321
40. Larson BL, Rolleri GD (1955) J Dairy Sci 38:351–357
41. Schmidt RH, Packard VS, Morris HA (1984) J Dairy Sci 67:2723–2733
42. Levieux D (1980) Ann Rech Vet 11:89–97
43. Manji B, Kakuda Y (1987) J Dairy Sci 70:1355–1361
44. Morr CV, Josephson RV (1968) J Dairy Sci 51:1349–1355
45. Garrett JM, Stairs RA, Annett RG (1988) J Dairy Sci 71:10–16
46. Bernal V, Jelen P (1985) J Dairy Sci 68:2847–2852
47. Harwalkar VR (1979) Milchwissenschaft 34:419–422
48. Reimerdes EH (1982) Changes in the proteins of raw milk during storage. In: Fox PF (ed) Developments in dairy chemistry-1. Applied Science Publishers, pp 271–288
49. Euber JR, Brunner JR (1982) J Dairy Sci 65:2384–2387
50. Harwalker VR (1982) Age gelatin of sterilized milks. In: Fox PF (ed) Developments in dairy chemistry-1. Applied Science Publishers, pp 229–269
51. Burton H (1984) J Dairy Res 51:341–363
52. Freeman NW, Mangino ME (1981) J Dairy Sci 64:1772–1780
53. Mehta RS (1980) J Food Protect 43:212–225
54. Masters PM, Friedman M (1980) Amino acid racemization in alkali-treated food proteins-chemistry, toxicology, and nutritional consequences. In: Whitaker JR, Fajimaki M (eds) Chemical deterioration of proteins. ACS Symposium Series 123. American Chemical Society, Washington, DC, Chapter 8
55. Finot PA (1983) Nutr Abstr Rev 53:67–80
56. Maga JA (1984) J Agric Food Chem 32:955–964
57. Adrian J (1982) The Maillard reaction. In: Rechcigl M Jr (ed) Handbook of nutritive value of processed foods. CRC Press, Boca Raton, FL, USA, pp 529–608
58. Liardon R, Friedman M (1986) J Agric Food Chem 35:557–565

59. Liardon R, Friedman M (1987) J Agric Food Chem 35:661–667
60. Chung S-Y, Swaisgood HE, Catignani GL (1986) J Agric Food Chem 34:579–584
61. Swaisgood HE, Catignani GL (1982) J Food Protect 45:1248–1256
62. Friedmann M (1977) Protein crosslinking: nutritional and medical consequences. Plenum Press, New York, pp 1–27
63. Patrick PS, Swaisgood HE (1976) J Dairy Sci 59:594–600
64. Andrews GR (1986) J Dairy Res 53:665–680
65. Moller AB, Andrews AT, Cheeseman GC (1977) J Dairy Res 44:267–275
66. Moller AB, Andrews AT, Cheeseman GC (1977) J Dairy Res 44:277–281
67. Renner E (1979) Nutritional and biochemical characteristics of UHT milk. In: Proceedings of the International Conference on UHT Processing and Aseptic Packaging of Milk and Milk Products, North Carolina State University. Raleigh, NC, USA, pp 21–52
68. Turner LG, Swaisgood HE, Hansen AP (1978) J Dairy Sci 61:384–392
69. Sheffner AL, Eckfeldt GA, Spector H (1956) J Nutr 60:105–120
70. Akeson WR, Stahmann MA (1964) J Nutr 83:257–261
71. Mauron J (1973) The analyses of food proteins, amino acid composition and nutritive value. In: Porter JWG, Rolls BA (eds) Proteins in human nutrition. Academic Press, New York, pp 139–154
72. Satterlee LD, Kendrick JG, Miller GA (1977) Food Technol 31:78–88
73. Pedersen B, Eggum EO (1983) Z Tierphysiol Tiernahrg Futtermittelkde 49:265–277
74. Thresher WC, Swaisgood HE, Catignani GL (In press) Plant Foods Human Nutr
75. Chang HI, Catignani GL, Swaisgood HE (1987) Abstr, Institute of Food Technologists, National Meeting
76. Paquet A, Thresher WC, Swaisgood HE, Catignani GL (1985) Nutr Res 5:891–901
77. Paquet A, Thresher WC, Swaisgood HE (1987) Nutr Res 7:581–588
78. Chang PhD (1987) Dissertation, North Carolina State University, Raleigh, NC
79. Culver CA, Swaisgood HE (1987) J Dairy Sci [Suppl] 70 1:62 (Abstr. D20)

Technological and Functional Aspects of Milk Proteins

J. N. de Wit and G. Klarenbeek

Netherlands Institute for Dairy Research, Ede, Netherlands

Introduction

Milk proteins, which are widely used as ingredients of food products, are well-known for their high nutritional quality and versatile functional properties. Their behavior during food processing, however, is very complex and is governed by the heat sensitivity of the whey proteins [4]. Many attempts have been made to predict the functional properties of milk proteins, and these may be summarized by the two working methods shown in Fig. 1. They are: the systematic approach of estimating functional properties in aqueous solutions (lefthand side), and the empirical approach of assessing protein functionality by using whey protein products directly in food products (right).

The term "functional properties" is generally used for the physico-chemical properties of proteins in aqueous solutions or in simple model systems [10]. These param-

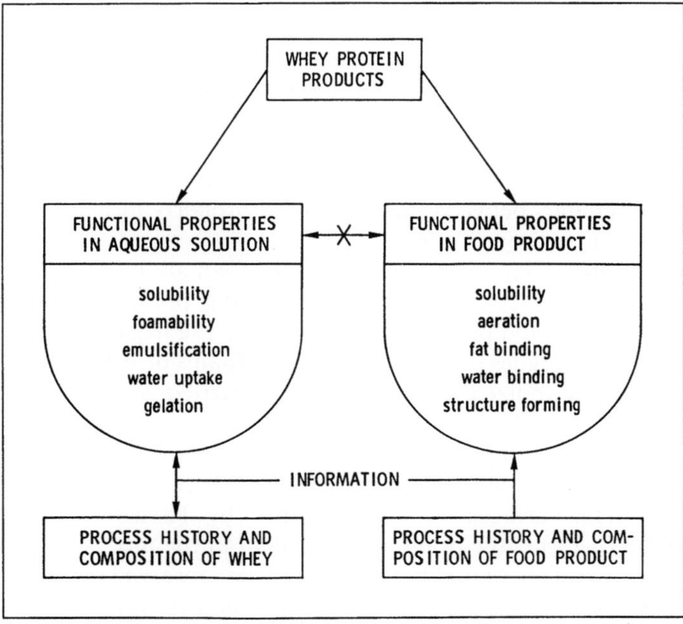

Fig. 1. Functional characterization of whey protein products

eters have the advantage of being primarily related to information on process history and composition of a given protein source, as is shown schematically in Fig. 1.

However, the usefulness of these parameters for predicting similar functional characteristics in food products, is limited or even absent. This is mainly due to the complex behavior of proteins in interactions with other ingredients of any food product during food processing. In particular the inevitable heat treatments used during processing and preservation of food products ensures that such properties as solubility, aeration, fat-binding, water-binding and structure-forming capacity in food products are not simply related to similar functional protein properties in aqueous solutions [6].

The heating of fluids containing milk proteins may also result in fouling of processing equipment. In the dairy industry, in particular, the whey proteins play a dominant role in the fouling of heat exchangers, because these proteins have a high heat sensitivity [7, 11, 14, 15]. A better understanding of the behavior of whey proteins during heat treatment is essential for the control of both their functional properties and their fouling ability.

In the present paper the kinetics of heat-induced changes of whey proteins are analyzed under various conditions and the results are related both to protein functionality in food products and to fouling problems in processing equipment.

The discussion will be focused on the difference between the rate of protein unfolding observed by differential scanning calorimetry (DSC), and the rate of protein aggregation observed by high-performance gel permeation chromatography (HP-GPC) during and after heat treatment.

Effects of heat treatment on the behavior of whey proteins

The solubility of whey proteins in aqueous solution may differ widely from that in a complete fruit drink, as is illustrated in Fig. 2. Addition of 1% whey proteins from a highly soluble whey protein concentrate (WPC) to the orange juice (center) caused a precipitate in the poorly soluble drink (right). This beverage was completely de-stabilized in the presence of whey proteins when pasteurized at 74 °C for 5 min, which clearly shows the discrepancy between this functional property of whey proteins in aqueous solution and in a food product.

But also solubility-related functional properties such as the foamability and emulsification, may be impaired by the commonly applied heating processes during processing and/or preservation of food products. Examples are the negative effects of baking on the properties of meringues prepared from sugar foams of WPC [4] and of sterilization of whey protein-stabilized whipped toppings carried out under improper conditions [16].

In addition to this, liquids containing whey proteins may cause fouling of heat exchangers during pasteurization or sterilization processes, as a consequence of a deposit being formed on metal surfaces. Results obtained by Tissier et al. [15] showed that β-lactoglobulin (β-Lg) is the dominating protein in the mechanism of deposit formation at the inner surface of a plate-sterilizer. They observed a correlation between the maximum denaturation rate of β-Lg and the maximum deposit formation.

212

Fig. 2. Destabilization of an orange juice due to precipitation of 1% whey proteins after pasteurization during 5 min at 74 °C

It is well-known [5] that the denaturation of whey proteins may be considered as a two-step process: 1) an unfolding, which may be reversible or irreversible, and 2) aggregation, which generally follows irreversible unfolding. Arnebrant et al. [1] studied the thermal unfolding of β-Lg by differential scanning calorimetry (DSC) and the adsorption of this protein on metal surfaces by ellipsometry in the same temperature range. By comparing these results they concluded that, in order to produce deposit formation, a metal surface has to be present when the unfolding of β-Lg takes place. Based on these results obtained in phosphate buffer at pH 6.0, these authors postulated that bulk denaturation of β-Lg controls its deposition on metal surfaces.

Both protein unfolding and protein aggregation are, however, influenced by environmental conditions such as pH, ionic strength, and protein concentration [9]. In order to be able to optimize heat treatments that ensure the proper functional use of milk proteins in food products and to reduce fouling of heating equipment, it is important to have more detailed information available on the kinetics of unfolding and aggregation of whey proteins under various environmental conditions.

Kinetics of protein unfolding

Unfolding of globular proteins is accompanied by an endothermal heat effect (heat uptake). This effect may be observed by DSC as a function of temperature or time. We used a Dupont 990 Thermal Analyzer with a so-called heat flux DSC-cell, operating on the measuring principle shown schematically in Fig. 3.

The cell consists of a small heating box (oven), in which hermetically sealed sample pans are placed on a constantan plate provided with chromel/alumel thermocouples fitted close to the samples. The sample and an inert reference sample with a similar heat capacity (usually water or solvent) are heated at a programmed heating rate of 5 °C/min. The observed differences in heat flow between sample and reference sample are recorded as a function of the sample temperature (or time, since constant heating rates are used), as shown in Fig. 3.

From the shape of the heat absorption curve, information on the kinetics of the unfolding process can be obtained by the well-known method of Borchardt & Daniels [2]. This method is based on the assumptions that 1) the rate of heat uptake (i.e., protein unfolding) is proportional to the base line deflection (dH/dt) at a given temperature and 2) the extent of the unfolding is proportional to the heat evolved up to that temperature.

As protein unfolding is a reaction of the first order, the rate constant can be calculated from the following equation [13]:

$$k_t = dH/dt * 1/\Delta Hr .$$

where k = the first-order rate constant at temperature T,
     dH/dt = the deflection from the baseline at temperature T (see Fig. 3).
     ΔHr = the total area below the curve (ΔH) minus the area up to temperature T,
     t = time.

The quantities dH/dt and ΔHr are taken from the DSC curve at different temperatures.

Figure 4 A shows a thermogram of a whey protein concentrate (WPC) and, drawn in the same plot, one of β-Lg in whey permeate. It is shown that both curves overlap in the temperature range between 71° and 80 °C, which indicates that kinetic data on the unfolding of β-Lg may be derived from this part of the WPC-curve. Figure 4 B shows first-order Arrhenius plots of the WPC curve analyzed between the onset (65 °C) and the transition temperature (76 °C). The Arrhenius plot becomes nonlinear above the transition temperature, which may be caused by heat effects evolved from protein aggregation expressed in this part of the curve. As shown in Fig. 4 B, the Arrhenius plot of WPC reflects the kinetic data of β-Lg from 71° to 76 °C, whereas a lower activation energy (130 kJ/mol) is shown from 65° to 71 °C. This lower activation energy is probably caused by the unfolding of other whey proteins (α-lactalbumin and serum albumin) [5]. Based on this observation we decided to select for the kinetic analysis of unfolding of β-Lg a temperature range of maximum 5 °C just before the transition temperature in thermograms of milk proteins.

214

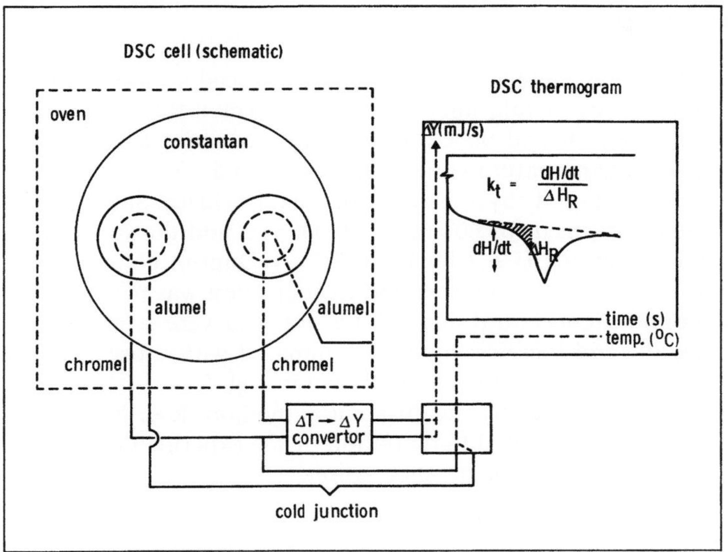

Fig. 3. Differential scanning calorimetric (DSC) procedure for the kinetic analysis of whey proteins

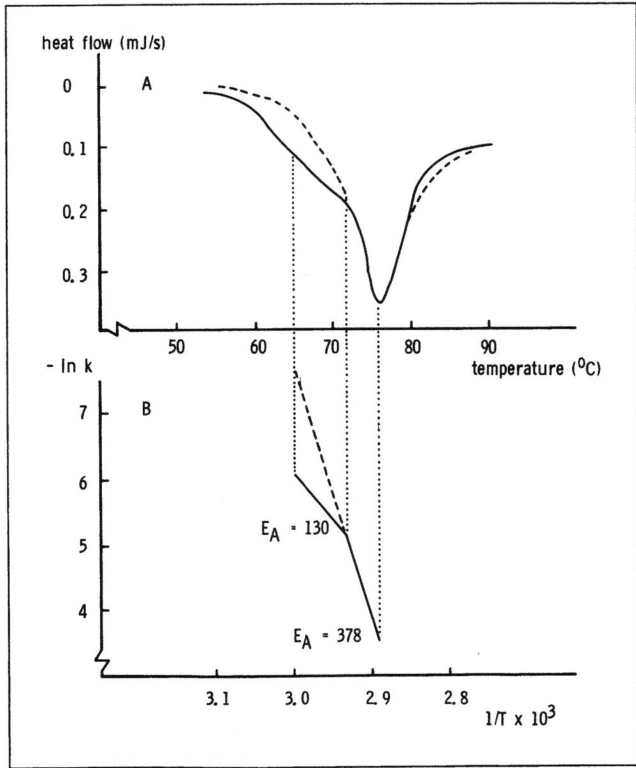

Fig. 4. Kinetic analysis of DSC thermograms of whey protein concentrate (—) and β-lactoglobulin in whey permeate (– –), pH = 6.5

Kinetics of protein aggregation

To determine the rate of protein aggregation the following procedure was followed. Aliquots (10 ml) of a 0.5% protein solution were heated in small tubes in a water bath at temperatures between 70 and 90 °C. The heating rate was 5 °C/min; the holding times at the selected temperature varied between 30 and 300 seconds, after which the samples were cooled rapidly by immersing of the tubes in ice. Subsequently the tubid solutions were centrifuged for 60 min at 90 000 × g, and the supernatant analyzed in 0.25 M phosphate at pH 6.7 by HP-GPC. A Dupont column, Type GF 250, was used at a flow rate of 1.5 ml/min, and the detection was performed at 280 nm. The residual monomer concentrations of β-Lg and α-La were determined by measuring the peak heights, and these results were compared with those obtained using non-heated solutions of β-Lg and α-La, as shown in Fig. 5.

The reaction order was determined by the differential method described by Hill and Grieger-Block [8]. This is based on differentiation of the experimental concentration-vs-time data, starting from the general equation:

$$-dC/dt = k\,C^n$$

where n = the order of the reaction, obtained by using the logarithmic form of this equation.

k = the rate constant derived from the integral of the general equation, giving:
$C_t/C_0 = \exp(-k\,t)$ for a first-order reaction and $C_0/C_t = [1 + k\,C_0\,t]$ for second-order reactions,

where C_0 = the initial monomer concentration and

C_t = the monomer concentration after a heating time t.

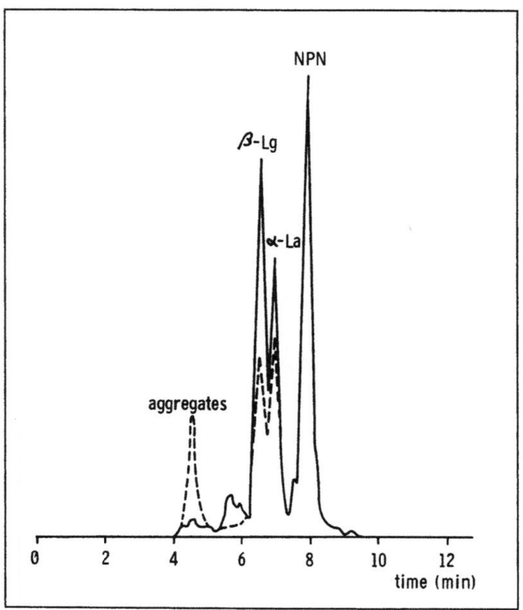

Fig. 5. High performance gel permeation chromatograms of whey before (—) and after a heat treatment for 3 min at 80 °C (---)

216

The temperature dependence of the rate constants was derived from the Arrhenius equation according to:

$$k = k_0 \cdot \exp\left(-E_A/RT\right),$$

where E_A = the energy of activation in J/mol,
R = the universal gas constant (8.314 J/mol/K), and
T = the absolute temperature.

Environmental effects on unfolding and aggregation of β-Lg

Some of the kinetic data of unfolding and aggregation of β-Lg in water, whey permeate, exhaustively desalted WPC, whey, skim milk and whole milk are summarized in Table 1. As mentioned earlier, unfolding was observed by DSC as a first-order reaction up to the transition temperature. The reaction order of the aggregation process varied, however, depending on composition, temperature and duration of heating. The most accurate and representative results were obtained at a temperature of 85 °C, at which the changes in the concentration of the native protein molecules are fairly large (i.e., a great deal of aggregation occurred).

Table 1. Kinetic data of β-lactoglobulin (0.5%) in various media at pH 6.5

Medium	Data of unfolding			Data of aggregation		
	Reaction order	k (75) (s^{-1})	E_A (70−75) (kJ/mol/K)	Reaction order	k (75) (l/g · s)	E_A (75−85) (kJ/mol/K)
Water	1.0	17.25	367	1.8	0.02	214
Permeate	1.0	21.53	338	1.0	0.59	275
WPC	1.0	12.72	346	2.4	0.08	185
Whey	1.0	13.64	294	1.8	0.12	300
Skim milk	1.0	13.65	261	1.8	0.19	307
Milk 3% fat	1.0	N.D.	N.D.	1.8	0.19	374

[a] Determined at 85 °C
[b] Derived from first-order reaction kinetics (k in s^{-1})

Most aggregation reactions appeared to follow (nearly) second-order kinetics, as was expected, except those taking place in desalted WPC and permeate. The higher reaction order in WPC may be explained by the participation of other protein species that are present. The aggregation of β-Lg in permeate, however, was found to follow a (pseudo) first-order reaction at all temperatures studied. This may be explained by the effect of a mechanism of two competing reactions (e.g., β-Lg and calcium for aggregate formation). Pseudo first-order conditions are only accomplished if the (molar) concentration of calcium or of other participating cations is much greater than that of β-Lg. Under those conditions the participating cation concentration remains approximately constant during aggregation [3].

217

Table 1 shows that at 75 °C the rate constants of unfolding are approximately two to three orders higher than those of aggregation. This implies that protein unfolding under these conditions is never a rate-determining process during denaturation of whey proteins. It is striking, however, that the aggregation of β-Lg in water (and desalted WPC) proceeds at a much lower rate than in salt-containing solutions. Also, the temperature coefficients (E_A) of the rate constants of β-Lg in these media are lower. This confirms the results achieved by Pantaloni [12] and Vreeman and Poll [17], who observed a time lag of several minutes in this temperature range before aggregation started. The lag phase was attributed to a slow, rate-determining, conformational change which precedes aggregation. This change is retarded by electrostatic repulsions which dominate in media of low ionic strength [17].

The large differences between the rates of unfolding and aggregation of β-Lg are shown more clearly in Fig. 6, showing the percentages of unfolding and aggregation in various media at 80 °C. The differences in the percentage of unfolding in the various media are small and fall within the curves shown for β-Lg in water and in whey. Within one minute 99% of this protein becomes unfolded, whereas, after a heat treatment of five minutes at 80 °C, the aggregation of β-Lg only reaches 10% in water and just over 50% in permeate. The aggregation of β-Lg in the other media shows a regular increase with the concentration of other protein species, including the (proteinuous surfaces of) fat globules.

The differences between the aggregation of β-Lg in water and in permeate are even more pronounced with increasing temperature at a given time, as shown in Fig. 7. This figure shows the percentage of aggregation of 0.5% β-Lg in water (dashed lines)

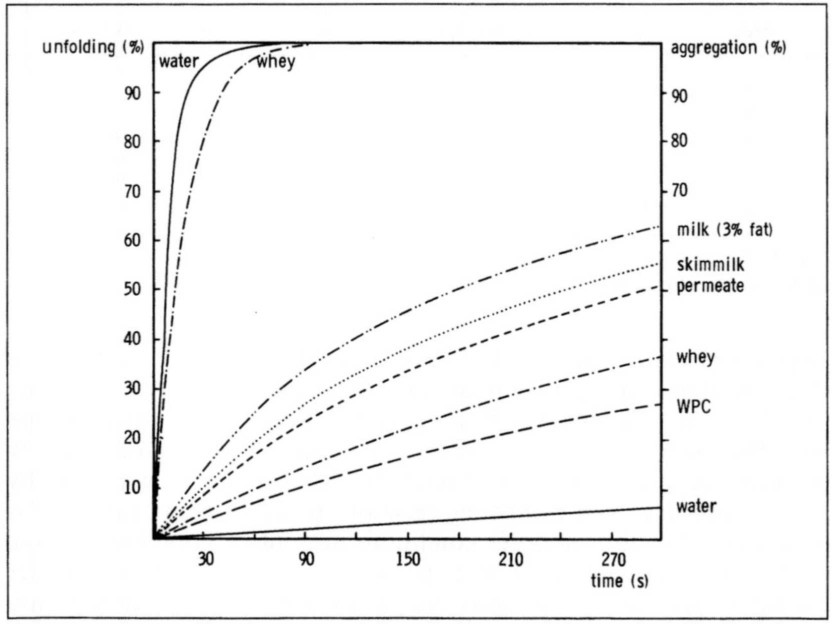

Fig. 6. Unfolding and aggregation of β-lactoglobulin in various media at 80 °C

218

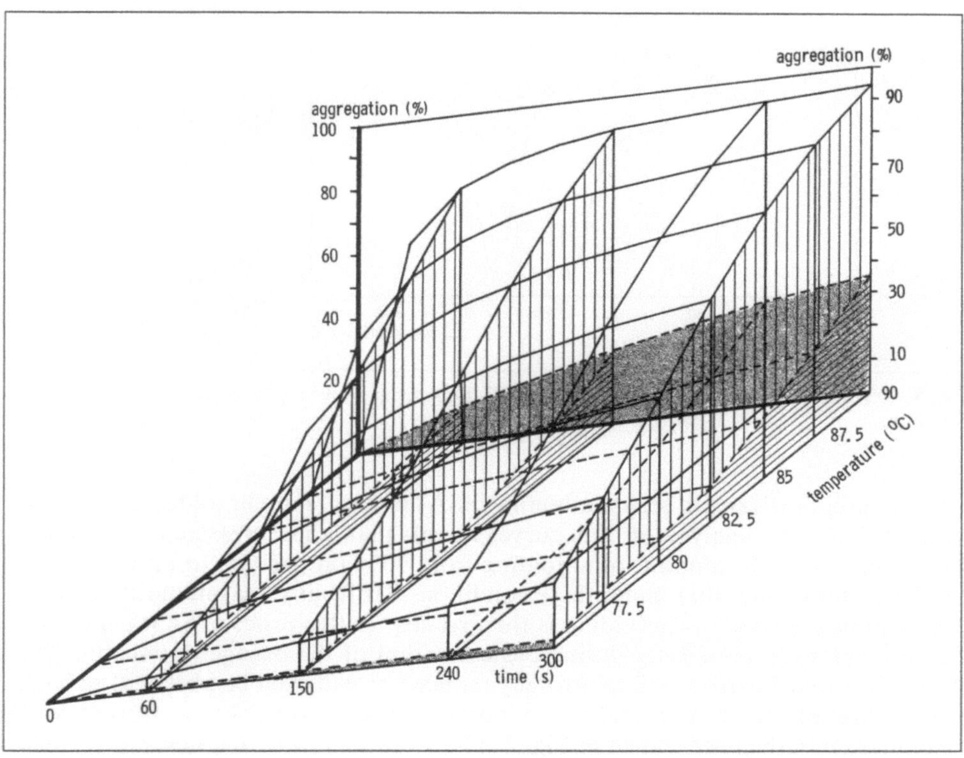

Fig. 7. Percentage of aggregation of β-lactoglobulin in water (– –) and permeate (—) as a function of temperature and time

and in permeate (solid lines) both as a function of temperature and time. It appears that the aggregation observed at 75 °C increases to only 3% in water and to 20% in permeate after 300 s (see dotted and white planes in front). A heat treatment at slightly increased temperature (300 s at 90 °C) results in 35% and 95% aggregation of β-Lg in water and permeate, respectively (see hatched planes on the right hand side). Moreover, it can be clearly seen that in permeate most of the aggregation is achieved within one minute.

Some functional and technological consequences

The kinetic information on unfolding and aggregation of whey proteins thus obtained under various conditions may explain a number of features of a functional and technological nature. An example of a functional nature is a DSC analysis of the mechanism responsible for the destabilization of the orange juice, discussed in a previous section. The thermograms of WPC shown in Fig. 8 reveal that reduction of the pH from 6.5 to 3.5 leads to an increase in transition temperature of nearly 10 °C. Moreover the WPC-curve at pH 3.5 resembles that of β-Lg, because the proteins

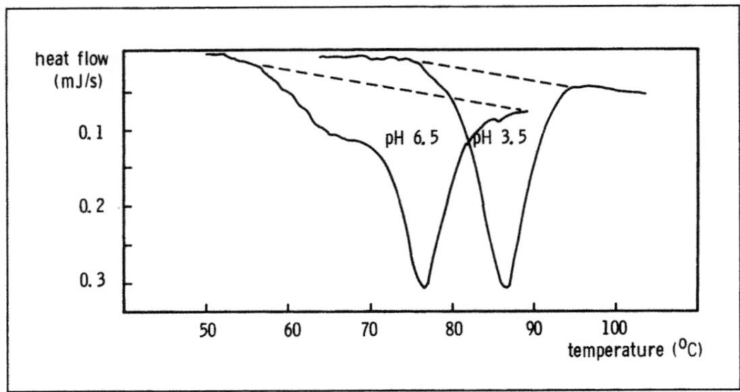

Fig. 8. DSC thermograms of 5% (m/v) protein solutions from WPC at pH 6.5 and 3.5. Heating rate 5 K/min

α-lactalbumin (α-La) and serum albumin (BSA) have already unfolded at this pH-value [14]. Kinetic analysis of both curves reveals that a heat treatment of 5 min at 74 °C will reduce the amount of unfolding from 96% (at pH 6.5) to only 3.5% (at pH 3.5). This implies that under these conditions hardly any heat-induced aggregation may be expected, as unfolding is the first step in this process. So it is obvious that the aggregation of β-Lg during pasteurization of the orange juice-containing WPC, mentioned earlier, will be strongly reduced or will even be absent at pH 3.5. A reduction of the pH from 4.2 to 3.5 indeed appeared to be sufficient to prevent precipitation in the juice shown in Fig. 2. The more acid taste that developed could be masked by adding sugar, provided that the buffer capacity of the beverage was not too high (low salt concentration).

An example of a technological nature that is connected with the differences in the kinetics of unfolding and aggregation of β-Lg is the improved control of the fouling of heat exchangers. Table 2 shows some results of numerical calculations adapted to a continuous heating process of skim milk at two extreme heating rates through the indicated temperature range. A heating rate (Hr) of 60 °C/min, may occur in the heating section of industrial pasteurizers, and a rate of 1 °C/min in the regenerative sections of heat exchangers with extensive facilities for heat recovery. C_u indicates the availability of unfolded β-Lg, which is the difference between the (relative) amount of unfolded and that of aggregated protein, going through the indicated temperature range. V_a is the aggregation rate, i.e., the change in the amount of aggregation of β-Lg per second through this temperature range. The data in Table 2 clearly illustrate that the availability of unfolded β-Lg is at maximum near 72.5 °C at a heating rate of 1 °C/min, and near 85 °C at a heating rate of 60 °C/min. This implies that, according to the hypothesis that unfolded and not yet aggregated β-Lg is preferentially adsorbed on metal surfaces, maximum adsorption may be expected near 85 °C and 72.5 °C at heating rates of 60 °C/min and 1 °C/min, respectively. The rate of aggregation at 85 °C is, however, over 40 times higher than that at 72.5 °C, indicating that increased aggregation and thus fouling may be expected at 85 °C and higher when the highest heating rate is used. At 80 °C the amount of unfolding and the rates of aggregation are practically identical for both heating rates. At $Hr = 1$ °C/min, how-

Table 2. Effects of two heating rates on the availability of unfolded β-Lg (C_u) and on its aggregation rate (V_a) in skim milk at various temperatures

Temperature (°C)	Heating rate			
	(60 °C/min)		(1 °C/min)	
	C_u (% net)	V_a (1/s)	C_u (% net)	V_a (1/s)
70.0	1.0	0.15	34.0	0.01
72.5	2.5	0.15	94.5	0.05
75.0	7.0	0.15	84.0	0.10
77.5	15.0	0.15	62.0	0.20
80.0	29.0	0.35	17.5	0.35
82.5	55.5	1.10	0	0
85.0	92.5	2.15	0	0
87.5	85.5	3.80	0	0
90.0	70.5	6.25	0	0

ever, most of the β-Lg has already unfolded and aggregated at 80 °C, so that hardly any fouling may be expected at higher temperatures or prolonged heating times.

Summarizing, our kinetic model predicts serious fouling problems in heat exchangers or holding sections at temperatures of 85 °C or higher, when high heating rates are used. This prediction is in accordance with the general experience of fouling in continuous-heating equipment, and stimulates the use of this kinetic approach for a more quantitative interpretation of these features.

Conclusion

– Some functional and technological aspects of milk proteins during and after heat treatments were studied on the basis of a new kinetic approach.
– In this approach the kinetics of whey protein denaturation is split into two successive processes: 1) unfolding and 2) aggregation.
– In the temperature range between 70° and 90 °C the rate of unfolding appears to be several orders higher than that of aggregation.
– The rate of unfolding of β-Lg (the most important whey protein) can be slowed down by reducing the pH to 3.5. It is shown that this is sufficient to permit pasteurization of a whey protein-enriched orange juice without encountering stability problems.
– The relative rates of unfolding and aggregation may be optimized by the selection of proper temperature/time relations during the heating processes. It is argued that all this may be used to predict and control fouling of heat exchangers.

References

1. Arnebrant T, Barton K, Nylander T (1987) Adsorption of α-lactalbumin and β-lactoglobulin on metal surfaces versus temperature. J Colloid and Interface Sci 119:383–390
2. Borchardt HJ, Daniels F (1957) The application of differential thermal analysis to the study of reaction kinetics. J Am Chem Soc 79:41–46

3. Capellos C, Bielski BHJ (1972) In: Capellos C, Bielski BHJ (eds) Kinetic systems, determination of order of reaction. Wiley-Intersci, London, p 4–57
4. De Wit JN (1984) Functional properties of whey proteins in food systems. Neth Milk Dairy J 38:71–89
5. De Wit JN (1984) Effects of various heat treatments on structure and solubility of whey proteins. J Dairy Sci 67:2701–2710
6. De Wit JN (1985) In: Galesloot TE, Tinbergen B (eds) "Milk Proteins '84". New approach to the functional characterization of whey proteins for use in food products. Pudoc, Wageningen, p 183–195
7. Hiddink J, Lalande M, Maas AJR, Streuper A (1986) Heat treatment of whipping cream. 1. Fouling of the pasteurization equipment. Milchwissenschaft 41:542–546
8. Hill CC, Grieger-Block RA (1980) Kinetic data: generation, interpretation and use. Food Technol 94:56–66
9. Joly M (1965) In: Horecker B, Kaplan NO (eds) A physico-chemical approach to the denaturation, phenomenological aspects of denaturation. Academic Press, London New York, 6B, p 9–189
10. Kinsella JE (1976) Functional properties in food proteins, a review. Crit Rev Food Sci Nutr 7:219–280
11. Lalande M, Tissier JP, Corrieu G (1985) Fouling of heat transfer surfaces related to β-lactoglobulin denaturation during heat processing of milk. Biotechnol. Progress 1:131–139
12. Pantaloni MD (1964) Etude de la transition R→S de la β-lactoglobuline par spectropolarimétrie de différences. Compt. Rend Acad Sci (Paris) 258:5753–5756
13. Sharp JH (1979) In: McKenzie RC (ed) Reaction kinetics, differential thermal analysis, Academic Press, New York, 2
14. Skudder PJ, Thomas EL, Pavey JA, Perkin AG (1981) Effects of adding potassium iodate to milk before U.H.T. treatment. 1. Reduction in the amount of deposit on the heated surface. J Dairy Res 48:99–113
15. Tissier JP, Lalande M, Corrieu G (1984) In: McKenna BM (ed) Engineering and food. Elsevier Applied Science Publ. London New York, 1, p 49
16. Towler C (1982) Cream products for the consumer. Rev NZ J Dairy Sci Technol 17:191–202
17. Vreeman HJ, Poll JK (1985) Het gedrag van β-lactoglobuline tijdens verhitten tot 150 °C zoals dat volgt uit spectrofluorescentiemetingen. Intern NIZO Report NOV-1077.

In Vitro Digestion of Bovine Milk Proteins by Trypsin Hydrolysis and pH-Stat Analysis

P. Antila

State Institute for Dairy Research, Jokioinen, Finland

Introduction

Knowledge of the proteolytic properties of milk proteins and the factors affecting them is becoming increasingly important, owing to the food industry's growing use of milk protein concentrates and fractions.

This study investigated the in vitro proteolysis of casein and various whey proteins, using both native and modified proteins.

Material and methods

Acid casein (CN) was prepared in our laboratory [1]. Serum albumin (SA), β-lactoglobulin (β-Lg), α-lactalbumin (α-La), trypsin, and pepsin were purchased from the Sigma Chemical Co.

Modification of proteins: the acid treatments, pepsin predigestions, and heat treatments were performed as shown in Figs. 1 and 2. The protein concentration of the substrate solutions was 0.3%.

In vitro digestion of proteins: the digestion was carried out by trypsin hydrolysis and pH-stat analysis as described by Antila et al. [1]. The following conditions were used: protein concentration 0.3%, protein-enzyme ratio 150:1, pH 8.0 and temperature 37°C (Mettler MemoTitrator).

Results and conclusions

Native proteins: the results indicated that trypsin hydrolyzed casein considerably more efficiently than whey proteins, the most endurant of which was α-lactalbumin (see Fig. 1).

Modified proteins: both the acid and pepsin treatments improved the digestion of whey proteins in trypsin hydrolysis (Fig. 1). Pepsin predigestion, however, had an opposite effect on casein.

Heat treatments enhanced the trypsin hydrolysis of β-lactoglobulin (Fig. 2). No change was observed for the other whey proteins or for casein.

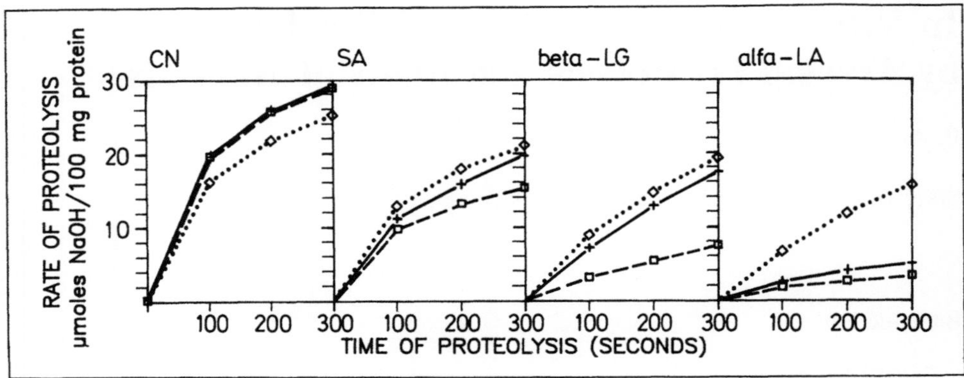

Fig. 1. In vitro digestion of bovine milk proteins by trypsin. Native and modified proteins pretreated with acid and pepsin. Key: □ = Protein without pretreatment, + = Acid treatment at pH 2 (HCl) and 38 °C for 30 min, ◇ = Pepsin treatment at pH 2 (HCl) and 38 °C for 30 min, Protein-enzyme ratio 100 : 1

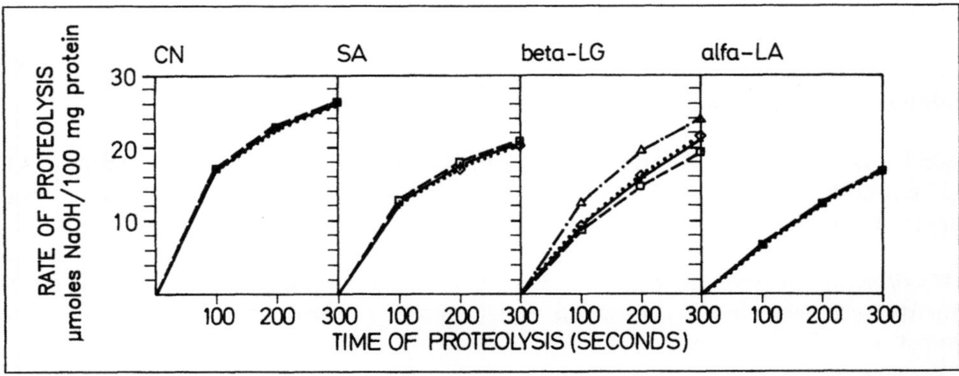

Fig. 2. In vitro digestion of bovine milk proteins by trypsin. Proteins modified by heat treatment and predigested with pepsin. Key: □ = Proteins not heat treated, + = heat treatment at 60 °C, pH 7.0, for 1 h, ◇ = heat treatment at 70 °C, pH 7.0, for 1 h, △ = heat treatment at 80 °C, pH 7.0, for 1 h

The different proteolytic behavior of milk proteins has also been shown in studies carried out in in vitro conditions with other proteolytic enzymes, e.g., pronase, pepsin, chymotrypsin, and pancreatin [2–5].

The differences in the initial rates of the proteolysis of native milk proteins correlated well with their structural properties. Casein with limited ordered structure was more susceptible to proteolytic enzymes than structurally compact whey proteins. The modifications promoted the digestion of whey proteins more than that of casein.

References

1. Antila P, Haque Z, Kinsella JE (1986) Milchwissenschaft 42:485–489
2. Church FC, Catignani GL, Swaisgood HE (1981) J Dairy Sci 64:724–731
3. Porter DH, Swaisgood HE, Catignani GL (1984) J Agric Food Chem 32:334–339
4. Reddy M, Kella ND, Kinsella JE (1988) J Agric Food Chem in Print (The author is indepted to these authors for the preprint of this publication)
5. Rothenbuhler E, Kinsella JE (1985) J Agric Food Chem 33:433–437

Degradation of β-casein by Mesophilic Starter Lactococci

W. Bockelmann, B. Kiefer, A. Geis, and M. Teuber

Federal Dairy Research Centre, Institute of Microbiology, Kiel, FRG

Introduction

During growth in milk lactococci depend on a cell wall-associated proteolytic system [1, 3]. For the rapid acid production required in the dairy industry fast growth to high cell densities is necessary. Because of the low level of free amino acids and peptides in milk, milk proteins have to be utilized by the cells for a sufficient supply of nitrogen for the cells' protein synthesis. The degradation of milk proteins is mediated by a protease and several peptidases localized in the cell wall. These enzymes are a potential source for the bitter flavor and β-casomorphins found in fermented milk [2, 4].

The aim of our work is to purify and characterize all components of the lactococcal cell wall-proteolytic system for a better understanding of the pathway of bacterial protein degradation in milk.

Materials and methods

- source of enzymes: Lc. lactis subsp. cremoris strains P8-2-47 and Ac1;
- growth: M17-Medium, 30 °C, harvest of cells: late log phase;
- purification from cellwall extracts:
  a) protease: anionexchange-chromatography (Q-Sepharose FF) in soluble buffer containing 20 mM $CaCl_2$; NaCl gradient
  b) peptidase: first step same as a); 2nd step: FPLC anionexchange-chromatography in soluble buffer (Mono Q); NaCl gradient;
- assays:
  a) protease: forming of radioactive TCA soluble products from ^{14}C-methylated total casein
  b) peptidase: splitting of oligopeptides, dipeptides, and aminoacid-nitroanilids;
- aminoacid analysis: "Biotronic" analyzer; sequencing: EDMAN

Results

The protease from cell wall extracts (Lc. lactis subsp. cremoris AC1 & P8-2-47) could be separated completely from peptidase activity and was purified in one step. A number of acid-soluble peptides is released from β-casein by the action of the purified protease (Fig. 1). One peptide contains the sequence of β-casomorphin-7, one is reported to have a bitter taste (C-terminus of β-casein, pos. 194–209).

```
ARG-GLU-LEU-GLU-GLU-LEU-ASN-VAL-PRO-GLY-GLU-ILE-VAL-GLU-SER-
                                                          P
LEU-SER-SER-SER-GLU-GLU-SER-ILE-THR-ARG-ILE-ASN-LYS-LYS-ILE-
        P    P    P
GLU-LYS-PHE-GLN-SER-GLU-GLU-GLN-GLN-GLN-THR-GLU-ASP-GLU-LEU-
                    P              ↓        ↓        ↓    ↓    ↓
GLN-ASP-LYS-ILE-HIS-PRO-PHE-ALA-GLN-THR-GLN-SER-LEU-VAL-TYR-
                                      ↓              ↓
PRO-PHE-PRO-GLY-PRO-ILE-PRO-ASN-SER-LEU-PRO-GLN-ASN-ILE-PRO-
PRO-LEU-THR-GLN-THR-PRO-VAL-VAL-VAL-PRO-PRO-PHE-LEU-GLN-PRO-
              ↓
GLU-VAL-MET-GLY-VAL-SER-LYS-VAL-LYS-GLU-ALA-MET-ALA-PRO-LYS-
HIS-LYS-GLU-MET-PRO-PHE-PRO-LYS-TYR-PRO-VAL-GLN-PRO-PHE-THR-
GLU-SER-GLN-SER-LEU-THR-LEU-THR-ASP-VAL-GLU-ASN-LEU-HIS-LEU-
PRO-PRO-LEU-LEU-LEU-GLN-SER-TRP-MET-HIS-GLN-PRO-HIS-GLN-PRO-
                                        ↓              ↓
LEU-PRO-PRO-THR-VAL-MET-PHE-PRO-PRO-GLN-SER-VAL-LEU-SER-LEU-
    ↓              ↓              ↓    ↓    ↓
SER-GLN-SER-LYS-VAL-LEU-PRO-VAL-PRO-GLU-LYS-ALA-VAL-PRO-TYR-
    ↓    ↓                                  ↓    ↓    ↓    ↓
PRO-GLN-ARG-ASP-MET-PRO-ILE-GLN-ALA-PHE-LEU-LEU-TYR-GLU-GLN-
                                                  ↓
PRO-VAL-LEU-GLY-PRO-VAL-ARG-GLY-PRO-PHE-PRO-ILE-ILE-VAL-OH
```

Fig. 1. Cleavage sites of the 'AC1' protease in β-casein

Table 1. Substrate specificity

	β-casein	β-casomorphin-7[c]	leu-gly	lys-NA	Pro-NA
Protease[a]	+	−	N. D.	−	−
Endopep.	N. D.	+	−	−	−
Dipep.	−	−	+	−	−
Aminopep.[b]	−	−	−	+	−
Prolidase	N. D.	N. D.	N. D.	N. D.	+

[a] Cleavage of a_{s1}-, a_{s2}-, and $\varkappa$-casein at a very slow rate
[b] Cleavage of Leu-NA and Ala-NA at a slower rate
[c] Peptide: tyr-pro-phe-pro-gly-pro-ile (60−66 in β-casein)

In preparations of the P8-2-47 aminopeptidase recently dipeptidase- and endopeptidase activities were detected. It could be shown by native PAGE (activity staining of proteinbands) that these peptidase activities belong to different enzymes (Table 1). Purification of these peptidases is still in progress. Low level prolidase activity was detected in cell wall extracts but not in any of the peptidase preparations after the first step of purification (Table 1).

Our in vitro experiments showed that during growth in milk the proteolytic system of lactococci is a potential source of bitter peptides and β-casomorphins. These peptides, however, can be further degraded by a set of at least four cell surface peptidases. The combined action of the described cell wall associated protease and peptidases seems to be capable of degrading β-casein to an extent that supplies the cells with all amino acids required for their nitrogen metabolism.

Conclusion

Five enzymes of a cell wall-associated proteolytic system were identified in *Lactococcus lactis* subsp. cremoris strains AC1 and P8-2-47 and partially characterized. The cell wall proteinases of strains AC1 and P8-2-47 were purified. β-Casein is the preferred substrate for the proteinases. Several small, acid soluble peptides (length 4–25 amino acids) are liberated from β-casein by the purified proteinase. At least one of these peptides contains the amino acid sequence of β-casomorphin-7, another is reported to have a bitter taste. All peptides are further hydrolyzed by cell wall peptidase extracts. Degradation stops at N-terminal proline residues, but it can be continued by a prolidase which has been detected in cellwall extracts of *Lactococcus lactis* subsp. cremoris AC1 & P8-2-47. The described five components of the cell wall-proteolytic system seem to be able to hydrolzye β-casein efficiently. Liberated aminoacids and small peptides can be transported through the cell membrane as a nitrogen source necessary for the growth of the highly auxotrophic bacteria in milk.

Acknowledgement

I wish to thank Dr. V. Monnet from INRA, Jouy-en-Josas, France, for her help in performing peptide sequencing.

References

1. Geis A, Bockelmann W, Teuber M (1985) Simultaneous extraction and purification of a cell wall-associated peptidase and β-casein specific protease from Streptococcus cremoris AC1. Appl Microbiol Biotechnol 23:79–84
2. Hamel U, Kielwein G, Teschemacher H (1985) β-Casomorphin immunoreactive materials in cows' milk incubated with various bacterial species. J Dairy Res 52:139–148
3. Thomas TD, Pritchard GG (1987) Proteolytic enzymes of dairy starter cultures. FEMS Microbiol Rev 46:245–268
4. Visser S, Hup G, Exterkate FA, Stadhouders J (1983) Bitter flavour in cheese. 2. Model studies on the formation and degradation of bitter peptides by proteolytic enzymes from calf rennet, starter cells and starter cell fractions. Neth Milk Dairy J 37:169–180

Determination of Furosine, Lysinoalanine (LAL) and 5-Hydroxymethylfurfural (HMF) as a Measure of Heat Intensity for UHT-Milk

B. Dehn-Müller, B. Müller, M. Lohmann and H. F. Erbersdobler

Institute of Human Nutrition and Food Science, University of Kiel, Kiel, FRG

Introduction

UHT treatment of milk results in such chemical alterations of the product as degradation of essential nutrients, or formation of undesired substances. To evaluate the intensity of the heating process it is possible either to measure losses of nutrients or to analyze the concentrations of new substances whose formation depends on different heating conditions.

One special reaction that takes place during the heat treatment of milk is the Maillard condensation, where the lactose reacts namely with the ε-amino group of lysine. The first main intermediate formed in this way, lactuloselysine, is degraded during acid hydrolysis of the protein, but can be estimated by analyzing for furosine which is formed during hydrolysis with strong HCl [1]. For the determination of 5-hydroxymethylfurfural (HMF), which also results from the Maillard condensation, precursors of browning products in milk are transformed into HMF after addition of oxalic acid and following heating. Principally, the HMF value of a milk can be used as an indicator for the heating process, but data from literature offer a wide range for this value. A comparison between the furosine and the HMF-method should give information about the usefulness of the HMF-method as a rapid and simple measure of heat damage caused by the UHT process.

Additionally, lysinoalanine (LAL) was determined in commercial UHT milks. LAL is formed throughout heat and/or alkali treatment of proteins by nucleophilic reaction of the lysyl-ε-amino-group with the activated double bond of dehydroalanine, which is formed by β-elimination of cystine and phosphoserine in the peptide chain. Like furosine, LAL creates unwanted nutritional effects like a reduction in protein digestibility and amino-acid availability.

Furosine, HMF, and LAL do not exist in raw milk. They represent, therefore, a good means to characterize the reactions during the heating process.

Methods

Furosine and HMF were measured in 270 model milk samples obtained by direct heating in a pilot plant with heating temperatures of 100–150 °C and heating times of 2–128 s. In 190 commercial UHT milk samples from 45 different dairies in W. Germany, furosine, HMF, and LAL were determined.

The furosine and LAL analyses were performed by ion-exchange-chromatography with a Liquimat III amino acid analyser (Kontron Instruments, Munich, FRG)

following the procedure of Erbersdobler et al. [2]. HMF was determined by the method of Keeney and Bassette [4] corresponding to Konietzko [5].

Results

The results of the determinations in the model milk samples show that there is a clear relationship between the severity of heat treatment in terms of heating time and temperature, and the furosine and HMF values of the milks. Figure 1 shows the effect of heating temperature and time on HMF values in directly heated UHT milks.

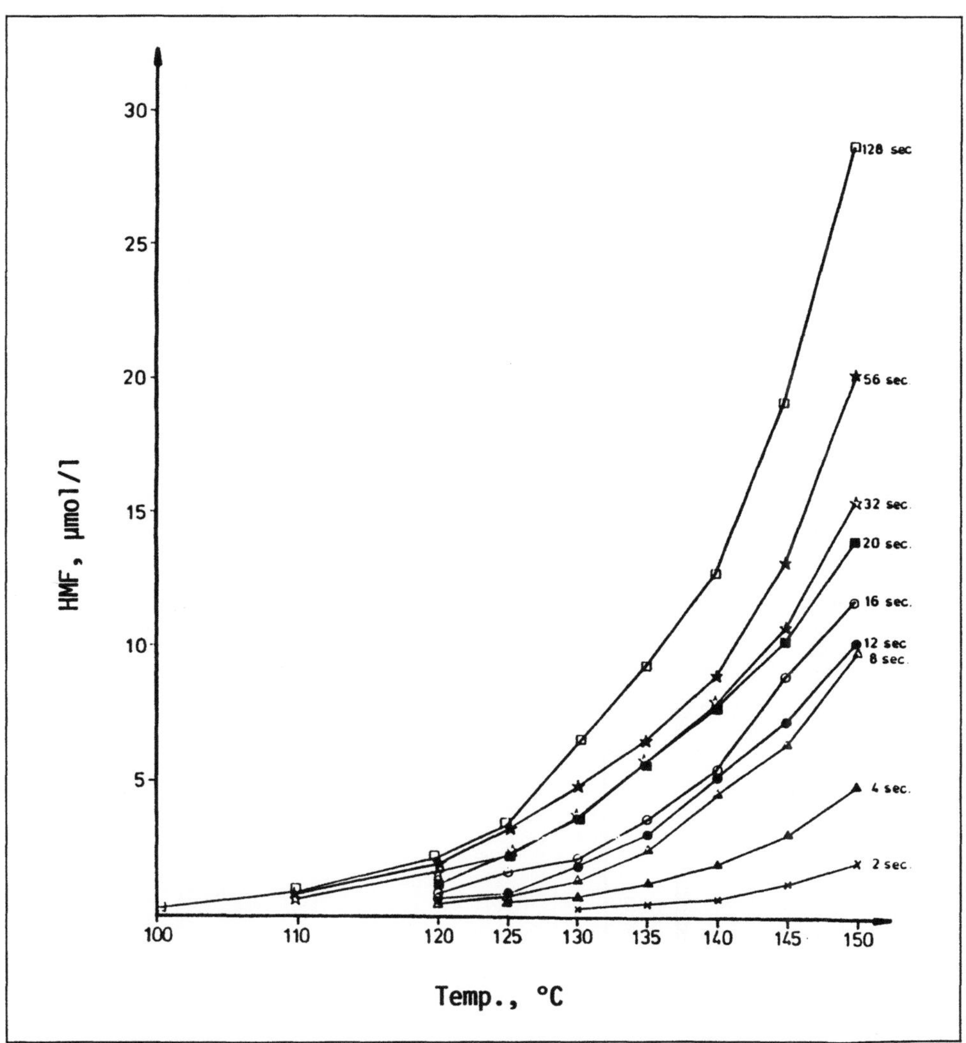

Fig. 1. Effect of heating temperature and time on HMF values in directly heated UHT milks

Table 1. Furosine, LAL, and HMF contents of UHT milk samples

FRG states from which milk samples originated (from various dairies)	n	Furosine contents in mg/l			LAL contents in mg/l			HMF contents in µmol/l		
		$\bar{x}$	$\pm s$	range	$\bar{x}$	$\pm s$	range	$\bar{x}$	$\pm s$	range
Baden Württemberg	11	42	5.8	31−51	1.9	1.0	0.9−4.3	14.7	2.2	10−17
Bayern	33	40	16.8	9−88	1.6	1.0	0.5−5.8	14.1	4.8	6−28
Hessen	13	46	8.4	32−60	2.3	1.4	1.0−5.1	17.0	1.7	14−19
Niedersachsen	22	38	7.2	30−58	1.9	0.8	0.9−4.7	13.6	2.4	9−20
Nordrhein-Westfalen	42	36	6.8	23−49	1.4	0.6	0.6−2.8	13.6	2.1	10−19
Rheinland-Pfalz	9	33	4.9	23−39	1.0	0.3	0.5−1.3	12.2	0.7	11−13
Schleswig-Holstein	60	35	7.3	23−59	1.9	1.0	0.5−5.2	13.2	2.2	9−18

A very similar picture was obtained for the furosine values as is given in [2]. The correlation of the furosine and the HMF values of $r = 0.961$ ($y = 1.76 + 2.34\,x$) was very good. According to this study furosine values between 5 and 15 mg/litre can be expected in commercially processed UHT milks with direct UHT treatment. Indirectly heated model milk samples showed somewhat higher values (15–35 mg/l, unpublished data). Directly heated milks normally show lower furosine and HMF values than indirectly heated milks; however, this depends on the adjustment of the UHT plant.

In the commercial UHT milks furosine ranged from 9 to 88 mg/l, while HMF values of 6–28 µmol/l (without subtracting the blank value) and LAL values of 0.5–5.8 mg/l were obtained. About 70% of the milks contained furosine in the range of 28–49 mg/l and HMF contents of 10–16 µmol/l. Higher values revealed excessive heating and confirm other findings that termal processes are often too severe (e.g. [3]).

Table 1 shows the furosine, LAL, and HMF results in the milk samples tested, given as the mean values.

With a correlation-coefficient of $r = 0.846$ for the commercial UHT milk samples the furosine and the HMF methods showed a fairly good agreement in estimating the heat intensity of processing.

Conclusion

The results of the present project show that furosine is a precise indicator of heat damage in UHT milks. Its determination, however, is expensive and difficult. The HMF-method, for these reasons, is a useful alternative, particularly for routine checking of milk quality. The LAL values on the other hand appear to be too low for precise testing but can be a useful means in cases of overprocessing or incorrect alkali treatment.

The losses in available lysine were calculated on the basis of the furosine levels, according to [1]. Supposing an appropriate UHT milk production (furosine values < 50 mg/l) the heat-induced lysine losses of 0.4–2.6% are of inferior nutritional significance for the consumer considering the high lysine values in raw milk – but the data may be an aid to good manufacturing practice and could assist in the improvement of processing equipment and in the study of the effect of new process techniques in general, e.g., sensorial and milk quality.

Acknowledgements

The authors thank Prof. Dr. H. Reuter and Dr. A. Nangpal (Institut für Verfahrenstechnik der Bundesanstalt für Milchforschung, Kiel, FRG) for providing the model milks from the UHT pilot plant.

Supported by a research grant from Deutsche Forschungsgemeinschaft.

References

1. Erbersdobler HF (1986) Twenty years of furosine – better knowledge about the biological significance of Maillard reaction in food and nutrition. In: Fujimaki M, Namiki M, Kato H (eds) Amino carbonyl reactions in food and biological systems. Amsterdam: Elsevier (Developments in Food Science 13), pp 481–491
2. Erbersdobler HF, Dehn B, Nangpal A, Reuter H (1987) Determination of furosine in heated milk as a measure of heat intensity during processing. J Dairy Res 54:147–151
3. Geier H, Klostermeyer H (1983) Formation of lactulose during heat treatment of milk. Milchwissenschaft 38:475–477
4. Keeney M, Bassette R (1959) Detection of intermediate compounds in the early stages of browning reactions in milk products. J Dairy Sci 42:945–960
5. Konietzko M (1981) Abtötung von Bacillus stearothermophilus und Bildung von Gesamt-Hydroxymethylfurfural während des Ultrahocherhitzens von Vollmilch. Agr Diss, Kiel

Analysis of Milk Proteins and Their Proteolytic Products by Use of a Modified OPA-method

H. Frister, H. Meisel, and E. Schlimme

Institute for Chemistry and Physics, Federal Dairy Research Centre, Kiel, FRG

The reaction of o-phthaldialdehyde (OPA) (A, Fig. 1) and mercaptoethanol (B1) with free α- and ε-amino groups (C) in amino acids, peptides, and proteins, as well as their hydrolytic and proteolytic products has recently attracted much attention because of the high sensitivity of the assay which can be carried out in aqueous solutions [1, 3, 10].

However, the use of mercaptoethanol as thiol component in the OPA-reaction has the disadvantage that the initially generated 1-alkylthio-2-alkylisoindole (D) often undergoes a matrix-dependent rearrangement to form 2,3-dehydro-1-H-isoindole-1-one (E, Fig. 2). This process is induced by an intramolecular nucleophilic attack on the C-1-atom of the pyrrole system by the hydroxyl group of the mercaptoethanol [9]. This leads to a decrease in absorbance at the monitoring wavelength of 340 nm. Substitution of mercaptoethanol by ethanethiol (B2) [2, 3, 8] prevents the rearrangement of alkylthioalkylisoindole mentioned above and gives stable absorbance values over a longer period of time. This is demonstrated in Fig. 3 for the reaction of human α-lactalbumin with OPA [3]. A disadvantage remains in that the plateau in the

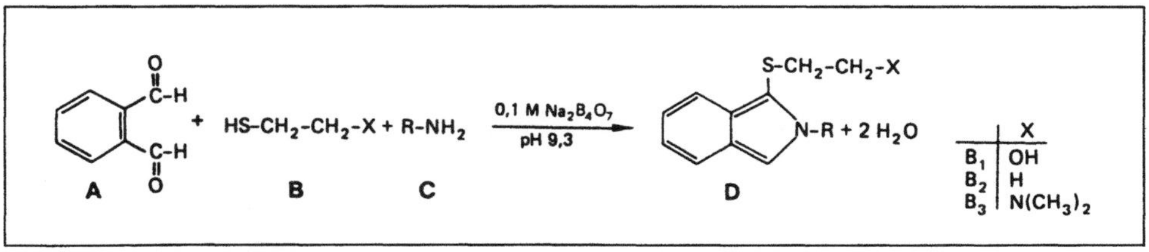

Fig. 1. Formation of 1-alkylthio-2-alkylisoindoles

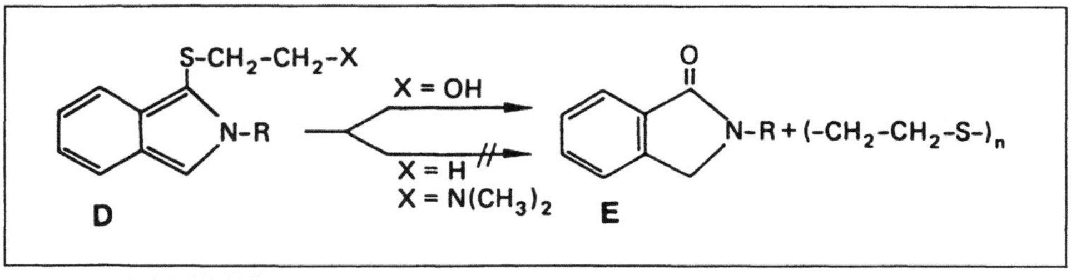

Fig. 2. Formation of isoindolones

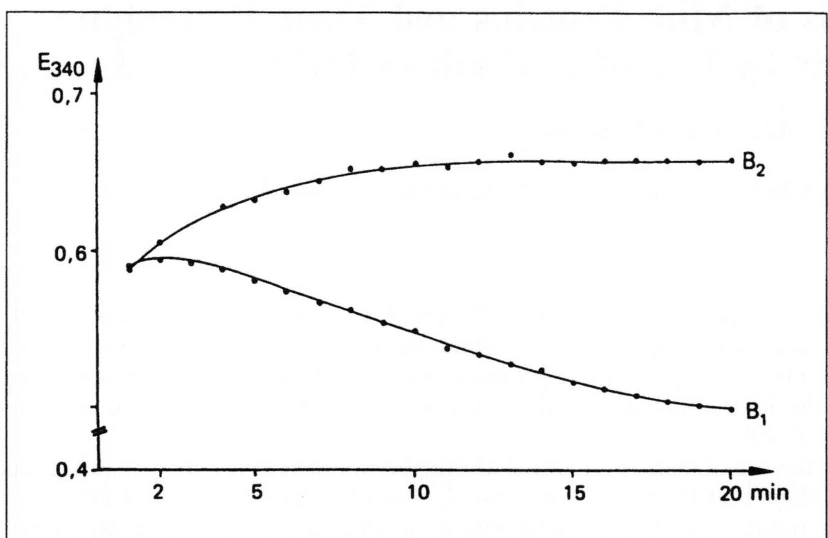

Fig. 3. Absorbance vs time diagram of α-lactalbumin (human) using mercaptoethanol (B₁) and ethanethiol (B₂)

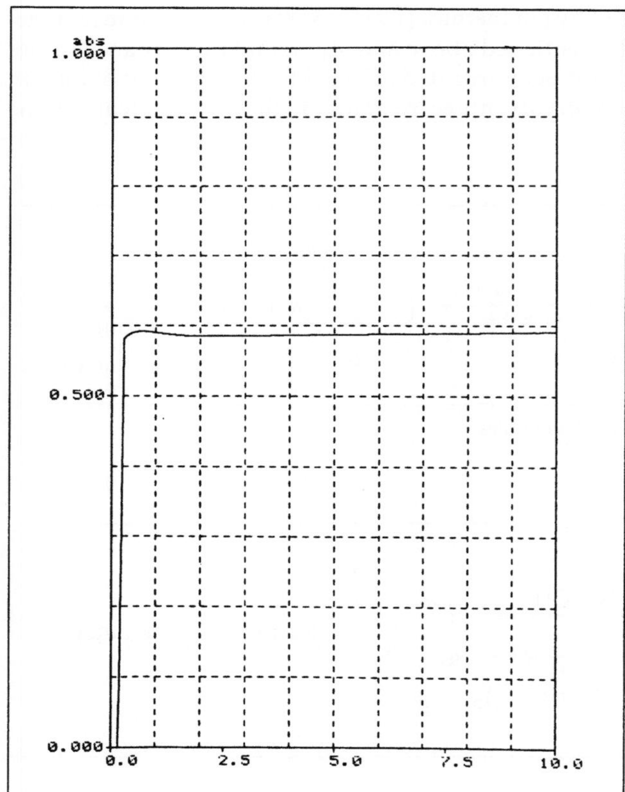

Fig. 4. Absorbance vs time diagram of α-lactalbumin (human) using N,N-dimethyl-2-mercaptoethylammonium chloride

absorbance vs time curve is not always constant. Furthermore, a strong, repulsive odor is generated.

We therefore modified the OPA-method by using N,N-dimethyl-2-mercapto-ethylammonium chloride which does not give off a repulsive odor in OPA-solution. The use of this new thiol component in the OPA-reaction results in long-term stability of the absorbance values [4]. This is shown in Fig. 4 for the reaction of human α-lactalbumin with OPA [5].

Spectroscopic measurements of 19 different amino acid-OPA reaction products gave extinction coefficients within the range of $\varepsilon = 5.98$ (glycine, one OPA-sensitive amino group) to $\varepsilon = 12.96$ (lysine, two OPA-sensitive amino groups) $\mathrm{mmol}^{-1} \times l \times \mathrm{cm}^{-1}$. For quantitation of amino acid mixtures and proteolytic products the mean extinction coefficient of an OPA-sensitive amino group was found to be $\varepsilon = 6.42 \pm 0.20 \ \mathrm{mmol}^{-1} \times l \times \mathrm{cm}^{-1}$ [4]. Based on this mean extinction coefficient the recovery of OPA-sensitive amino groups was investigated for several peptides, whey proteins, and caseins before and after HCl hydrolysis (Table 1) [4, 5].

Furthermore, the quotients (F-values) calculated from OPA-sensitive amino groups after and before HCl-hydrolysis of the peptides and milk proteins have been determined (Table 2).

Using this method it was possible to determine average chain-lengths of protein-derived soluble peptides in the intestinal chyme of minipigs fed a casein or soya diet. The F-values (five different animals) ranged from 2.8 to 3.9 after the casein diet and from 2.8 to 5.4 after the soya diet [5]. The results support previous findings [6] that the soluble part of chyme contains mainly small peptides as well as amino acids.

Table 1. OPA-reaction of peptides and milk proteins before and after HCl hydrolysis

	OPA-sensitive NH_2-groups before hydrolysis[b]	Validation $\pm s$ % of the calculated value	OPA-sensitive NH_2-groups after hydrolysis[c]	Validation $\pm s$ % of the calculated value
Glu-Lys	2	91.2 ± 1.3	2.85	93.1 ± 0.6
Thr-Lys-Tyr	2	87.1 ± 0.8	3.80	87.9 ± 0.8
Ser-Ser-Ser	1	94.6 ± 2.1	2.85	95.2 ± 0.9
Thr-Lys-Pro-Arg	2	98.9 ± 1.3	3.80	107.8 ± 1.1
β-Lactoglobulin A (bovine)	16	97.7 ± 1.2	158.65	93.7 ± 1.3
β-Lactoglobulin B (bovine)	16	101.8 ± 0.9	158.65	100.0 ± 2.0
α-Lactalbumin (bovine)	13	95.5 ± 1.2	122.55	95.6 ± 0.7
α-Lactalbumin (human)	13	96.0 ± 0.6	123.50	93.3 ± 0.9
Whole casein (bovine)	13.7[a]	110.0 ± 1.0	178.30[a]	100.2 ± 1.7
α-Casein 70% (bovine)	14.1[a]	91.3 ± 0.6	181.20[a]	93.0 ± 1.4
α-Casein 90% (bovine)	15.7[a]	95.5 ± 1.6	185.90[a]	92.7 ± 1.4
α_{s1}-Casein (bovine)	15.5[a]	86.8 ± 1.6	185.40[a]	86.5 ± 0.5
$\varkappa$-Casein 80% (bovine)	11[a]	96.2 ± 1.7	149.20[a]	99.4 ± 1.3

[a] Calculated according to the protein pattern obtained by PAGE
[b] Calculation: N-terminal α-NH_2-group + ε-terminal NH_2-group of lysine
[c] Calculation: α-NH_2-groups of aminoacids + ε-terminal NH_2-groups of lysine -Try-Pro- 5% destruction after HCl hydrolysis

Table 2. Relative chain-lengths F of the substrates

	$F_{calc.}$	$F_{exp.}$
Glu-Lys	1.4	1.4 ± 0.1
Thr-Lys-Tyr	1.9	1.9 ± 0.1
Ser-Ser-Ser	2.9	2.9 ± 0.1
Thr-Lys-Pro-Arg	1.9	2.0 ± 0.1
β-Lactoglobulin A (bovine)	9.9	9.5 ± 0.2
β-Lactoglobulin B (bovine)	9.9	9.8 ± 0.4
α-Lactalbumin (bovine)	9.4	9.4 ± 0.3
α-Lactalbumin (human)	9.5	9.2 ± 0.2
Whole casein (bovine)	13.0[a]	13.3 ± 0.4
α-Casein 70% (bovine)	12.9[a]	13.1 ± 0.4
α-Casein 90% (bovine)	11.8[a]	11.5 ± 0.5
α_{s1}-Casein (bovine)	12.0[a]	11.9 ± 0.4
$\varkappa$-Casein 80% (bovine)	13.6[a]	14.0 ± 0.5

[a] Calculated according to the protein pattern obtained by PAGE

The chain-lengths of the casein derived phosphopeptide Ser(P)-Ser(P)-Ser(P)-Glu-Glu-Ile-Val-Pro-Asn isolated from intestinal chyme of minipigs given a casein diet was determined with F = 9.3 [7].

In conclusion, the application of N,N-dimethylmercaptoethylammonium chloride as thiol component in the OPA-reaction in the analysis of free amino groups, both in milk proteins and their proteolytic/hydrolytic products, is characterized by good repeatability and recovery. Furthermore, the stability of the absorbance values makes the reaction particularly suitable for pre- and post-column derivatization in HPLC.

References

1. Church FC, Swaisgood HE, Porter DH, Gatignani GL (1983) Spectrophotometric assay using o-phthaldialdehyde for determination of proteolysis in milk and isolated milk proteins. J Dairy Sci 66:1219–1227
2. Fleury MO, Ashley DV (1983) High-performance liquid chromatographic analysis of amino acids in physiological fluids. On-line precolumn derivatization with o-phthaldialdehyde. Anal Biochem 133:330–335
3. Frister H, Meisel H, Schlimme E (1986) Modifizierte OPA-Methode zur Charaterisierung von Proteolyse-Produkten. Milchwissenschaft 41:483–487
4. Frister H, Meisel H, Schlimme E (1988) OPA method modified by use of N,N-dimethyl-2-mercaptoethylammonium chloride as thiol component. Fresenius Z Anal Chem 330:631–633
5. Frister H, Meisel H, Schlimme E (1988) Unpublished results
6. Meisel H, Hagemeister H (1987) Postprandiale Proteolyse von Casein und Sojaprotein. Milchwissenschaft 42:153–158
7. Meisel H, Frister H (1988) Chemical characterization of a caseinphosphopeptide isolated from in vivo digests of a casein diet. Biol Chem Hoppe-Seyler, in press
8. Simons SS, Johnson DF (1977) Ethanethiol: A thiol conveying improved properties to the fluorescent product of o-phthalaldehyde and thiols with amines. Anal Biochem 82:250–254
9. Simons SS, Johnson DF (1978) Reaction of o-phthalaldehyde and thiols with primary amines: formation of 1-alkyl (and aryl) thio-2-alkylisoindoles. J Org Chem 43:2286–2891
10. Roth M (1971) Fluorescence reaction for amino acids. Anal Chem 43:880–882

Effect of Technological Treatments of Milk on Gastric Digestion

P. Garnier, B. Savalle, G. Miranda, and J-P. Pélissier

Station de Recherches Laitières INRA, Jouy-en-Josas, France

Introduction

Technological treatments of milk modify its digestibility. Coagulation plays an important role during the first part of digestion which takes place in the stomach. To compare the effects of technological treatments of milk on the gastric emptying of proteins and peptides, three diets were studied: crude skim milk, pasteurized skim milk (95 °C, 45 s), and skim yoghurt. In the preruminant calf (monogastric) all effluents leaving the stomach during 12 h were collected and analyzed for amino acid composition, N emptying, NPN level, and identification of proteins and peptides (electrophoresis, HPLC).

Results

For the three diets, α-lactalbumin was degraded when the pH value was under 3.5; β-lactoglobulin is not proteolyzed. Casein emptying depends on the ability of diets to coagulate. With crude skim milk casein coagulation is almost immediate in the stomach and caseins are evacuated in the form of peptides. With pasteurized skim milk, casein coagulation is slower. Caseins are also evacuated in the form of peptides. With yoghurt, caseins are evacuated during the whole digestion process and gradually degraded and there is no coagulation. This result is confirmed by proline emptying (proline may be considered as a marker of caseins). Therefore, the proportions of each amino acid vary according to the different diets. This especially may have consequences for essential amino acids.

The change of the NPN level during digestion occurred in two steps. During the first hours of digestion, the NPN level increased considerably. After four hours the proportion of large peptides increased. This is not observed with yoghurt – the NPN level is always lower which indicates that yoghurt is a less coagulable diet.

Amino acid compositions of the effluents are more homogeneous with pasteurized skim milk than with crude skim milk. With yoghurt, these compositions are almost identical during the whole digestion process.

Conclusion

Technological treatments of milk do not seem to modify the whey protein emptying but they do modify casein evacuation. With crude skim milk, caseins coagulate immediately in the stomach. Thus, whey proteins are emptied at first, and caseins are degraded and afterwards evacuated as peptides. With yoghurt, caseins are continually emptied; at the same time they are gradually degraded. Pasteurization of milk induces an intermediate behavior.

Molecular Weight Determination of Protein Hydrolysates (FPLC)

G. Georgi, G. Sawatzki

Milupa AG Research Department, Friedrichsdorf, FRG

Introduction

Food allergy, especially with cow's milk proteins, is a problem in infant nutrition. The prevention of such allergic reactions can be managed by using structural-modified proteins derived by heat denaturation or enzymatic hydrolysis of the proteins. Recently there is an increased use of the partial enzymatic hydrolysis of proteins [2, 3], since it is known that decreasing the length of the resulting peptides is parallel to a decrease in immunogenicity. The length of the peptides can be determined by gel electrophoresis or by gel chromatography. The use of fast protein liquid chromatography (FPLC) with a Superose-12-column (Pharmacia) for determining the molecular weight of protein hydrolysates was investigated.

Materials and methods

Gel filtration chromatography was performed with FPLC by using a Superose-12-column, HR 10/30 (Pharmacia); 1 mg of each standard (Table 1) and 30 mg of the samples (Pregomin, Milupa AG; Nutramigen, Mead Johnson) were dissolved in 1 ml of eluent I (10 mM Tris, 100 mM NaCl, 0.1% SDS; pH 6.8) or eluent II (6 M Guanidine-HCl, 50 mM Tris, 5 mM EDTA; pH 8.6), centrifuged at $4\,000 \times g$ for 15 min (samples), and membrane-filtered (0.45 µm); 100–200 µl of each filtrate was injected. The peptides were separated using a flow rate of 0.4 ml/min (eluent I) or 0.3 ml/min (eluent II) and detected at 214 nm and/or 280 nm.

Both infant formulas, Pregomin and Nutramigen, contain protein hydrolysates.

Table 1. Proteins and peptides for the calibration of the Superose-12-column

Protein	Molecular weight	Source
Bacitracin	1 450	Serva
Insulin, B chain	3 400	Serva
Glucagon	3 500	Serva
Aprotinin	6 000	Sigma
α-Lactalbumin	14 200	Sigma
α-Chymotrypsinogen A	25 700	Serva
Myoglobin and fragments	2 500–16 900	Sigma

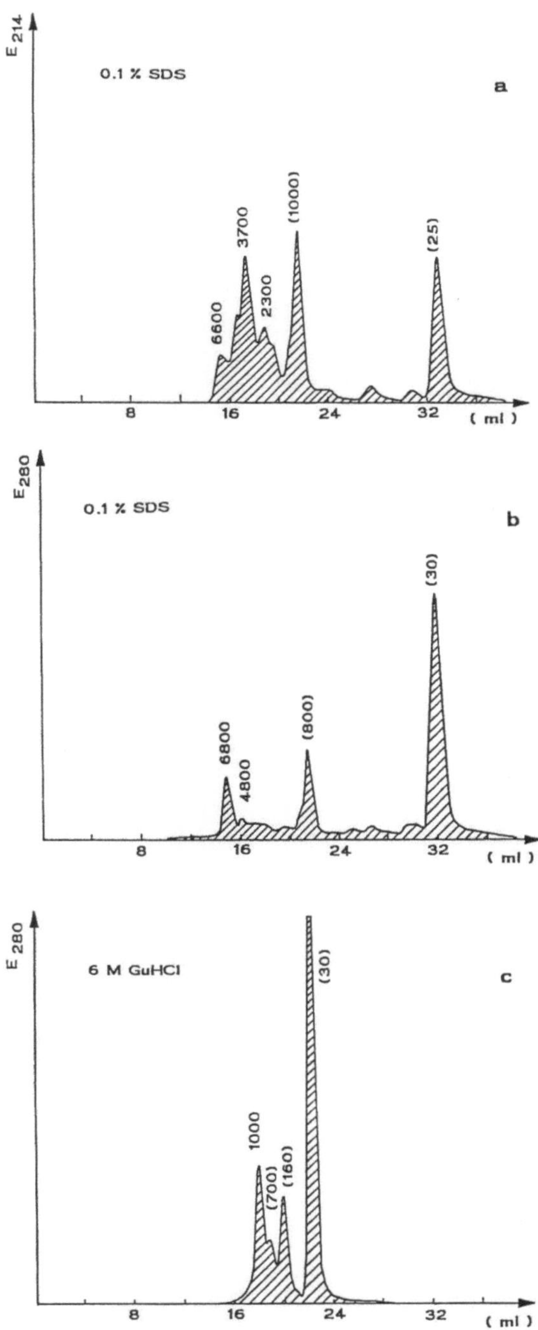

Fig. 1 a–c. FPLC elution patterns of Nutramigen using Superose-12-column. Different buffers are given. (Numbers in brackets are apparent molecular weights)

239

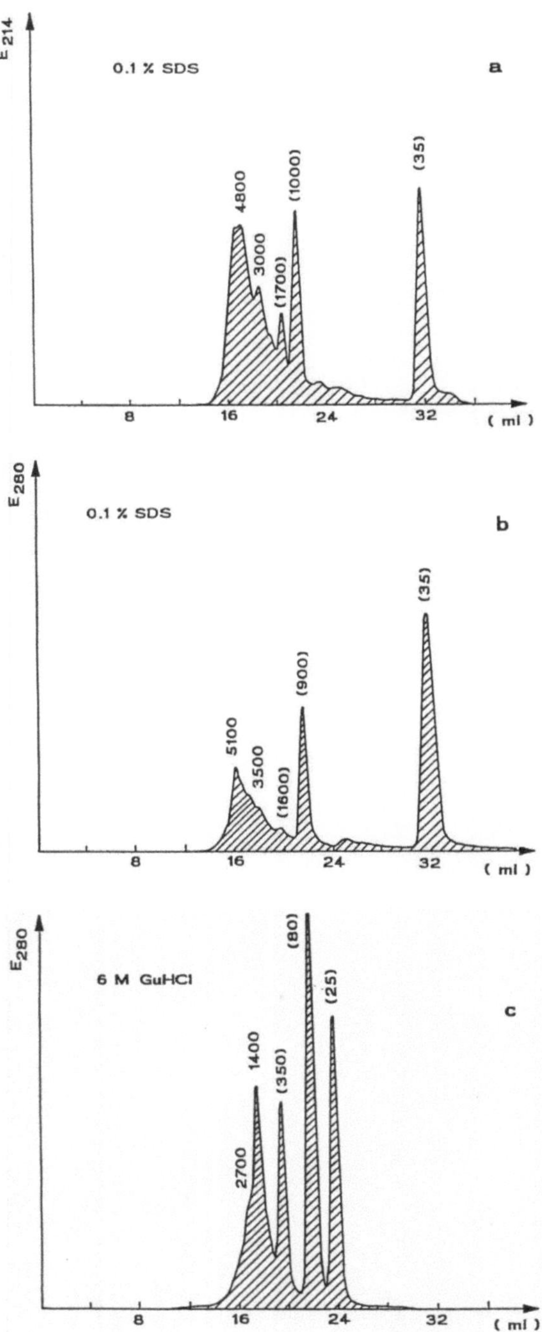

Fig. 2a–c. FPLC elution patterns of Pregomin using Superose-12-column. Different buffers are given. (Numbers in brackets are apparent molecular weights)

240

Results and discussion

Due to difficulties with SDS-PAGE during fixation and staining of small peptides from cow's milk protein hydrolysates, we used liquid chromatography (FPLC) with Superose-12-column for the determination of the molecular weights. This is less time consuming and it allows to isolate definite peptides. For our experiments we used two different eluents (containing SDS or GuHCl) for the separation of the peptides, and compared both methods. The calibration curves for both eluents were linear in the selected range. Figures 1 a–c show the molecular weight patterns for the peptides of Nutramigen and Figs. 2 a–c show those for Pregomin. The elution with 0.1% SDS was monitored not only at 280 nm but also at 214 nm to guarantee that peptides without aromatic amino acids could be detected as well. For both Pregomin and Nutramigen the calculated molecular weights were smaller using GuHCl compared to SDS. Using SDS for the elution of peptides the calculated molecular weights are higher than the actual [1, 4]. Since the calibration of our column was performed under the same conditions as those used for the molecular weight determination of the hydrolysates it can be assumed that the molecular weights are comparable. Nevertheless we observed different molecular weight patterns and found different calculated molecular weights dependent on the eluent (SDS or GuHCl). This effect can be considered as an interaction between the separation matrix (Superose-12) and the peptides and/or interaction between the matrix and the eluent. These interactions seem to be protein specific. The verification of both methods for the determination of the real molecular weights is under investigation.

Conclusion

The determination of molecular weights of hydrolysate peptides using FPLC (Superose-12) is reproducible and easily performed. The hydrolysates analyzed showed apparent higher molecular weights when SDS was used as eluent compared to GuHCl. Due to these observations it is inevitable for discussing molecular weight data of protein hydrolysates to give the method used.

References

1. Fish WW, Reynolds A, Tanford C (1970) J Biol Chem 245:5166
2. Pahud J-J, Monti JC, Jost R (1985) J Pediatr Gastroenterol Nutr 4:408
3. Takase M, Fukuwatari Y, Kawase K et al. (1979) J Dairy Sci 62:1570
4. Wong P, Barbeau A, Roses AD (1985) Anal Biochem 146:191

Heat Coagulability of Whey Proteins in Acidic Conditions

P. Jelen

Department of Food Science, University of Alberta, Edmonton, Alberta, Canada

Introduction

While whey proteins are highly susceptible to heat-induced changes, heating of casein-containing liquid dairy systems usually does not result in visible coagulation unless the whey-protein and calcium content are increased [1]. However, when caseins are removed by ultracentrifugation or by the action of rennet or acid as in production of various cheeses, heating of the remaining sweet or acid wheys normally results in protein aggregate formation and often in precipitation. The size and the shape of the aggregates formed in sweet whey systems may be such that the heating effect is observed as a change from translucent to milky white appearance with no apparent coagulation.

For certain food applications, the susceptibility of whey protein to heat-induced precipitation is a defect limiting the whey protein use. In development of whey-based beverages, the heat-coagulability of whey protein is especially important as it may result in an undesirable cloud and/or sediment formation.

The objective of this brief communication is to review some of the recent information on the heat stability of natural and isolated whey protein systems in acidic conditions suitable for whey beverage manufacture.

Theoretical considerations

Information about thermal behavior of whey protein systems in acidic conditions below pH 4.0 is scant as shown by a recent literature review [2]. What little information is available pertains mostly to isolated individual whey proteins in pH regions as low as pH 2.0–2.5. For whey beverage manufacture, the pH region of 3.0–4.0 is of particular interest since this is the typical range of acidity for most thirst-quenching beverages such as fruit juices, cola-type products or other carbonated soft drinks.

Both of the major whey proteins, β-Lg and α-La, are known to undergo conformational changes at this pH range. While β-Lg merely dissociates from its dimer configuration to a monomeric state at approximately pH 3.5, the α-La is said to undergo an acid denaturation-type conformation at about the same pH range, probably due to the removal of the stabilizing calcium from the molecular structure [3]. However, neither of these changes appears to result in coagulation in isolated protein systems; in fact, the monomerization of the β-Lg may be responsible for the observed resistance to heat-induced coagulation [4].

242

The effects of heating in acidic conditions on isolated serum albumin (SA) have been studied to even a lesser degree. This protein is known to be effective in fatty acid binding and the different heat-related stability data reported in literature [5] may be due to the type of the isolated protein used in the experiments.

Behavior of the complex whey protein systems in heated acidified wheys and modified whey systems is also subject to uncertainties. A coagulative interaction between α-La and β-Lg has been described for the pH 4.5–6.5 range [2]. Whether the same occurs in acidic conditions has not been established. While it is recognized that the coagulative behavior of whey protein systems changes with pH, the limiting pH for these changes and the consequences of heating with respect to visible sediment formation have been established only recently.

Effect of pH on heat-induced precipitation of whey proteins

Investigations of problems associated with the various whey beverage technologies available today indicated [6] that pH 3.8–3.9 is critical for the change in heat stability of proteins in whey. The transition from heat-unstable to heat-stable systems occurs in a relatively narrow pH range of about 0.2 pH units. The heat stability of acidified wheys has been demonstrated by the lack of any visible change quantified as the Absorbance Index (A.I.) upon heating at 92 °C for up to 30 min (Fig. 1).

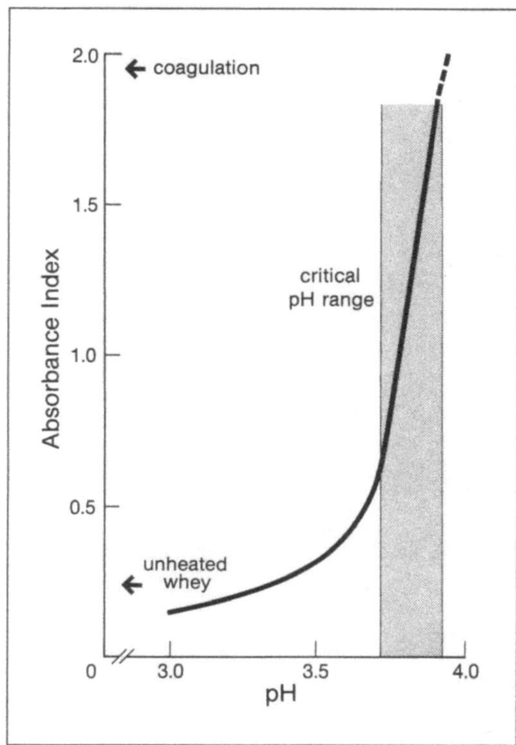

Fig. 1. Effect of pH on heat stability of whey protein in acid whey heated to 92 °C for 15 min (adapted from [9])

243

The calcium content of the particular whey system plays a major role in the stabilization of the whey protein fraction towards heat-induced precipitation. Sweet (rennet) whey systems and wheys in which Ca was chelated by citric acid, or EDTA, or removed by ion exchange, behaved similarly with [6] the critical pH range for transition from heat-unstable to heat stable conditions, shifting by about 0.2 pH units from 3.9 – 3.8 to 3.7 – 3.5.

When investigated separately in a UF permeate from cottage cheese whey, the individual major whey protein fractions (β-Lg, α-La, SA) responded differently to heat treatment. The precipitation patterns of the individual proteins were illustrated [4] by sodium dodecyl sulphate polyacrylamide gel electrophoresis (SDS-PAGE) of the model protein solutions before and after heating to 93 °C for 20 min. The behavior of the β-Lg generally followed the trends observed in whole wheys, with the protein being resistant to heat-induced coagulation at pH 3.7 in the normal permeate but precipitating at this pH in decalcified permeate. However, the isolated SA and α-La, in the absence of β-Lg but with the Ca present, appeared quite resistant to heat coagulation. In particular, the α-La showed no signs of any precipitation up to pH 4.5, typical for regular acid wheys. In decalcified permeates, slight increases in turbidity in both α-La and SA solutions were observed at the onset of the heating experiments but no further change was caused by the heating. No removal by precipitation was seen on the PAGE patterns for α-La up to pH 4.5, while the SA band disappeared at pH 4.2. However, in a mixed model system containing all three proteins, some of the α-La and SA appeared to co-precipitate with the β-Lg (Fig. 2)

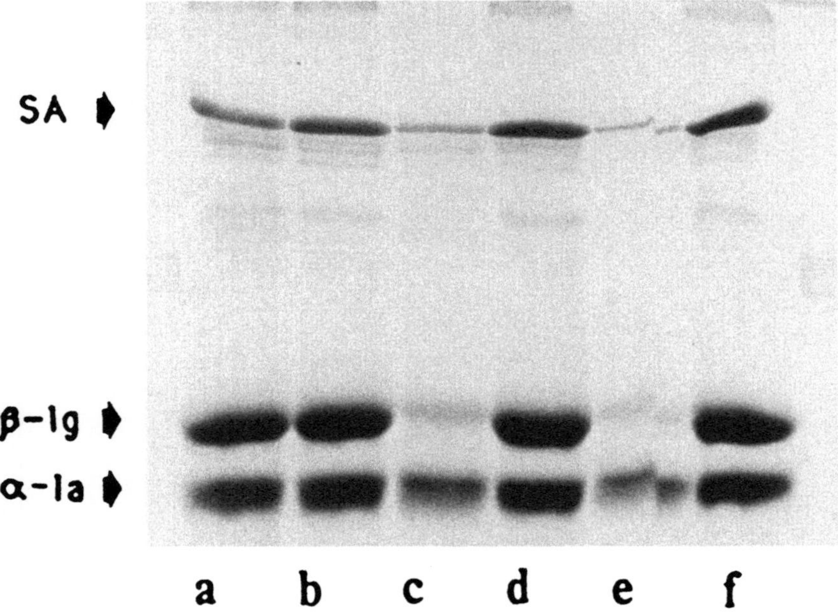

Fig. 2. Effect of heating (92 °C, 20 min) on whey protein precipitation in model acid whey solutions acidified to pH 3.7 (a), 4.0 (c), or 4.2 (e) in comparison to corresponding unheated controls (b, d, f, resp.); (from [4]).

244

above the critical pH range of 3.7–3.8. Similar patterns were observed in regular and decalcified cottage cheese wheys above the critical pH ranges established earlier.

Whey protein denaturation

Although the phenomena of protein coagulation and denaturation are related, they are not synonymous. The resistance of the whey protein fraction to heat-induced coagulation below the critical pH range does not mean that the whey proteins are not denatured by heating in this pH range. Investigations with differential scanning calorimetry (DSC) showed that temperatures of denaturation for 10% solutions of β-Lg and SA heated in the simulated milk ultrafiltrate (SMUF) were not substantially different at pH 3.5 from values obtained for sweet and acid whey pH conditions (6.5 and 4.5; Table 1). However, α-La showed a definite effect of the acidic pH, mainly evidenced by the 50% reduction in the enthalpy of denaturation. Because of the presumed acid denaturation, it is not surprising that at pH 2.5 no heat denaturation signal was obtained while at pH 3.5 the experimental observations showed great variability, particularly with respect to the reversibility of the denaturation phenomenon.

Recent experimental evidence [7] confirmed that α-La in pH conditions of sweet or acid wheys is stabilized by calcium. As a result, the heat denaturation reaction is reversible at pH 4.5 or higher when the calcium remains available in the solution. However, when the Ca is chelated by EDTA addition, the heat denaturation is irreversible and the observed denaturation temperature is much lower. The reversibility of the denaturation reaction and the resulting limited coagulability, may have been the reasons why α-La was earlier considered to be the most heat-stable whey protein, while in fact its denaturation temperature is lower than for any of the other main whey proteins in the acid or rennet whey pH range.

The same is also true for the highly acidic range (pH 3.5), however, the reversibility of the denaturation reaction is reduced and the temperature of denaturation is lower by about 3 °C as compared to pH 4.5 or higher; β-Lg and SA showed no effect of the acidic pH on denaturation temperature as compared to the regular whey pH range of 4.5–6.5, and also no evidence of renaturability.

Table 1. Heat denaturation temperatures for main whey protein fraction in SMUF[a]

Isolated protein	pH		
	3.5	4.5	6.5
α-La	58.6	61.5	61.0
β-Lg	81.9	81.2	75.9
SA	73.5	74.0	71.9
SA – defatted	n. d.	65.5	n. d.

[a] Adapted from [8]; SMUF = simulated milk ultrafiltrate; n. d. = not determined

Heat-induced co-precipitation of whey proteins

Binding of β-Lg with casein and possibly with α-La are the best examples of interactive heat-induced whey protein reactions. The limited co-precipitation of α-La and especially SA with β-Lg above the critical pH range was recently illustrated (Fig. 2). Similarly, in the mixed component systems such as whey beverages containing fruit juices, additional heat-induced whey protein co-precipitation with certain fruit components (pectins, tannins) may occur below the critical pH range. In our recent experience, a whey beverage containing berry fruit juices was formulated for final pH below the critical whey protein coagulation range. Upon UHT processing, the product showed evidence of strong cloud formation resulting in substantially reduced appearance acceptability. Whether this heat-induced change was due to interaction of any specific whey protein fraction with the fruit component has not yet been determined. Evidence in literature for heat-induced co-precipitation of whey proteins with fruit components in the acidic pH range is not available, and much additional work to identify the causes of the various co-precipitation reactions of whey proteins is required.

References

1. Jelen P, Patocka J (1988) This volume, pp 158–161
2. Hill AR (1986) Thermal precipitation of whey proteins. Ph D Thesis, Univ of Guelph, Canada
3. Kronman MJ, Singh SK, Brews K (1981) J Biol Chem 256:8582
4. Patocka J, Drathen M, Jelen P (1987) Milchwissenschaft 42:700–705
5. Mulvihill DM, Donovan M (1987) Ir J Food Sci Technol 11:43–75
6. Patocka J, Renz-Schauen A, Jelen P (1986) Milchwissenschaft 41:490–494
7. Bernal V, Jelen P (1984) J Dairy Sci 67:2452–2454
8. Bernal V, Jelen P (1985) J Dairy Sci 68:2847–2852
9. Jelen P, Buchheim W (1984) Milchwissenschaft 39:215–218

Selected Gelation Properties of Beta-Lactoglobulin in Comparison with Whey Protein Concentrate

P. Chr. Lorenzen and M. Grzinia

MILEI GmbH, Stuttgart, FRG

Introduction

The functional properties of whey protein concentrates (WPCs) are the sum of the properties of their fractions. Therefore the ultimate performance of the whey protein complex will be influenced by either the synergistic or antagonistic properties of these fractions.

Some authors [e.g., 1, 3] suggest that the functionality of WPCs is essentially determined by its beta-lactoglobulin-fraction (β-Lg).

The object of this paper is to examine this suggestion with regard to some selected gelation properties of β-Lg and WPC.

For better comparison of the results of the gelation properties, the materials used were defatted WPC and β-Lg (which was prepared from defatted WPC) and gel strength was used as the indicator for gelation.

Materials and method

The WPC was obtained by ultrafiltration (UF) of Emmental whey. Prior to UF the whey was defatted according to the procedure described by de Wit et al. [5]. UF was carried out at 20 °C in a pilot plant unit (De Danske Sukkerfabrikker, Denmark). After concentration of the protein up to 80% (in dry matter) the WPC was spray-dried.

β-Lg was prepared from the defatted WPC according to Pierre and Fauquant [4].

Solutions (protein content = 11.25%, pH 7.0) of WPC and β-Lg were prepared in demineralized water. 150 g of the solutions were weighed into 250 ml beakers. The beakers were sealed with aluminum foil and heated in a water bath at 75 °C for 45 min. Afterwards the samples were cooled for 30 min under running water and stored for 90 min in a refrigerator (6–8 °C).

The determination of the gel strength was performed by means of a texture analyzer (Stevens, UK).

Technical Data. Probe area, 1 cm^2; penetration rate, 0.5 mm/sec; penetration distance, 20 mm; gel strength, maximum load during 20 mm penetration expressed as penetration load (g) (see Fig. 1). The only parameters to be changed during these tests were temperature and protein content.

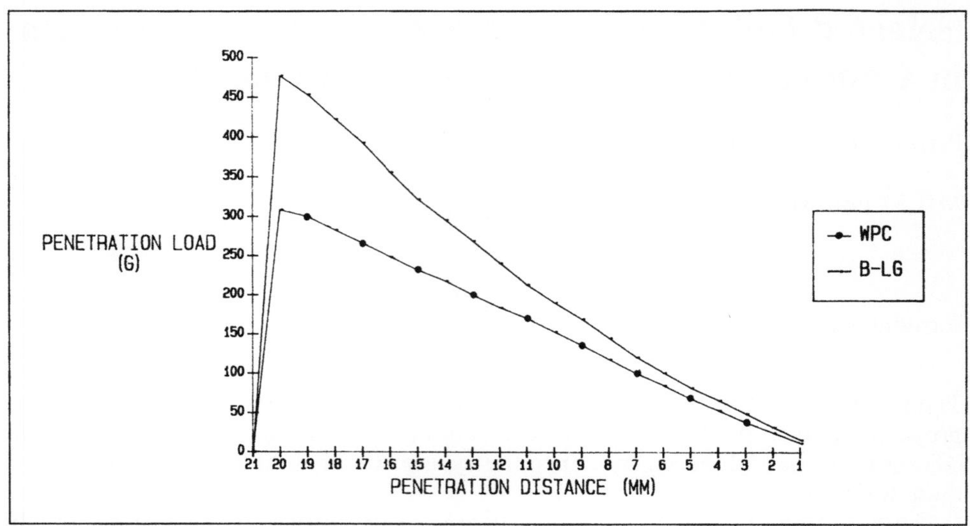

Fig. 1. Gel texture profiles of WPC and β-Lg

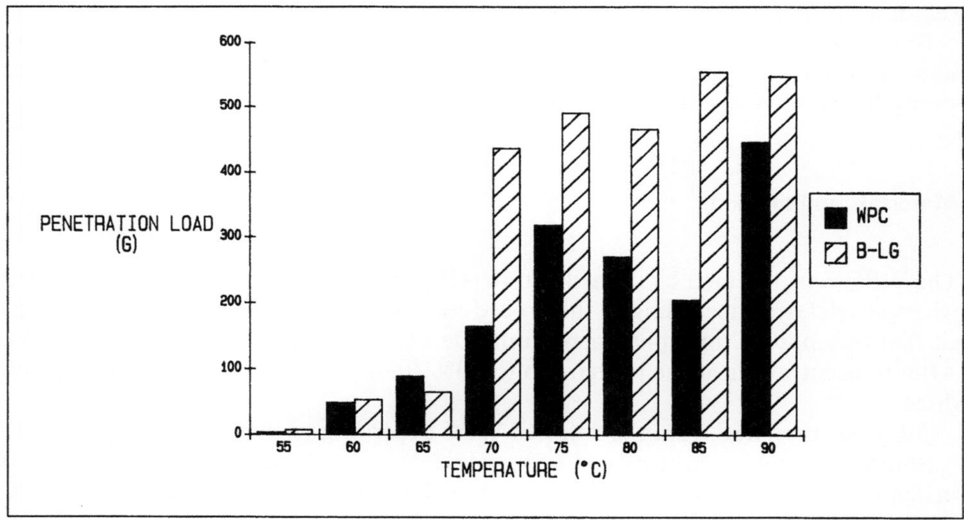

Fig. 2. Influence of the heating temperature

Results

Figure 1 illustrates the typical gel texture profiles for penetration load (g) against penetration distance (mm) obtained with WPC and β-Lg.

Both products achieve their maximum value after 20 mm penetration (end of measurement). Thereby the profile of WPC increases linearly whereby the curve of β-Lg is slightly concave. Both gels were milky in appearance and showed slightly brittle and aggregated structures.

248

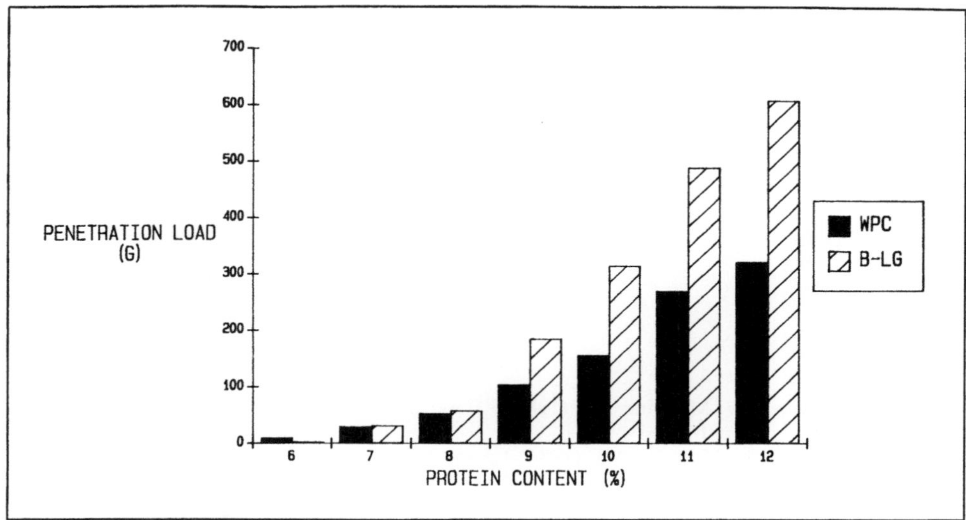

Fig. 3. Influence of the protein content

Figure 2 shows the effect of varying heating temperatures on the gel strengths of WPC and β-Lg.

Prior to heating the solution of WPC was translucent while that of the β-Lg was milky-white. After heating to 55 °C for 45 min, β-Lg showed evidence of slight gelation in the form of flakes while with WPC no gelation appeared and the solution remained translucent. At a temperature of 60–65 °C both products reached comparable levels of the gel strength. The gels of WPC took on a milky-white appearance similar to that of β-Lg. From a temperature of 70 °C and above the gel strength of the β-Lg became much higher than that of the WPC.

Figure 3 shows the influence of varying protein levels on gel strength.

At 6% protein the WPC solution showed clear evidence of gelation which had taken on a milky appearance. However, the β-Lg solution remained unchanged with no indication of any gelation. At protein levels of 7% and 8% both products had similar gel strengths. As the protein content increased (9% up to 12%) the β-Lg achieved twice the gel strength of WPC.

Discussion

The gel texture profiles illustrated in Fig. 1 indicate that both WPC and β-Lg form plastic gels as opposed to elastic gels which would result in a slightly convex curve and a clearly visible gel break point [2].

A surprising result of these investigations was the appearance of a slight gelation at 55 °C in the β-Lg because previously it had been determined that the denaturation temperature was 74 °C [6]. At 70–75 °C the gel strength of β-Lg is increased enormously, which is most probably caused by the unfolding of the globular structure of this protein.

At 6% protein the WPC showed clear signs of gelation while there were none with the β-Lg. This result clearly indicates that the β-Lg-fraction is not the only fraction which will influence gelation in the whey protein complex.

Acknowledgements

The authors thank Helga Meswarb and Michael Smallwood for their skillful assistance.

References

1. Kim YA, Chism GW, Mangino ME (1987) J Food Sci 52:124–127
2. Lorenzen PC, to be published
3. Peltonen-Shalaby R, Mangino ME (1986) J Food Sci 51:91–95
4. Pierre A, Fauquant J (1986) Le Lait 66:405–419
5. De Wit JN, Klarenbeek G, de Boer R (1978) Proc. 20th int Dairy Congr, Paris, p 996–997
6. De Wit JN, Klarenbeek G, Hontelez-Back E (1983) Neth Milk Dairy J 37:37–49

Study on Heat Stability and Coagulation Properties of Milks with Different Protein Genotype

M. Rampilli, T. Cattaneo, A. Caroli*, P. Bolla*, A. San Martino, and A. Sciocco Saita

Istituto sperimentale lattiero caseario, Lodi, Italy
* Istituto di Zootecnica Veterinaria, Milano, Italy

Introduction

The genetic polymorphism of caseins and β-lactoglobulin can heavily affect the composition and technological properties of milk [11, 12, 15]. While the effect of genetic variants on rennet coagulation has been extensively investigated, few and conflicting reports have been concerned with the relationships among protein polymorphism and heat stability of milk [9, 3, 13].

Several factors influence heat stability, like pH, composition of milk, stage of lactation, and processing (see [5] for review).

The aim of this work was to investigate heat and rennet stability of milks different for $\varkappa$-casein ($\varkappa$-CN) and β-lactoglobulin (β-Lg) genotypes.

Experimental

Selection of samples

Individual milks, previously typed for casein and β-Lg polymorphisms [1], were collected from one Friesian herd. Six samples were pooled from 97 mastitis-free cows, selected in a limited range (1 : 4) for parity and stage of lactation (30 : 180 days). Samples were homogeneous for α_s1-CN BB and β-CN AA, differing for $\varkappa$-CN (AA, AB) and β-Lg (AA, AB, BB). No pooled milk from $\varkappa$-CN BB was tested because of the few individual samples available.

Analytical determinations

Each pooled sample was submitted to the following measurements:
- total protein, casein and whey protein contents were determined by Kjeldahl method as described on IDF Standards [6, 7];
- proportions (%) of casein fractions and soluble whey proteins were calculated from the densitometer peak areas (Ultroscan XL-LKB) after polyacrylamide agarose gel electrophoresis, according to Uriel [17];
- urea concentration was determined by an enzymic method using a Boehringer testing kit (Boehringer Mannheim GmbH, FRG);

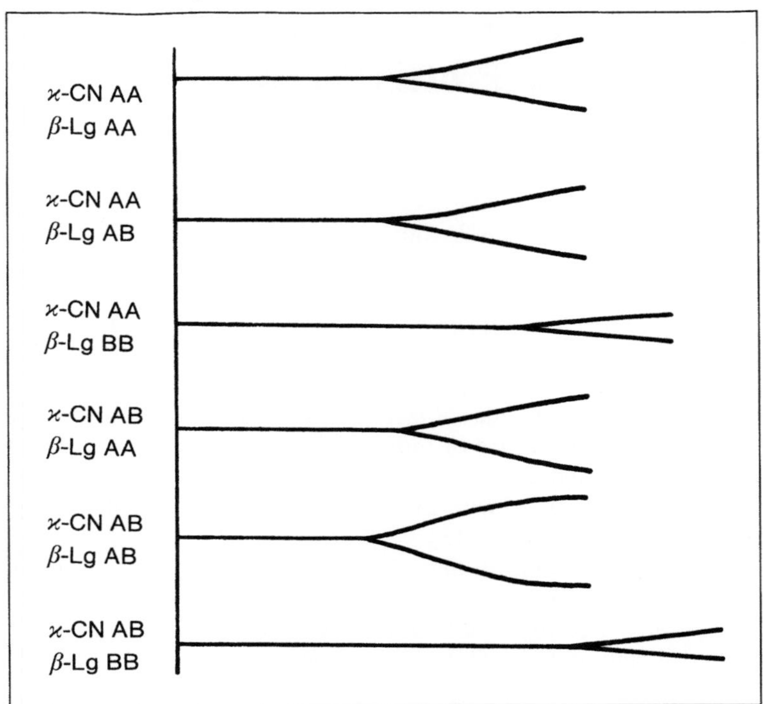

Fig. 1. Lactodynamographic patterns for each typed milk. (FORMAGRAPH Apparatus – Foss & Co., Denmark)

– Ca and P content were determined, respectively, by atomic absorption on ashes and a colorimetric method according to IDF Standard [8]. The colloidal amounts were calculated by subtracting the soluble amounts determined in the rennet whey from the total Ca and P contents of milk;
– rennet coagulation (Fig. 1) was investigated by lactodynamographic analysis using a Formagraph apparatus (Foss & Co., Denmark);
– heat coagulation time (HCT) was determined at natural and modified pH (Fig. 2) on 2 ml sample sealed in an airtight glass tube heated in an oil bath at 140 °C, as described by Feagan et al. [3].

Results and discussion

No remarkable differences were found between $\varkappa$-CN AA and AB milks for rennet and heat coagulation, while milk stability was higher in the two samples from β-Lg BB cows (Figs. 1 and 2). These samples showed the highest IG and BSA level, while the contents in β-Lg and colloidal calcium phosphate were the lowest (Tables 1 and 2). Moreover, β-Lg BB samples presented a higher casein content, either in absolute and relative terms, but lower level in $\varkappa$-CN fraction.

252

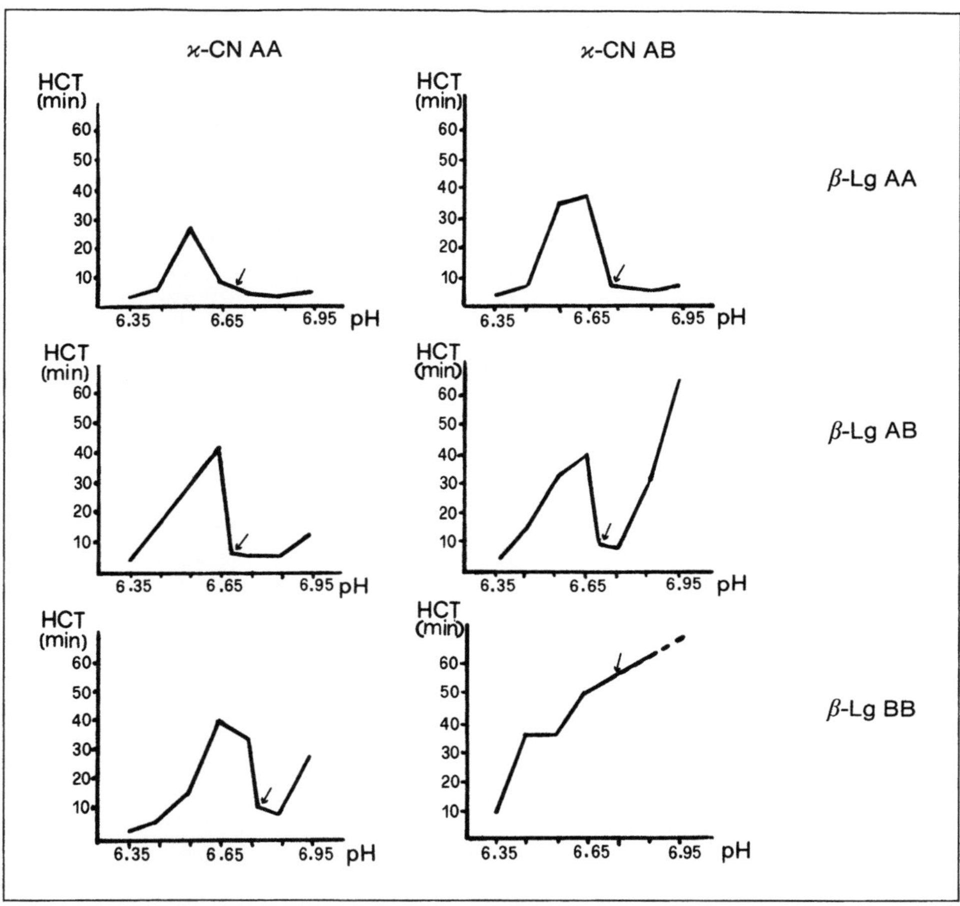

Fig. 2. Heat coagulation time-pH curves of the different samples at 140 °C. Arrows indicate the HCT at natural pH

The effect of natural pH on either rennet and heat coagulation was remarkable – higher pH values being associated with a higher stability of milk.

The present findings agree with literature for relationships between β-Lg genotype and composition of milk (casein, β-Lg content in particular; [12]). As regards heat stability of milk, these results confirm McLean's findings on concentrated milk [13], while they conflict with previous reports about genotype influence on heat coagulation, which indicated that β-Lg AA is associated with a higher stability of milk [3].

Moreover, although Rose [14] found heat stability positively related to β-Lg concentration, this relationship has not been confirmed in any other studies [13]. In the present work β-Lg content seems to be negatively associated with heat stability.

An important effect could be ascribed to the different β-Lg/x-CN ratio, which we found to be lower in milks with higher heat stability, as previously suggested by Feagan [3]. It should also be noted that heat coagulation time-pH curves (Fig. 2)

Table 1. Physico-chemical parameters of the six pooled milks. In brackets: number of cows selected for each sample

	$\varkappa$-CN AA			$\varkappa$-CN AB		
	β-Lg AA (13)	β-Lg AB (36)	β-Lg BB (16)	β-Lg AA (10)	β-Lg AB (22)	β-Lg BB (10)
pH	6.71	6.69	6.78	6.73	6.69	6.76
°SH/100	6.6	6.8	6.2	6.1	6.6	6.2
Ashes (g/100 g)	0.77	0.75	0.76	0.87	0.72	0.65
Fat (g/100 ml)	3.10	3.40	3.20	3.00	3.60	3.10
Protein (g/100 ml)	3.26	3.21	3.27	3.10	3.29	3.28
Casein (g/100 ml)	2.41	2.44	2.47	2.29	2.53	2.50
Casein number	74.0	76.0	75.4	73.9	76.9	76.3
Sol. Whey Prot. (g/100 ml)	0.64	0.58	0.62	0.61	0.57	0.57
Urea (mg/100 ml)	37.2	33.7	29.9	36.1	32.4	39.9
Ca/P	1.40	1.39	1.19	1.36	1.52	1.47
Colloidal Ca + P (g/100 ml)	0.122	0.136	0.109	0.163	0.131	0.107
Colloidal Ca + P (g/100 cas)	5.06	5.57	4.41	7.11	5.17	4.28

Table 2. Quantitative distribution (%) of casein and whey protein fractions

	$\varkappa$-CN AA			$\varkappa$-CN AB		
	β-Lg AA	β-Lg AB	β-Lg BB	β-Lg AA	β-Lg AB	β-Lg BB
α_s-casein	52.4	54.3	55.2	55.0	54.1	54.1
β-casein	35.5	35.4	34.2	35.0	37.5	37.0
$\varkappa$-casein	6.7	5.1	5.1	6.7	4.7	5.3
Minor caseins	5.4	5.2	5.5	3.3	3.7	3.6
β-lactoglobulin	64.8	58.7	46.8	63.7	60.8	52.8
α-lactoalbumin	22.5	23.8	24.4	19.1	21.8	21.2
BSA	4.4	4.8	6.7	5.0	5.8	7.1
Ig + PP	8.3	12.7	22.1	12.2	11.6	18.9

presented in five samples of A-type profile [16] in which the natural pH is close to HCT minimum, with a value included between HCT_{max} and HCT_{min}. Only one pooled milk (β-Lg BB – $\varkappa$-CN AB) showed the B-type curve (no minimum HCT), probably due to higher urea [10] and lower colloidal calcium phosphate contents [4].

References

1. Aschaffenburg R, Michalak W (1968) Simultaneous phenotyping procedure for milk proteins: improved resolution of the β-lactoglobulins. J Dairy Sci 51:1849
2. Feagan JT, Griffin AT, Lloyd GT (1966) Effects of subclinical mastitis on heat stability of fluid milk. J Dairy Sci 49:933–939
3. Feagan T, Bailey LF, Hehir AF, McLean DM, Ellis NJS (1972) Coagulation of milk proteins. 1. Effect of genetic variants of milk proteins on rennet coagulation and heat stability of normal milk. Aust J Dairy Technol 27:129–134

4. Fox PF, Hoynes MCT (1975) Heat stability of milk: influence of colloidal calcium phosphate and β-lactoglobulin. J Dairy Res 42:427–435
5. Fox PF (1982) Heat-induced coagulation of milk. In: Fox PF (ed) Developments in Dairy Chemistry – 1. Proteins. Applied Sciences Publishers LTD, London, pp 189–228
6. IDF Standard 20:1962 Determination of the total nitrogen content of milk by the Kjeldahl method
7. IDF Standard 29:1964 Determination of the casein content of milk
8. IDF Standard 42:1967 Determination of the phosphorus content of milk
9. Jenness R, Parkash S (1967) Heat stability of milks containing different casein polymorphs. J Dairy Sci 50:952
10. Kelly PM (1982) The effect of preheat temperature and urea addition on the seasonal variation in the heat stability of skim-milk powder. J Dairy Res 49:187–196
11. McLean DM, Graham ERB, Ponzoni RW, McKenzie HA (1982) Heat stability of preheated concentrated skim milk from individual cows: association with milk protein genotype and composition. Australian Dairy Technology Review Conference. Canberra, pp 190–191
12. McLean DM, Graham ERB, Ponzoni RW, McKenzie HA (1984) Effects of milk protein genetic variants on milk yield and composition. J Dairy Res 51:531–546
13. McLean DM, Graham ERB, Ponzoni RW, McKenzie HA (1987) Effects of milk protein genetic variants and composition on heat stability of milk. J Dairy Res 54:219–235
14. Rose D (1962) Factors affecting the heat stability of milk. J Dairy Sci 45:1305–1311
15. Schaar J (1984) Effects of k-casein genetic variants and lactation number on the renneting properties of individual milks. J Dairy Res 51:397–406
16. Tessier H, Rose D (1964) Influence of k-casein and β-lactoglobulin on the heat stability of skimmilk. J Dairy Sci 47:1047–1051
17. Uriel J (1966) Méthode d'électrophorèse dans des gels d'acrylamide-agarose. Bull Soc Chim Biol 48:969–982

Modelization of Gastric Digestion of Milk Proteins

B. Savalle, G. Miranda, and J-P. Pélissier

Station de Recherches Laitières, INRA, Jouy-en-Josas, France

In vivo digestion of proteins is initiated in the stomach by pepsins, and in some young animals by chymosin. It has been shown that milk coagulation in the stomach contributes to slowing down the degradation and it retains some peptides; casein degradation into small peptides takes place afterwards. β-lactoglobulin appears to be resistant to gastric proteolysis, but α-lactalbumin is hydrolyzed when stomach pH is about 3.5.

In vivo experiments for the study of protein digestion are time consuming and expensive. Experiments on animals and man are time consuming and difficult to realize. The aim of the present work was to simulate in vitro the most important phenomena observed in vivo in the stomach during milk protein digestion. The development of the model was achieved by control of different parameters: acidification curve of the medium, enzyme distribution (rennet), shaking mode, and emptying of peptides.

Samples recovered during in vivo and in vitro digestion were compared by SDS-electrophoresis, N measurement, amino acid composition, and the kinetic of liberation of characteristic peptides.

Results

SDS electrophoresis showed the rapid coagulation of caseins. These proteins are evacuated forwards as small peptides (evolution of the proline content confirms this result). β-lactoglobulin is not proteolyzed by gastric enzymes; α-lactalbumin is degraded when pH decreased below 4.0.

In vivo as in vitro, two groups were detected by multivariate analysis of the amino acid analysis. They correspond to sediment and supernatant in 12% TCA. The "small" peptides, which are probably absorbed faster, have essential amino acid contents that are different from "large" peptides. NPN level increased up to 50% during digestion and probably decreased a little after. No amine N (difference between Kjeldahl and amino acids) is evacuated earlier.

During digestion some peptides appeared rapidly and then disappeared, in particular the CMP (sequence 106–169 of $\varkappa$-casein) and sequence 1–23 from α_s1-casein. Quantities of other peptides increased slowly, such as 193–209 of β-casein.

Gastric emptying of proteins and peptides is reproducible. N and NPN level as well as HPLC profiles are similar to results obtained in vivo.

Conclusion

The comparison of the results obtained in vivo with those obtained in vitro using the "artificial stomach" (gastric emptying of N and NPN, NPN level, amino acid compositions, electrophoresis, HPLC profiles, and emptying of characteristic peptides) confirms the validity of our model. The described procedure is simple, does not require sophisticated material, and is much less expensive than in vivo experiments. Moreover, in vivo reactions that take place in six hours can be simulated in three hours.

General Discussion:
Milk Proteins – Technological and Functional Aspects

E. H. Reimerdes

Research and Development, Meggle Milk Industry GmbH & Co KG, Wasserburg, FRG

Milk constituents and milk products are traditionally recommended raw materials for the food industry. Due to their high nutrient density and specific functional properties they are designated to improve the nutritional and technological quality of food products.

Among milk constituents the proteins are of great importance for modern food production. The development of custom milk protein ingredients is directly related to improvements in modern food processing. Among modern processing technologies, fractionation, physical-chemical modifications, and recombination should be mentioned – by these methods the milk industry meets the requirements for optimization of their protein products with respect to a highly sophisticated food technology.

Due to the complex composition of milk proteins and progress made in protein technology the development of specific milk protein products by numerous modifications is a challenge to research and practice.

Functional properties of food ingredients, especially of milk proteins, are physical-chemical parameters which depend on structure and conformation as well as on interactions with other components of the product matrix. With respect to milk proteins these parameters are higly dependent on composition and processing conditions. The development of functional food ingredients on the basis of milk proteins is related to solubility, dispersibility, water-binding, fat-binding, viscosity, and the formation and stabilization of disperse systems such as foams, emulsions, and gels. In many food systems different types of disperse systems are integrated, e.g., in sausages, fat emulsification, and gel formation are of importance.

Another example is ice-cream where dispersed gases must be stabilized in destabilized oil-water emulsion matrix systems.

Besides the surface activity the viscosity of protein solutions or suspensions have to be considered in optimal food production. Each protein and protein product, as well as the complex food matrix, has its own characteristics and functionalities. Large differences can be found among the various types of proteins, e.g., caseins, whey proteins, fat globule membrane proteins, and even with the non-protein nitrogen (NPN) components.

There are large differences in structure/function relationships among caseins, whey proteins, fat globule membrane proteins, and the peptides of the NPN-fraction, dependent on the amino acid composition, sequence, and conformation.

In milk the caseins are organized in so-called micelles. After isolation the caseins show typical ion-exchange and detergent properties. This means the caseins can be

258

modified by monovalent and divalent cations and anions. Due to the fact that caseins have a flexible open-chain structure they show a high heat stability.

With whey proteins an entirely different situation has to be accounted for. According to their amino acid composition whey proteins have a typical globular structure which is further stabilized by disulfide-bridges. Due to the fact that globular proteins show related denaturation pattern upon heating, any type of heat treatment above a critical temperature/time combination will change structure and functional properties of whey protein products. In addition, denaturation and aggregation are dependent on pH, ions, and other food matrix constituents. Evaluating the kinetics of heat influence on whey proteins by sophisticated research techniques is crucial for the development of optimized milk and especially whey protein products. Skim-milk powders are already produced in five different types according to whey protein denaturation. Considering the application of these different products the significant different functionalities become apparent.

Further development is related to techniques such as ultrafiltration, ion-exchange chromatography, molecular-sieve chromatography, etc., which allow the production of a range of products with different protein contents, mineral compositions, and lactose contents. Including the above mentioned parameters, a multitude of products with different functionalities can be envisioned [Kinsella].

Structure/function relationships of proteins include the interactions with other food components, e.g., lipids, carbohydrates, minerals, etc. [Swaisgood]. This means that for the successful incorporation of a protein the recipe has to be evaluated with respect to the composition and processing parameters and the processing conditions have to be carefully evaluated in order to avoid unwanted reactions [Erbersdobler et al.] which can reduce the biological value of the food products. Typical indicators for suboptimal processing conditions are elevated, e.g. furosine, lysino alanine, or HMF. These indicators are useful for the evaluation of proper processing parameters, especially those concerning heat treatment.

During recent years special modifications of milk proteins have been developed. Among these are biotechnological procedures, e.g., proteolyses [Antila et al., Teuber et al.]. Functional properties like solubility and foam stabilization are directly related to the molecular weight of the proteins, which means by specific proteolytic degradation these parameters can be optimized with respect to food production. Other opportunities are the dephosphorylation of typical phosphor proteins like the caseins, or changes in the composition due to fermentation processes.

The fractionation of milk proteins to obtain single components with very specific biological or functional properties, e.g., immunoglobulins, is another interesting field of milk protein utilization [Klostermeyer et al.].

From traditional food product processing with raw materials, modern foods have been developed for high nutritional and techno-functional properties. Modern protein products are essential for quality parameters like structure, texture, and matrix consistency in many foods. For these purposes detailed information about ultrastructure and interfacial protein layers in foams and emulsions is essential. The fat globule membrane is the model system for such stabilized interfacial surfaces [Buchheim].

In summary, the inherent technological and functional properties of milk-protein products and the various possibilities for modification allow us to optimize the quality of food systems with respect to structure, texture, and stability. By integration of basic research results with practical considerations the development of new milk-protein products with specific functionalities for the improvement of food quality can be realized.

Milk Protein Allergy: Clinical Features, Pathogenesis, and Therapeutic Implications

A. Ferguson and K. C. Watret

Gastrointestinal Unit, Western General Hospital and University of Edinburgh, Edinburgh, UK

Introduction

The subject of food intolerance is of great interest to the general public as well as to the medical profession, the food industry, government and community regulatory bodies. There is considerable current research, particularly into the mechanisms underlying clinical food intolerance. It is clear that allergy, immunologically mediated intolerance, is relatively rare even when food intolerance has an organic cause, as opposed to the psychologically-based food aversion.

Food intolerance – definitions and clinical spectrum

Adverse reactions to ingested food vary enormously in severity as well as in pathogenesis. Food intolerance may be merely a nuisance, for example producing a transient rash or abdominal discomfort or there may be features strongly suggestive of disease. Many aspects of this subject are discussed in the report of the joint working party on food intolerance of the Royal College of Physicians and the British Nutrition Foundation, published in 1984. One of the most important sections of this report deals with definitions of diseases and with mechanisms. Two main conditions can be identified, and it is recommended that the terms food intolerance and food aversion are used, and defined as follows:

Food intolerance itself, which is a reproducible, unpleasant (i.e., adverse) reaction to a specific food or food ingredient and is neither immunologically nor psychologically based. This occurs even when the affected person cannot identify the type of food which has been given (for example, when it is disguised by flavoring and given as a puree).

Food allergy is a form of food intolerance where there is both a reproducible food intolerance and evidence of an abnormal immunological reaction to the food.

Food aversion comprises both psychological avoidance – when the subject avoids food for psychological reasons – and psychological intolerance, which is an unpleasant bodily reaction caused by emotions associated with the food rather than the

261

food itself and which does not occur when the food is given in an unrecognizable form.

Food aversion, intolerance, and allergy are very difficult to distinguish from one another. Whether objective changes are present or not, in most instances the clinical diagnosis of food intolerance can only be established if the symptoms and signs disappear with an elimination diet and if a controlled challenge then leads either to a recurrence of symptoms or to some other clearly identified change – for example, in a jejunal biopsy. Although much publicity has been given to a wide range of additional tests, both clinical and laboratory-based, these vary in their validity and interpretation and cannot reduce the importance of the challenge test.

True food allergy is predominantly a problem in infants and young children, and in the occasional highly atopic adult with multisystem disease. Cow milk allergy, similarly, is predominantly manifest in infants; however milk intolerance due to lactose malabsorption is, globally, by far the most common of all types of food intolerance.

In childhood, a wide range of conditions has been associated with food allergy, including eczema, wheeze, urticaria, mood alterations, angio-edema, diarrhea, vomiting, malabsorption, colitis, and gastrointestinal blood loss. The evidence linking hyperactivity to food intolerance is poor, despite some claims to the contrary.

In adults, classical allergic symptoms – urticaria, asthma or anaphylaxis – may occur and clinical effects on the gastrointestinal tract include nausea, vomiting, abdominal pain and diarrhea, anaphylaxis and angio-edema and in such patients the basis for the symptoms is truly allergic.

Milk allergy within the spectrum of food intolerance and disease

It is very important to clearly define the relevance of immunological mechanisms, and of the particular foodstuff, in the context of the symptom or disease under consideration. In relation to milk allergy, the pattern may be:
a) all cases of the disease are caused by milk allergy (for example milk-sensitive enteropathy). However, usually there are other single foods which can produce an identical condition, such as soya;
b) several substances, milk, other foods and other allergens particularly inhalants, are implicated in disease pathogenesis (for example atopic eczema);
c) allergy (including food allergy) is only one of several possible causes of a symptom or disease (for example diarrhea, wheeze).
In the diagnosis of milk allergy, the pioneer work by Goldman has now been superceded, partly because of recognition that challenge may be hazardous in atopic infants. Goldman's criteria required that symptoms subside after dietary elimination of milk; symptoms recur within 48 h after milk challenge and reactions to three such challenges must be positive and have a similar onset, duration, and clinical features [11]. The increasing use of double-blind testing of foods and placebo in older children and adults has greatly strengthened the clinical information to be obtained from a challenge test. Walker-Smith [29] has recommended that histological criteria substitute for the clinically-based "Goldman" criteria for diagnosing cow milk allergy.

Diseases caused by milk allergy in some patients

Atopic eczema: the incidence of atopic eczema is rising in Britain and foods are among the many environmental factors which contribute to exacerbation of this distressing skin disease [1]. The strongest evidence of a role for food in the pathogenesis of atopic eczema derives from a study in which 36 children with atopic eczema entered a double-blind controlled crossover trial of an egg and cow milk exclusion diet [2]. Twenty children finished the study of whom 14 responded to the antigen avoidance diet whereas only one responded to the control diet which contained egg and cow milk. Eczema can occur in exclusively breast-fed infants but this can be due to the carry over of food antigens in the mother's milk. Cant et al. [7] found that 11 of 34 breast-fed infants with atopic eczema had strongly positive skin test reactions to egg, and five of the 11 to cow milk, whereas none of the 34 controls had such skin-test reactivity.

Food intolerance and enhanced immune responsiveness to foods is also a feature of atopic eczema in adults. However, the antigens concerned are usually fish, shellfish, eggs, and nuts and milk sensitivity does not seem to be important [3, 4].

Asthma: an excellent and thoughtful appraisal of food sensitivity in childhood asthma was published recently [30]. In a group of hospital-based asthmatic children, one or more foods was thought to exacerbate symptoms in 70%, but eggs and milk were incriminated in less than 10% and this would support clinical observations that milk intolerance does not appear to be an important factor in asthma in childhood.

Malabsorption syndrome with cow milk intolerance: the classical description is that of Kuitunen et al. [19] who reported on 54 patients. All had diarrhea and failure to thrive; in the majority vomiting was also a feature and some 20% had atopic eczema and recurrent respiratory infections. Gastrointestinal investigations demonstrated malabsorption, and jejunal biopsy revealed abnormalities of the jejunal mucosa, ranging from moderate villus atrophy to a pathology indistinguishable from celiac disease. These Finnish pediatricians treated their patients with human milk and once clinical recovery was complete, patients were challenged with cow milk. Clinical effects of milk provocation varied: 28 of the infants had a rapid reaction (gastrointestinal symptoms occurring with one day of the challenge), however, in 26 infants the response was much slower. All of the children were clinically tolerant of cow milk by the age of one year, although the proximal jejunal mucosa often showed persistent but minor abnormalities.

In clinical practice, now that the disease is fully recognized by pediatricians, the need for jejunal biopsy in diagnosis and to confirm the existence of the condition by challenge when it is normally self-limiting, is being questioned [28, 8].

Lactose intolerance overlaps the syndrome of cow milk protein sensitive enteropathy. Where there is extensive villus atrophy, loss of disaccharidase-containing mature enterocytes leads to a relative reduction in the disaccharidase activity of the small bowel mucosa and thus there may be a reversible lactose intolerance consequence upon the enteropathy of cow milk protein intolerance [13, 16]. The most important practical consequence of this is that lactose intolerance should not be accepted as a primary explanation of malabsorption, failure to thrive, etc., but it should be appreciated that it can accompany a number of other more serious enteropathies. Additionally, when milk challenge is to be considered, it is often advisable first to carry out a lactose challenge in the healthy infant on a milk free diet. When this produces no clinically adverse effect, reactions to cows' milk challenge can be attributed to the protein constituents rather than the lactose.

Cow milk sensitive colitis: rectal bleeding accompanying other features of cow milk allergy in infants was recognized many years ago. However, it was only when the bimodal age distribution of children with colitis was recognized that the specific entity of probable allergic infantile colitis was formally documented [18, 23]. Typically, an infant with food sensitive colitis presents before the age of one year with loose, mucousy, bloody stools. In a series of 16 infants, 11 were less than four months old, of whom seven were taking cow milk protein only and the remaining four were exclusively breast-fed. Elimination diet and challenge procedures, with clinical observation and rectal biopsy have given similar findings to jejunal biopsy in cow milk sensitive malabsorption syndrome. Rectal biopsy pathology differs from classical ulcerative colitis in that there is preservation of crypt architecture with no crypt abscess formation and no depletion of goblet cell mucus. Additionally, there are substantial numbers of eosinophils and plasma cells in the lamina propria infiltrate. These infants respond well to elimination of cows' milk from their diet or from that of the mother. Severe clinical colitis was induced by challenge in a small number and it is recommended that challenge should not be used to confirm a diagnosis of food-sensitive colitis. As is the case for food-sensitive enteropathy, most children can tolerate cow milk by the age of two.

Infantile colic: in 1978, Jakobsson and Lindberg suggested that exposure to antigens of cow milk might cause infantile colic in breast-fed infants. They reported that when breast feeding mothers of 19 infants with infantile colic eliminated cow milk protein from their diets, the colic disappeared promptly in 13 of 19 infants. In 12 infants, milk-protein-induced colic was confirmed by dietary challenge of the mother. Subsequently, five of the infants were fed cow milk and four reacted promptly with colic. It is of considerable interest that three of these infants developed other symptoms of cow milk protein intolerance (skin rash, vomiting, and diarrhea).

Immunological aspects of the proteins of cow milk

Antigens of cow milk which induce hypersensitivity reactions are confined to the protein components. Digestion of milk proteins may reveal new immunogenic epitopes which are not present on the intact molecule. This has been studied in man for IgE responses to β-lactoglobulin (β-Lg) degradation products [25], and in an animal model new amino acid sequences were revealed during digestion of bovine serum albumine (BSA), and these induced quite distinct patterns of immune responses when compared with the native protein [22].

   In clinical studies particular attention has been paid to β-Lg, but reactions to all of the milk proteins may occur. For example Goldman et al. [11] orally challenged a group of children with cow milk allergy: 62% reacted with β-Lg but, in addition, 60% reacted with casein, 53% with α-lactalbumine (α-La), and 52% with BSA.

264

Lebenthal [20] reviewed the results of five studies and pointed out that sensitivity to β-Lg occurred in 82% of patients, to casein in 43%, to α-La in 41%, to BGG in 27%, and to BSA in 18%. Kuitunen et al. [19], studying enteropathy and malabsorption, challenged eight children with two or more individual cow milk proteins. Five of six responded to casein, one of four to α-La, six of seven to β-Lg, one of seven to BSA, and one of five to BGG.

Immunogenicity, as reflected by serum antibodies, of cow milk protein in man: there have been many reports, using a wide range of techniques, of the titres and patterns of antibodies to cow milk proteins in the serum of human infants and children. For example May et al. [21] used a protein-binding technique to detect and measure antibodies to BSA, casein, α-La, β-Lg, and BGG in infants and children who were not hypersensitive to foods. They found antibodies in the serum of a proportion of children up to age 15 years, to all of these antigens, but antigen-binding-capacity of serum was distinctly higher in the first year of life. In the 1970s we used passive hemagglutination and precipitin tests in the clinical immunology laboratory in Glasgow, and found that serum antibodies to all five major proteins were present in more than 50% of children aged under four years. Children with diffuse small bowel disease and enhanced intestinal permeability, such as in celiac disease, tend to have high titres of serum antibody to many foods, but we found that the distribution of antibody to the various milk proteins was similar in celiac disease and other children (unpublished).

IgE responses to food proteins are of greater relevance than other isotypes as the mechanism of food allergic disease. The development of such antibodies in a group of 86 normal female infants followed from birth to age seven years has been reported [14, 15]. Transient IgE antibodies to food proteins were detected in a proportion of the children and at the age of seven, seven of the 86 children had detectable IgE antibodies to cow milk in serum.

Immune responses to milk proteins encountered via the gut: when antigen is introduced into the tissues of an animal a variety of immune responses may be evoked (Fig. 1), and these are not mutually exclusive. In the case of antigen normally encountered via the gut it is relatively unusual for there to be induction of a serum antibody response (the guinea pig being an exception to this rule), and the most striking effects are induction of mucosal IgA antibodies, and suppression of systemic immune responsiveness and oral tolerance. Active immune responses can readily be detected and measured in humans as well as in animals whereas the phenomenon of immunological tolerance to ingested protein is best demonstrated in experimental animals. We are currently examining immune responses to β-Lg in mice, and find that, as with most other proteins, this is immunogenic when administered parenterally and tolerogenic when fed. Our experiments with the egg protein ovalbumin have shown that both immunoregulatory T cells and "processing" of antigen by the gut are involved in tolerance induction [5, 6, 10, 26, 27].

The fact that cow milk proteins are immunogenic in man and, indeed, in experimental animals, does not necessarily indicate that the immune responses which ensue are harmful. In the absence of induction of oral tolerance to a fed antigen, systemic immunity to foods, as with serum IgM and IgG antibodies, is usually harmless. IgE

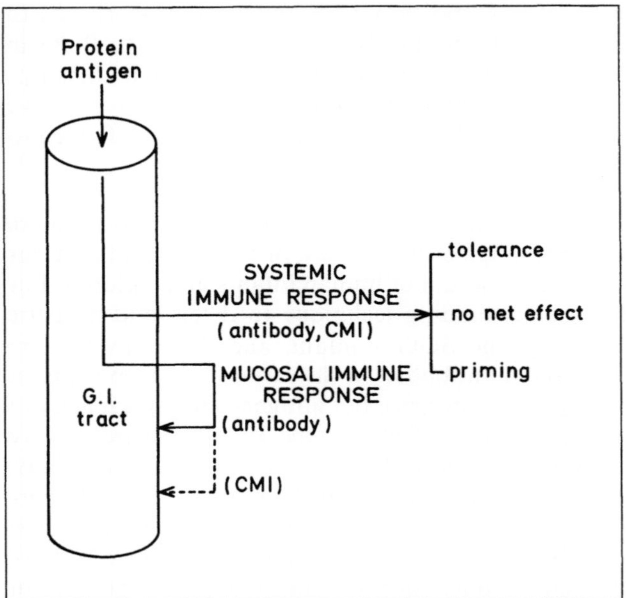

Fig. 1. Range of immune responses which may occur when antigen is fed

antibodies are evidence of an atopic state and often are of no clinical relevance being not associated with clinical food allergy. However, occasionally the presence of specific IgE to a defined antigenic determinant correlates very well with clinical reactions to the foodstuff concerned. Induction of cell-mediated immunity can be demonstrated by positive in vitro tests with blood lymphocytes and, theoretically, should be demonstrable by delayed skin test reactions. Although these are useful in animals they have not found any clinical role in man.

Mucosal hypersensitivity reactions: when oral tolerance fails and a state of active systemic immunity is induced, re-exposure to antigen is likely to produce intestinal and systemic hypersensitivity reactions analogous to those experienced in clinical food allergy. The cow milk protein BSA and the chicken egg protein ovalbumin (OVA) have been extensively used in investigation of the induction and expression of intestinal anaphylaxis (reviewed in [9]). Granato and Piguet [12] succeeded in producing a monoclonal IgE antibody directed against bovine β-Lg. They found that 1 ng of this purified antibody was capable of eliciting cutaneous anaphylaxis and demonstrated very clearly the development of an immediate hypersensitivity reaction to BLG within the gut in passively immunized mice fed β-Lg. Fluid accumulation within the small intestinal lumen, increased permeability of the gut and edema within the villi were present at 30 min after the challenge feed. Studies in this and similar animal models fully support the descriptions of the clinical and pathological effects of intestinal anaphylaxis in half a century of biomedical literature.

IgE mediated hypersensitivity reactions produce edema, diarrhea, and functional changes in epithelial fluid and electrolyte transport without any histological changes

266

or malabsorption. There is a wealth of evidence, based on animal work, that it is the T-cell mediated reactions in small bowel mucosa which cause immune mediated villus atrophy, crypt hyperplasia with malabsorption (reviewed in [9]). To our knowledge there have been no studies of animal models of cell-mediated immune responses to cow milk proteins in the gut mucosa.

Conceptually, it is essential to separate the induction phase of the immune response which relates to the dose and nature of antigen, the route of first encounter, and which may produce active immunity or tolerance, and expression of immunity after antigen re-exposure of an immune animal. This involves, in the first instance, antigen-specific, antigen-antibody, or antigen-cell reactions followed by non-antigen specific recruitment of cells and mediator release, and finally, effects on the function of tissue which may be detrimental, thus by definition: hypersensitivity.

Implications for prevention and treatment

Regulation of the induction of the immune response to an antigen which all of us ingest daily in large amounts, is the critical point at which any food allergy and food allergic disease must be considered. In the case of atopic individuals there is a defect in the regulatory mechanisms involved in help and suppression for the IgE B cell system. As far as cellular immunity is concerned, it is our hypothesis, based on experimental animals, that both the balance between help and suppression, and some sort of antigen processing by the gut, are necessary to induce oral tolerance for cell-mediated immunity to fed antigen and that there are many circumstances in which oral tolerance fails to develop. This puts the experimental animal, and by analogy the patient, at risk of developing food hypersensitivity.

There are a number of nutritional and therapeutic implications if one accepts that the problem in milk protein allergy is more likely to be in the infant's response rather than the food itself. Infants are particularly at risk of milk allergy because of immaturity of their immunoregulatory mechanisms and also because of the limited selection of foods which they encounter as antigens via the gut. There is a wealth of evidence in animals and also in human allergy that if extremely small doses of antigen are administered this tends to induce IgE antibodies, whereas large amounts of antigen suppress this isotype. Thus, there are theoretical reasons why so called hypoallergenic diets, which contain only trace amounts of antigen such as beta-lactoglobulin, are contraindicated even in the children of atopic families. One could argue that large amounts of antigen, administered frequently, are theoretically much more advisable as mechanisms of preventing allergy.

Once an individual is actively immunized and expresses the immune response as a hypersensitivity reaction either in the gut or in the skin, the best treatment is by elimination diet and this is not too difficult to achieve for cow milk protein. However, where mast cells and IgE antibodies are involved, drugs which modify mast cell functions such as sodium cromoglycate, may have a role. T-cell mediated immunity is probably the underlying mechanism of enteropathy and colitis in the majority of infants with these rare syndromes and exciting new developments on immunomodulatory regimes affecting T cells either by the use of monoclonal antibodies or pharmacologically, may have scope for treatment of the expression of the harmful immune response in the gut of these infants. Ultimately, one would hope to develop protocols and regimes which recreate the healthy state of immunological tolerance.

Acknowledgement

Our research on food intolerance and immunogenicity of milk proteins has been supported by grants from the Scottish Hospitals Endowment Research Trust, the Nutritional Consultative Panel of the UK Dairy Industry, and the Priory of Scotland of the Order of St. John of Jerusalem.

References

1. Atherton DJ (1983) Breast feeding and atopic eczema. Br Medical J 287:775–776
2. Atherton DJ, Sewell M, Soothill JF, Wells RS (1978) A double-blind controlled crossover trial of an antigen-avoidance diet in atopic eczema. Lancet I:401–403
3. Barnetson RStC (1980) Hyperimmunoglobulinaemia E in atopic eczema (atopic dermatitis) is associated with "food allergy". Acta Dermato (Stockh) Supplement 92:94–96
4. Barnetson RStC, Merrett TG, Ferguson A (1981) Studies on hyperimmunoglobulinaemia E in atopic diseases with particular reference to food allergens. Clin Exp Immunol 46:54–60
5. Bruce MG, Ferguson A (1986) Oral tolerance to ovalbumin in mice: studies of chemically modified and "biologically filtered" antigen. Immunology 57:627–630
6. Bruce MG, Ferguson A (1986) The influence of intestinal processing on the immunogenicity and molecular size of absorbed, circulating ovalbumin in mice. Immunology 59:295–300
7. Cant A, Marsden RA, Kilshaw PJ (1985) Egg and cows' milk hypersensitivity in exclusively breast fed infants with eczema, and detection of egg protein in breast milk. Br Med J 291:932–935
8. Cucchiara S, De Ritis G, Follo D, Jaccarino E, Romaniello G, Latte F, Staiano A, Auricchio S (1983) Cow's milk protein intolerance (CMPI): evaluation of clinical, histological, endoscopic and laboratory criteria for postchallenge diagnosis. Ital J Gastroenterol 15:231–235
9. Ferguson A (1987) Models of immunologically driven small intestinal damage. In: Marsh MN (ed) Immunopathology of the small intestine. John Wiley & Co Ltd, Chichester, pp 225–252
10. Ferguson A, Bruce MG, Strobel S (1988) "Processing" of antigen by the gut. In: Hanson LA, Svanborg Eden C (eds) Cellularmolecular interactions in the mucosal immune system. Switzerland: Karger, pp 253–255
11. Goldman AS, Anderson DW Jr, Sellars WA et al. (1963) Milk allergy. 1. Oral challenge with milk and isolated milk proteins in allergic children. Pediatrics 32:425–443
12. Granato DA, Piguet PF (1986) A mouse monoclonal IgE antibody anti bovine milk B-lactoglobulin allows studies of allergy in the gastrointestinal tract. Clin Exp Immunol 63:703–710
13. Harrison M, Kilby A, Walker-Smith JA, France NE, Wood CBS (1976) Cows' milk protein intolerance: a possible association with gastroenteritis, lactose intolerance, and IgA deficiency. Br Med J 1:1501–1504
14. Hattevig G, Kjellman B, Bjorksten B (1987) Common food proteins and inhalants in the first 7 years of life. Clin Allergy 17:571–578
15. Hattevig G, Kjellman B, Johansson SGO, Bjorksten B (1984) Clinical symptoms and IgE responses to common food proteins in atopic and healthy children. Clin Allergy 14:551–559
16. Iyngkaran N, Davis K, Robinson MJ, Boey CG, Sumithran E, Yadav M, Lam SK, Puthucheary SD (1979) Cows' milk protein-sensitive enteropathy. Arch Dis Child 54:39–43
17. Jakobsson I, Lindberg T (1978) Cow's milk as a cause of infantile colic in breast-fed infants. Lancet II:437–440
18. Jenkins HR, Pincott JR, Soothill JF, Milla PJ, Harries JT (1984) Food allergy: the major cause of infantile colitis. Arch Dis Child 59:326–329
19. Kuitunen P, Visakorpi JK, Savilahti E, Pelkonen P (1975) Malabsorption syndrome with cow's milk intolerance. Clinical findings and course in 54 cases. Arch Dis Child 50:351–356

20. Lebenthal E (1975) Cow's milk protein allergy. Pediatr Clin North Am 22:827–833
21. May CD, Remigio L, Feldman J, Bock SA, Carr RI (1977) A study of serum antibodies to isolated milk proteins and ovalbumin in infants and children. Clin Allergy 7:583–595
22. Michael JG, Muckerheide A, Dosa S, Pesce AJ (1981) Modulation of immune responses by antigenic fragments. In: Steinberg CM, Lefkowitz I (eds) The Immune System, Volume 2. Karger, Basel, p 284
23. Milla PJ (1984) Milk sensitive colitis in infants. In: Ferguson A (ed) Advanced Medicine 20. Pitman, London, pp 39–47
24. Royal College of Physicians and the British Nutrition Foundation (Joint report) (1984) J R Coll Physic (Lond) 18:83–123
25. Schwartz HR, Nerurkar LS, Spies JR, Scanlon RT (1980) Milk hypersensitivity: RAST studies using new antigens generated by pepsin hydrolysis of betalactoglobulin. Ann Allergy 45:242–245
26. Strobel S, Ferguson A (1986) Modulation of intestinal and systemic immune responses to a fed protein antigen in mice. Gut 27:829–837
27. Strobel S, Mowat AMcI, Drummond HE, Pickering MG, Ferguson A (1983) Immunological responses to fed protein antigens in mice. II. Oral tolerance for CMI is due to activation of cyclophosphamide-sensitive cells by gut-processed antigen. Immunology 49:451–456
28. Sumithran E, Iyngkaran N (1977) Is jejunal biopsy really necessary in cow's milk protein intolerance? Lancet II:1122–1124
29. Walker-Smith J (1975) Cow's milk protein intolerance. Transient food intolerance of infancy. Arch Dis Child 50:347–350
30. Wilson N, Silverman M (1985) Diagnosis of food sensitivity in childhood asthma. J R Soc Med [Suppl] 5:11–16

Milk Protein and Enteral and Parenteral Feeding in Disease

G. K. Grimble and D. B. A. Silk

Department of Gastroenterology & Nutrition, Central Middlesex Hospital, London, UK

Introduction

The incidence of nutritional deficiencies in hospitalized patients is often high [20, 21, 62, 90] but the indications for instituting nutritional support are both controversial and difficult to define. In general, patients requiring nutrition support fall into three groups. The first group comprises patients with preexisting weight loss, a low plasma albumin level, muscle wasting, and peripheral edema. Patients in the second group present with no overt malnutrition but have a dietary history of reduced intake for 2–4 weeks preceding admission. Finally, there are patients with normal nutritional status but whose underlying pathology results in malnutrition if nutritional support is withheld. These aspects are discussed in detail elsewhere [115].

Having made the decision to institute nutrition support, the choice lies between the parenteral route of nutrient delivery (through a central or peripheral catheter) or the enteral route (via a nasoenteral tube or by oral supplementation). Developments in enteral nutrition have, to some extent, been overshadowed by advances in total parenteral nutrition (TPN), since Dudrick first reported long-term support of patients by TPN [36]. Although enteral nutrition is not a new technique [133] its widespread acceptance has only come about following formal evaluation of the benefits and risks associated with each mode of nutrient delivery [61, 112]. It is now clear that enteral nutrition is more physiological than, considerably cheaper than, and has fewer associated risks than TPN. There are situations where TPN alone is indicated but many patients who are inappropriately prescribed TPN would benefit from enteral nutrition. In the light of this change towards greater use of the enteral route, it is worth considering the protein source of the diet and the best form in which it should be given in different disease states. There is considerable controversy as to whether free amino acids, partially hydrolyzed protein, or whole protein is "best." The aim of this paper is twofold. Firstly, this question will be examined in relation to the physiology of nitrogen absorption in the normal gastrointestinal tract and in situations where gut absorptive capacity is impaired. Secondly, the use of hydrolyzed proteins for *parenteral* use will be examined in the light of the considerable body of experience with intravenous casein hydrolysates and more recent studies with short-chain peptides.

Physiology of protein nitrogen absorption

Sources of digested protein and site of absorption in the gastrointestinal tract: Although dietary nitrogen intake forms the major part of protein digested and ab-

270

sorbed in the intestine, endogenous sources (gastric, biliary and intestinal secretions) represent, in man, about one-third of an adequate dietary protein intake [65, 98, 99] and comprise about 20–30 g from direct secretions [43], the remainder being desquamated epithelial cells [44] and plasma proteins secreted into the intestinal tract [43]. The majority of ingested protein appears to be absorbed in the proximal jejunum [22, 65, 98, 99, 123] although ileal protein assimilation does not appear to be complete [50, 84]. Animal studies implicate the colon as the major site of assimilation of endogenous protein [33].

Luminal protein digestion: The initial phase of protein digestion occurs in the stomach by the dual mechanism of acid denaturation and the action of several pepsins with an acid pH optimum and with varying substrate specificities [33, 138]. The primary cleavage products are large oligopeptides, only small amounts of free amino acids are liberated in the process [26] and further hydrolysis is effected by pancreatic proteases in the small intestine [44, 51].

In addition to pancreatic enzymes, low levels of solubilized intestinal brush-border and cytoplasmic intestinal mucosal aminooligopeptidases can be detected in intestinal contents, implying only a small role in luminal digestion [68, 122]. However, digestion of large oligopeptides by brush-border membrane bound neutral metalloendopeptidases [127] may have quantitative significance in view of studies showing significant digestion and absorption of dietary protein in the absence of pancreatic secretions [16, 32, 34, 132, 136]. The high level of solubilized membrane peptidases in the ileal lumen suggests significant luminal hydrolysis of peptides. Luminal proteolysis produces both peptides of chain-length 2–6 [9, 26, 98 99] and a smaller quantity of free amino acids (ca. 30% [9, 28]).

Free amino acid uptake: Free amino acid transport, studied in vitro is dependent on a gradient of Na^+ across the brush-border membrane of intestinal epithelial cells [110] and in man is saturable, consistent with the existence of carrier mediated mechanisms for amino acid transport [4, 5, 7, 13, 60]. To date, the absolute Na^+ dependency of free amino acid transport has not been demonstrated in vivo in man [116] but competition studies in animals indicate the likely existence of four major group-specific active transport systems [82, 84].

Peptide uptake: Although early studies, showed that small amounts of unhydrolyzed dipeptides crossed the intestinal wall [15, 92, 139], or could be taken up intact [93, 94], it was concluded that free amino acid uptake was the major route for nitrogen assimilation and that peptide transport was quantitatively insignificant.

The oral-load experiments of Matthews and colleagues in man [31] challenged this concept decisively and to date, there are four lines of evidence which favor peptide transport in the small intestine:
1) In patients with Hartnup disease and cystinuria, those free amino acids whose uptake is blocked are absorbed normally when presented to the mucosa in the form of homologous or mixed dipeptides [17–19, 59, 89, 118, 120];
2) Competition for uptake, in vivo, between free amino acids is relieved when they are presented as heterologous dipeptides [6, 58, 113, 117];

3) Faster rates of uptake of at least one of the constituent residues has been observed from the dipeptide rather than the corresponding free amino acid solution [30, 58, 117];
4) Unhydrolyzed dipeptides (resistant to brush-border hydrolysis) have been detected in peripheral plasma samples during intestinal perfusion in vivo [6, 48, 49] and in peripheral circulation during oral feeding experiments in man [63, 103, 104]. This evidence and studies on the substrate specificity of mucosal and cytoplasmic peptidases of the enterocyte [44, 71, 73, 96, 97, 101, 109, 134] suggest that peptide uptake and mucosal hydrolysis are closely related phenomena (see [52]). As a result, a "dual hypothesis" for peptide assimilation has been proposed. In this model, there are two routes of uptake of peptide nitrogen [80, 83, 113, 114]:
   a) Intact absorption of di- and tripeptides and subsequent cytoplasmic hydrolysis;
   b) Brush-border hydrolysis and uptake of free amino acids and smaller peptide fragments.

Given the enormous number of possible small peptide sequences generated during luminal digestion, for any particular dipeptide the proportion of peptide-bound amino acid absorbed by each route would depend on the relative affinity for intact transport $[K_t]$ or brush-border hydrolysis $[K_m]$.

Tripeptide and tetrapeptide transport: All the evidence now supports the view that tripeptides are assimilated by the small intestine in a similar fashion to dipeptides [2, 3, 14, 119]. In contrast, with one exception [29], all experimental studies suggest that prior brush-border hydrolysis of tetrapeptides has to occur before uptake of peptide nitrogen can occur [3, 11, 24, 111, 126]. The appearance of luminal free amino acids is consistent with the affinity of brush-border amino oligopeptidases for tetrapeptides [72, 73, 97, 134] and the absence of significant enterocyte cytoplasmic tetrapeptidase activity [97].

General characteristics of dipeptide and tripeptide transport: Peptide assimilation in the small intestine appears to have the following characteristics (see [52]):
1) There is at least one carrier system not shared with free amino acids [10];
2) The in vivo and in vitro evidence is consistent with the existence of a single common carrier [35, 125], single or multiple transport systems which may be allosterically activated by more hydrophobic dipeptides [25, 85, 86], or multiple transport systems [1, 39, 76, 107];
3) Transport is energy dependent with a Na^+ dependent component and is a saturable process with Michaelis-Menten kinetics. The partial Na^+ dependence appears to be a reflection of the activity of the $Na^+/K^+/ATPase$ system at the basolateral membrane of the enterocyte. Recent studies have indicated the cotransport of H^+ with di- and tripeptides, the process being "driven" by a proton gradient adjacent to the mucosal surface [46].

Intestinal assimilation of modified peptides: So far, the discussion of the assimilation of dietary nitrogen has assumed that all protein amino acids arrive in the portal vein in the form of free amino acids. This is a gross oversimplification because some amino acid residues of casein and egg protein are modified by covalent addition of phosphate or glycosyl moieties and will be released in peptide bound form during the

272

process of luminal or brush-border digestion. Similarly, certain amino acid sequences (e.g., those rich in proline moieties [103, 104]) are particularly resistant to brush-border and cytoplasmic hydrolysis by the enterocyte and may enter the blood stream intact.

Estimates of the proportion of alpha-amino nitrogen present in peptide form in portal blood following a protein meal vary, but the evidence reviewed by Gardner [48, 49], suggests that it is more than 10%. In the light of this observation, some revision of current concepts of nitrogen absorption may be necessary. Since aminoaciduria in man is rarely more than 2% of dietary intake, it may be concluded that pathways for degradation of significant quantitities of intravenous peptides exist. The reason for this mechanism may be the removal of peptides with pharmacological activity from the circulation. In relation to milk protein, the best studied peptide is a fragment of beta-casein liberated by pancreatic and intestinal brush-border hydrolysis, beta-casomorphin [102] which can be detected in the plasma of calves ingesting milk [74]. The major site of hydrolysis of beta-casomorphin (Tyr-Pro-Phe-Pro-Gly) appears to be the renal tubule from which it is reabsorbed as Phe-Pro-Gly and Phe-Pro [87] by a carrier mediated peptide transport system [47]. Thus it can be seen that there is provision for the lack of absolute barrier function of the small intestine against translocation of intact peptides. The significance of this in relation to intravenous preparations of hydrolyzed egg and milk proteins in parenteral therapy will be considered later.

Nutritional significance of amino acid and peptide transport

Intestinal perfusion studies: In order to quantitate the uptake of nitrogen from different dietary proteins in man we, and others, have studied the absorption of partial enzymic hydrolysates of protein which simulate the post-prandial milieu of the small intestine. In each case, an intestinal perfusion system has been used which excludes pancreatic secretions from the perfused test segment.

Effect of concentration on relative uptake of amino acids and peptides: The rates of absorption of amino acid residues have been shown to be faster, and more even, from peptide based than the free amino acid based mixtures [40, 117, 121, 124] and, as a result, there may be a quantitative [rate of absorption] and qualitative [evenness of absorption] "kinetic advantage" conferred by peptide based hydrolysates on nitrogen uptake. Although this effect occurred at perfused loads approximately two to three times greater than those experienced during continuous nasogastric enteral nutrition, the evenness of absorption of amino acid residues from the hydrolysate remained constant at all concentrations [54, 57] and this improvement in absorption pattern may have a bearing on the clinical situation.

Effect of peptide chain-length on uptake: Other factors which may influence uptake of amino acid residues from protein hydrolysates include differences in starter protein amino acid composition, primary sequence, enzymatic system used, and chain length of the peptides [70]. In particular, the effect of peptide chain length on absorption of hydrolysates of two milk proteins and one egg protein has been studied.

The absorption of nitrogen from a lactalbumin hydrolysate containing mainly peptides of chain length 10–15 was significantly less than from a lactalbumin hydrolysate of chain length 2–5, and less than from the equivalent free amino acid mixture [53]. A series of partial enzymic hydrolysates of ovalbumin were therefore prepared [54] and two contained a predominance of di- and tripeptides while the third contained mainly tetra- and pentapeptides. During perfusion at 30 mmol/L and 100 mmol/L, we noted that the subtle increase in chain-length profile reduced nitrogen uptake at both concentrations perfused. Similar studies with hydrolysates of casein (containing either 70% of di- and tripeptides or 64% tetra- and pentapeptides) confirmed this observation [105] and it was concluded that in man, brush-border hydrolysis may be a rate limiting step in assimilation of nitrogen from mixtures which simulate the products of post-prandial luminal digestion. The optimum chain-length for absorption under these experimental conditions would appear to be two to three amino acid residues.

Thus, where the absorptive capacity of the small intestine is severely limited, there may be good grounds for presenting dietary nitrogen in the form in which it is best absorbed, namely as di- and tripeptides, rather than as tetra- and higher peptides.

Clinical applications of amino acids, partially hydrolyzed milk protein or whole milk protein in enteral nutrion

Where intestinal function is normal: Initial studies in animals [64] indicated that whole protein or free amino acid-based diets produced similar growth rates and nitrogen balance. Comparison of the plasma appearance of amino acids in subjects ingesting a fish protein hydrolysate-based diet or equivalent free amino acid-based meal [123] indicated more rapid absorption, although by the time the meal had reached the distal small intestine, total absorption of the two diets was identical. Similarly, no difference was observed in nitrogen balance in healthy subjects fed isonitrogenous diets based on lactalbumin, a lactalbumin hydrolysate, or an equivalent amino acid mixture [88]. In light of these studies, one would expect that where gut function is normal, an enteral diet based on whole protein would be as efficiently assimilated as one based on a protein hydrolysate or free amino acid mixture.

One difficulty in assessing the metabolic utilization of comparable nitrogen sources has been the confounding effect of larger clinical variables such as the poor performance of feeding tubes, the use of diluted starter regimes, the use of small volume diet reservoirs and the slowing of infusion rates due to the development of gastrointestinal side-effects including diarrhea (see [115]). However, controlled clinical trials of enteral diets based on whole-protein or free amino acids have been unable to detect differences in nitrogen balance or clinical benefit or complications [66, 67]. There is thus no basis for using anything other than whole protein based diets in patients with normal gastrointestinal function.

Where gastrointestinal function is impaired: There may, however, be a place for easily assimilated forms of dietary nitrogen where some impairment of digestive capacity (e.g., pancreatic insufficiency, pancreatectomy, intestinal resection, or more rarely, severe active celiac or Crohn's disease) exists. The issue in this case is whether the capacity of the remaining small intestine is sufficient to digest and absorb the load of nitrogen given by 24 h nasoenteral infusion.

With moderate impairment of function, we were unable to detect any differences in nutritional parameters or diet absorption in patients with moderately impaired gastrointestinal function [106] who were fed either a milk protein hydrolysate or a whole milk protein-based diet. As far as the protein moiety of the diet is concerned, there is probably no basis for feeding protein hydrolysate based diets where there is slight impairment of gastrointestinal function.

However, where absorptive capacity (e.g., the inadequate short-bowel syndrome) or digestive capacity (e.g., total pancreatectomy) is severely impaired it is unlikely that whole protein would be efficiently assimilated. Thus in enterally-fed pancreatectomized patients, a lactalbumin hydrolysate was well absorbed (ca. 90%) compared to only 48% absorption of lactalbumin itself [132]. Whether protein hydrolysate or free amino acid based diets should be used is still open to question. In the severe inadequate short bowel syndrome (<100 cm small intestine), we were unable to detect any difference in the efficiency of absorption or rates of whole-body protein turnover when patients fed diets containing either a partial hydrolysate of casein (chain length 4 to 5 amino acids) or an equivalent free amino acid mixture [55]. In view of the observation that absorption of peptides in this chain-length range is rate-limited by brush-border hydrolysis [53, 54, 105], it seems likely that a hydrolysate diet containing mainly di- and tripeptides will give maximal absorption.

We would therefore conclude that where there is some impairment of gastrointestinal function, a whole protein diet is indicated; more severe impairment will require the use of partial hydrolysates of protein. There do not appear to be any indications, on purely nutritional grounds, for the more expensive amino acid-based elemental diets.

Clinical use of partially hydrolzyed milk proteins in parenteral nutrition

Total parenteral nutrition, as such, was first introduced by Elman who succeeded in infusing mixtures of electrolytes, glucose, and amino-acids (as a partial hydrolysate of casein) in sufficient quantities to reverse the edema of long-term malnutrition seen in patients with intestinal obstruction who were unable to eat normally [37, 38]. These results were often repeated with both acid and enzymatic hydrolysates of casein or fibrin given parenterally [23, 41, 78, 100, 129, 135, 141] and these preparations appeared to be an efficacious source of nitrogen for Total Parenteral Nutrition. It was noted that there was some peptiduria but in early preparations this could be accounted for by the formation of Maillard reaction products when the hydrolysate and glucose were heat sterilized together [27]. When administered to healthy subjects, hydrolysates were well utilized [78], and in patients on TPN, appeared to be as well utilized as free amino acids [135].

Following the subsequent use of defined mixtures of crystalline amino acids in TPN, protein hydrolysates fell into disuse. The reasons for this change seemed to be the development of central vein parenteral nutrition enabling hyperosmolar amino acid solutions to be infused, the ability to manipulate the amino acid composition to give a balanced mixture for therapeutic reasons, and finally, the successful large scale economic production of L-amino acids suitable for TPN. However, it should be noted that amino acid formulations have theoretical and practical drawbacks, namely the poor solubility and stability of certain amino acids (Tyrosine, Cysteine, Tyrptophan, and Glutamine), and the high osmolarity of amino acid solutions which would prevent administration of adequate nitrogen intakes via a peripheral vein.

Interest in the use of peptides as an nitrogen source in TPN has been revived by the work of Furst, Adibi and others. Recent work from Furst's group has indicated that synthetic peptides containing Glutamine [45], or Tyrosine [91] are soluble, stable, and will replete the tissue Glutamine pool. Similarly [8, 12, 75], synthetic peptides (based on Glycine at the N-terminal) are well tolerated and appear to be as effectively utilized as free amino acids in model of complete TPN in baboons [131]. One feature of this work was that with glycine oligomers [12] there appeared to be efficient utilization of the injected peptides up to a chain length of four glycine residues. With increasing peptide chain length, the pattern of peptide metabolism altered and in the case of tetraglycine, renal tubular brush-border hydrolysis and reabsorption of smaller peptides or free amino acids predominated. In view of the similarity in the enzymology of the small intestinal and renal brush border, it may be tentatively predicted that highly purified protein hydrolysates containing di- and tripeptides or tetra- and pentapeptides would be similarly handled when given intravenously. Certainly the intestinal perfusion studies discussed above suggest that brush-border limitation of nitrogen uptake only occurred with milk or egg protein hydrolysates containing a predominance of tetra- and pentapeptides, rather than di- and tripeptides.

We therefore investigated the intravenous utilization of an ovalbumin hydrolysate which differed from earlier intravenous hydrolysates in one important respect. Whereas these contained less than 30% as peptides of undetermined length [79, 108], the ovalbumin hydrolysate contained approximately two-thirds of amino acids as di- and tripeptides [54]. In healthy volunteers infusion of a complete TPN mixture based on the ovalbumin hydrolysate or the equivalent free amino acid mixture indicated good utilization of the hydrolysate although peptiduria (6% of excreted nitrogen) was slightly higher than that observed during amino acid infusion or following overnight fasting [56]. Due to the nature of the ovalbumin hydrolysate used, it is likely that the increase in excretion of peptide species was due, in part, to the presence of brush-border peptidase resistant glycopeptides which could be detected in the hydrolysate, by GC/MS [42]. Further studies are planned to answer this point. In summary therefore, there may be several benefits from the use of highly purified and defined short-chain peptide mixtures prepared from enzymically hydrolyzed protein. Firstly, the lower osmolarity of such preparations (cf. free amino acids) may allow peripheral infusion of adequate nitrogen. Secondly, since the glutamine and asparagine content of the original protein is preserved in stable form, such hydrolysates may be a suitable carrier for glutamine, which appears to be a specific nitrogen and energy source for the gut and cells of the immune system [95, 128, 140]. Infusion of exogenous glutamine may spare the efflux which occurs from peripheral tissues in states of stress or sepsis. Finally, it is likely that such preparations will be an economic alternative to the use of crystalline free amino acids or synthetic dipeptides [130] as a TPN nitrogen source.

References

1. Addison JM, Matthews DM, Burston D (1974) Competition between carnosine and other peptides for transport by hamster jeunum in vitro. Clin Sci Mol Med 46:707–714
2. Addison JM, Burston D, Payne JW, Wilkinson S, Matthews DM (1975) Evidence for active transport of tripeptides by hamster jejunum in vitro. Clin Sci Mol Med 49:305–312
3. Addison JM, Burston D, Dalrymple JA, Matthews DM, Payne JW, Sleisinger MH, Wilkinson S (1975) A common mechanism for transport of di- and tripeptides by hamster jejunum in vitro. Clin Sci Mol Med 49:313–322
4. Adibi SA (1969) The influence of molecular structure of neutral amino acids on their absorption in the jejunum and ileum of human intestine in vivo. Gastroenterology 56:903–913
5. Adibi SA (1970) Leucine absorption rate and net movements of sodium and water in human jejunum. J Appl Physiol 28:753–757
6. Adibi SA (1971) Intestinal transport of dipeptides in man: Relative importance of hydrolysis and intact absorption. J Clin Invest 50:2266–2275
7. Adibi SA, Gray SJ (1967) Intestinal absorption of essential amino acids in man. Gastroenterology 52:837–845
8. Adibi SA, Johns BA (1984) Partial substitution of amino acids of parenteral solutions with tripeptides: effects on parameters of protein nutrition in baboons. Metabolism 33:420–424
9. Adibi SA, Mercer DW (1973) Protein digestion in human intestine as reflected in human mucosal and plasma amino acid concentrations after meals. J Clin Invest 52:1586–1594
10. Adibi SA, Soleimanpur MR (1974) Functional characterisation of the dipeptide transport system in human jejunum. J Clin Invest 53:1368–1374
11. Adibi SA, Morse EL (1977) The number of glycine residues which limits intact absorption of glycine oligopeptides in human jejunum. J Clin Invest 60:1008–1016
12. Adibi SA, Morse EL (1982) Enrichment of glycine pool in plasma and tissues by glycine, di-, tri-, and tetraglycine. Am J Physiol 243:413–417
13. Adibi SA, Gray SJ, Menden E (1967) The kinetics of amino acid absorption and alteration of plasma composition of free amino acids after intestinal perfusion of amino acid mixtures. Am J Clin Nutr 20:24–33
14. Adibi SA, Morse EL, Masilamani SS, Amin PF (1975) Triglycine absorption in human intestine: Evidence for a common carrier for dipeptide and tripeptide transport. J Clin Invest 56:1355–1363
15. Agar WT, Hird EJ, Sidhu GS (1954) The uptake of amino acids by the intestine. Biochim Biophys Acta 14:80–84
16. Anderson DM, Ash RW (1971) The effect of ligating the pancreatic duct on digestion in the pig. Proc Nutr Soc 30:34–35
17. Asatoor AM, Bandoh JK, Lant AF, Milne MD, Navab F (1970) Intestinal absorption of carnosine and its constituent amino acids in man. Gut 11:250–254
18. Asatoor AM, Cheng B, Edwards KDG, Lant AF, Matthews DM, Milne MD, Navab F, Richards AJ (1970) Intestinal absorption of two dipeptides in Hartnup disease. Gut 11:380–387
19. Asatoor AM, Harrison BDW, Milne MD, Prosser DI (1972) Intestinal absorption of an arginine-containing peptide in cystinuria. Gut 13:95–98
20. Bistrian BR Balckburn GL, Hallowell E, Heddle R (1974) Protein status of general surgical patients. J Am Med Assoc 230:858–860
21. Bistrian BR, Blackburn GL, Vitale J, Cochran D, Naylor J (1976) Prevalence of malnutrition in general medical patients. J Am Med Assoc 235:1567–1570
22. Borgstrom B, Dahlqvist A, Lundh G, Sjovall J (1957) Studies of intestinal digestion and absorption in the human. J Clin Invest 36:1521–1536
23. Brunschwig A, Clark DE, Corbin N (1942) Postoperative nitrogen loss and studies on parenteral nitrogen nutrition by means of casein digest. Ann Surg 115:1091–1105
24. Burston D, Taylor E, Matthews DM (1979) Intestinal handling of two tetra-peptides by rodent small intestine. Biochim Biophys Acta 553:175–178
25. Burston D, Wapnir RA, Taylor E, Matthews DM (1982) Uptake of L-valyl-L-valine and glycyl-sarcosine by hamster jejunum in vitro. Clin Sci 62:617–626

26. Chen ML, Rogers QR, Harper AE (1962) Observations on protein digestion in vivo. IV Further observation of the gastrointestinal contents of rats fed different dietary proteins. J Nutr 76:235–241

27. Christensen HN, Wilber PB, Coyne BA, Fisher JH (1955) Effects of simultaneous or prior infusion of sugars on the fate of infused protein hydrolysates. J Clin Invest 34:86–94

28. Chung YC, Kim YS, Shadchehr A, Garrido M, MacGregor IL, Sleisinger MH (1979) Protein digestion and absorption in human small intestine. Gastroenterology 76:1415–1421

29. Chung YC, Silk DBA, Kim YS (1979) Intestinal transport of a tetrapeptide, L-leucylgylcyclglycyclglycine, in rat small intestine in vivo. Clin Sci 57:1–11

30. Cook GC (1973) Independent jejunal mechanisms for glycine and glycylgylcine transfer in man in vivo. Br J Nutr 30:13–19

31. Craft IL, Geddes D, Hyde CW, Wise IJ, Matthews DM (1968) Absorption and malabsorption of glycine and glycine peptides in man. Gut 9:425–437

32. Crane CW (1964) Studies on the absorption of ^{15}N labelled yeast protein in normal subjects and patients with malabsorption. In: v. Munro HN (ed) The role of the gastrointestinal tract in protein metabolism. FA Davis, Philadelphia, pp 333–347

33. Curtis KJ, Gaines HD, Kim YS (1978) Protein digestion and absorption in rats with pancreatic duct occlusion. Gastroenterology 74:1271–1276

34. Curtis KJ, Kim YS, Perdomo JM, Silk DBA, Whitehead JS (1978) Protein digestion and absorption in the rat. J Physiol 274:409–419

35. Das M, Radhakrishnan AN (1975) Studies on a wide-spectrum intestinal dipeptide uptake system in the monkey and in the human. Biochem J 146:133–139

36. Dudrick SJ, Wilmore DW, Vars HM, Rhoads JE (1969) Can intravenous feeding as the sole means of nutrition support growth in the child and restore weight loss in the adult? Ann Surg 169:974–981

37. Elman R, Weiner DO (1939) Intravenous alimentation. With special reference to protein (amino acid) metabolism. J Am Med Assoc 112:796–802

38. Elman R (1940) Parenteral replacement of protein with the amino acids of hydrolysed casein. Ann Surg 112:594–602

39. Fairclough PD, Silk DBA, Clark ML, Matthews DM, Marrs TC, Burston D, Clegg KM (1977) Effect of glycylglycine on absorption from human jejunum of an amino acid mixture simulating casein and a partial enzymic hydrolysate of casein containing small peptides. Clin Sci Mol Med 53:27–33

40. Fairclough PD, Hegarty JE, Silk DBA, Clark ML (1980) A comparison of the absorption of two protein hydrolysates and their effects on water and electrolyte movements in the human jejunum. Gut 21:829–834

41. Farr LE, Emerson K Jr, Futcher PH (1940) The comparative nutritive efficiency of intravenous amino acids and dietary protein in children with the nephrotic syndrome. J Pediatr 16:595–614

42. Ford C, Grimble GK, Halliday D, Silk DBA (1986) GC-MS analysis of dipeptides in nutritionally significant enzyme hydrolysates of ovalbumin and casein. Biochem Soc Trans 14:1291–1293

43. Freeman HJ, Kim YS (1978) Digestion and absorption of protein. Ann Rev Med 29:99–116

44. Freeman HJ, Sleisinger MH, Kim YS (1983) Human protein digestion and absorption: Normal mechanisms and protein energy malnutrition. In: v. Sleisinger MH (ed/Hrsg) Clinics in Gastroenterology, vol 12 number 2. WB Saunders, London Philadelphia Toronto, pp/S 357–378

45. Furst P (1985) Peptides in parenteral nutrition. Clin Nutr 4 (Special Supplement):105–115

46. Ganapathy V, Leibach FK (1985) Is intestinal transport energised by a proton gradient? Am J Physiol 249:G153–G160

47. Ganapathy V, Leibach FK (1986) Carrier mediated reabsorption of small peptides in renal proximal tubule. Am J Physiol 251:F945–F953

48. Gardner MLG (1983) Evidence for, and implications of, passage of intact peptides across the intestinal mucosa. Biochem Soc Trans 11:810–813

49. Gardner MLG (1984) Intestinal assimilation of intact peptides and proteins from the diet – a neglected field? Biol Rev 59:289–331

50. Gibson JA, Sladen GE, Dawson AM (1976) Protein absorption and ammonia production: The effects of dietary protein and removal of the colon. Br J Nutr 35:61–65

51. Gray GM, Cooper HL (1971) Protein digestion and absorption. Gastroenterology 61:535–544

52. Grimble GK, Silk DBA (1986) The optimum form of dietary nitrogen in gastrointestinal disease: Proteins, peptides or amino acids. Verhandlungen der Deutschen Gesellschaft für Innere Medizin 92:674–685. JF Bergmann, München

53. Grimble GK, Keohane PP, Higgins BE, Kaminski MV, Silk DBA (1986) Effect of peptide chain-length on amino acid and nitrogen absorption from two lactalbumin hydrolysates in the normal human jejunum. Clin Sci 71:65–69

54. Grimble GK, Rees RG, Keohane PP, Cartwright T, Desreumaux M, Silk DBA (1987) The effect of peptide chain-length on absorption of egg-protein hydrolysates in the normal human jejunum. Gastroenterology 92:136–142

55. Grimble GK, Rees RG, Halliday D, Ford C, Silk DBA (1986) Are enterally fed peptides better utilised than free amino acids in the short bowel syndrome? Clin Nutr [Suppl] 5:50

56. Grimble GK, Raimundo AH, Rees RG, Hunjan MK, Silk DBA (1988) Parenteral utilisation of a purified short-chain enzymic hydrolysate of ovalbumin in man. J Parent Ent Nutr [Suppl] 12:15S

57. Hegarty JE, Fairclough PD, Moriarty KJ, Kelly MJ, Clark ML (1982) Effects of concentration on in vivo absorption of a peptide containing protein hydrolysate. Gut 23:304–309

58. Hellier MD, Holdsworth CD, Perrett D, Thirumalai CD (1972) Intestinal dipeptide transport in normal and cystinuric subjects. Clin Sci Mol Med 43:659–668

59. Hellier MD, Holdsworth CD, McColl I, Perrett D (1972) Dipeptide absorption in man. Gut 13:965–969

60. Hellier MD, Holdsworth CD, Perrett D (1973) Dibasic amino acid absorption in man. Gastroenterology 65:613–618

61. Heymsfield SB, Bethel RA, Ansley JD, Nixon DW, Rudman D (1979) Enteral hyperalimentation: An alternative to central venous hyperalimentation. Ann Int Med 90:63–71

62. Hill GL, Blackett RL, Pickfort I, Young GA (1977) Malnutrition in surgical patients. An unrecognised problem. Lancet I:689–692

63. Hueckel HJ, Rogers QR (1970) Urinary excretion of hydroxyproline-containing peptides in man, rat, hamster, dog and monkey after feeding gelatin. Comp Biochem Physiol 32:7–16

64. Itoh H, Kishi T, Chibata I (1973) Comparative effects of casein and amino acid mixture simulating casein on growth and food intake in rats. J Nutr 103:1709–1715

65. Johannson C (1975) Studies of gastrointestinal interactions: VII. Characteristics of the absorption pattern of sugar, fat and protein from composite meals in man: A quantitative study. Scand J Gastroenterol 10:33–42

66. Jones DC, Rich AJ, Wright PD, Johnston IDA (1980) Comparison of proprietary elemental and whole-protein diets in unconscious patients with head injury. Br Med J I:493–495

67. Jones BJM, Lees R, Andrews J, Frost P, Silk DBA (1983) Comparison of an elemental and polymeric diet in patients with normal gastrointestinal function. Gut 24:78–84

68. Josefsson L, Lindberg T (1967) Intestinal dipeptidases: IX. Studies on dipeptidases of human intestinal mucosa. Acta Chem Scand 21:1965–1966

69. Kaplan MS, Mares A, Quintana P, Strauss J, Huxtable RF, Brennan P, Hays DM (1969) High caloric glucose-nitrogen infusions. Postoperative management of neonatal infants. Arch Surg 99:567–571

70. Keohane PP, Grimble GK, Brown B, Spiller RC, Silk DBA (1985) Influence of protein composition and hydrolysis method on intestinal absorption of protein in man. Gut 26:907–913

71. Kim YS, Birtwhistle W, Kim YW (1972) Peptide hydrolases in the brush border and soluble fractions of small intestinal mucosa of rat and man. J Clin Invest 51:1419–1430

72. Kim YS, Kim YW, Sleisinger MH (1974) Studies on the properties of peptide hydrolases in the brush-border and soluble fractions of small intestinal mucosa of rat and man. Biochim Biophys Acta 370:283–296

73. Kim YS, Brophy EJ, Nicholson JA (1976) Rat intestinal brush border membrane peptidases 2. Enzymatic properties, immunochemistry and interactions with lectins of two different forms of the enzyme. J Biol Chem 251:3206–3212

74. Kreil G, Umbach M, Brantl V, Teschemacher H (1983) Studies on the enzymatic degradation of beta-casomorphins. Life Sci [Suppl] 33:137–140
75. Krzysik BA, Adibi SA (1977) Cytoplasmic dipeptidase activities of kidney, ileum, jeunum, liver, muscle and blood. Am J Physiol 233:E450–E456
76. Lane AE, Silk DBA, Clark ML (1975) Absorption of two proline containing peptides by rat small intestine in vivo. J Physiol (Lond) 248:143–149
77. Levenson SM, Smith Hopkins B, Waldron M, Canham JE, Seifter E (1984) Early history of parenteral nutrition. Fed Proc 43:1391–1406
78. Lidstrom F, Wretlind KAJ (1952) The effect of intravenous administration of a dialysed, enzymic casein hydrolysate (Aminosol) on the serum concentration and on the urinary excretion of amino acids. Scand J Clin Lab Invest 4:167–178
79. Long CL, Zikria JM, Kinney JM, Geiger JW (1974) Comparison of fibrin hydrolysates and crystalline amino acid solutions in parenteral nutrition. Am J Clin Nutr 27:163–174
80. Matthews DM (1975) Intestinal absorption of peptides. Physiol Rev 55:537–608
81. Matthews DM (1983) Intestinal absorption of peptides. Biochem Soc Trans 11:808–810
82. Matthews DM (1984) Absorption of peptides, amino acids and their methylated derivatives. In: v Stegnik LD, Filer LJ Jr (eds) Aspartame: Physiology and Biochemistry. Marcel Dekker, New York Basel, pp 29–46
83. Matthews DM, Adibi SA (1976) Peptide absorption. Gastroenterology 71:151–161
84. Matthews DM, Payne JW (1980) Transmembrane transport of small peptides. Current Topics In Membranes and Transport, 14:331–425. Academic Press, New York
85. Matthews DM, Burston D (1983) Uptake of L-leucyl-L-leucine and glycylsarcosine by hamster jejunum in vitro. Clin Sci 65:177–184
86. Matthews DM, Burston D (1984) Uptake of a series of neutral dipeptides including L-alanyl-L-alanine, glycylglycine and glycyclsarcosine by hamster jejunum in vitro. Clin Sci 67:541–549
87. Miyamoto Y, Ganapathy V, Barlas A, Neubert K, Barth A, Leibach FK (1987) Role of dipeptidyl peptidase IV in uptake of peptide nitrogen from beta-casomorphin in rabbit renal BBMV. Am J Physiol 252:F670–F677
88. Moriarty KJ, Hegarty JE, Fairclough PD, Kelly MJ, Clark ML, Dawson AM (1985) Relative nutritional value of whole protein, hydrolysed protein and free amino acids in man. Gut 26:694–699
89. Navab F, Asatoor AM (1970) Studies on intestinal absorption of amino acids and a dipeptide in a case of Hartnup disease. Gut 11:373–379
90. Nethercut WD, Smith ADS, McAllister JM, La Ferla GA (1987) Nutritional survey of patients in a general surgical ward: Is there an effective predictor of malnutrition? J Clin Pathol 40:803–807
91. Neuhauser M (1985) Utilisation of glycyl-L-tyrosine during long-term parenteral nutrition in the rat. Clin Nutr [Special Supplement] 4:124–130
92. Newey H, Smyth DH (1959) The intestinal absorption of some dipeptides. J Physiol (Lond) 145:48–56
93. Newey H, Smyth DH (1960) Intracellular hydrolysis of dipeptides during intestinal absorption. J Physiol (Lond) 152:367–380
94. Newey H, Smyth DH (1962) Cellular mechanisms in intestinal transfer of amino acids. J Physiol (Lond) 164:527–551
95. Newsholme EA, Crabtree B, Ardawi MSM (1985) Glutamine metabolism in lymphocytes: Its biochemical, physiological and clinical importance. Quart J Exp Physiol 70:473–489
96. Nicholson JA, Peters TJ (1978) Subcellular distribution of hydrolase activities for glycine and leucine homopeptides in human jejunum. Clin Sci Mol Med 54:205–207
97. Nicholson JA, Peters TJ (1979) Subcellular localisation of peptidase activity in the human jejunum. Eur J Clin Invest 9:349–354
98. Nixon SE, Mawer GE (1970) The digestion and absorption of protein in man. 1. The site of absorption. Br J Nutr 24:227–240
99. Nixon SE, Mawer GE (1970) The digestion and absorption of protein in man. 2. The form in which digested protein is absorbed. Br J Nutr 24:241–258
100. Peaston MJT (1968) A comparison of hydrolysed L- and synthesised DL-amino acids for complete parenteral nutrition. Clin Pharm Therap 9:61–66

280

101. Peters TJ (1970) The subcellular localisation of di- and tripeptide hydrolase activity in guinea pig small intestine. Biochem J 120:195–203
102. Petrilli P, Pucci P, Pelissier P, Addeo F (1987) Digestion by pancreatic juice of a beta-casomorphin-containing fragment of buffalo beta-casein. Int J Peptide Res 29:504–507
103. Prockop DJ, Sjoerdsma A (1961) Significance of urinary hydroxyproline in man. J Clin Invest 40:843–849
104. Prockop DJ, Keiser HR, Sjoerdsma A (1962) Gastrointestinal absorption and renal excretion of hydroxyproline peptides. Lancet II:527–528
105. Rees RG, Raimundo AH, Grimble GK, Hunjan MK, Silk DBA (1988) Peptide based nitrogen source of enteral diets: Studies with casein hydrolysates in man. J Parent Ent Nutr [Suppl] 12:21S
106. Rees RG, Payne-James JJ, Grimble GK, Silk DBA (1988) Requirement of peptides versus whole protein in patients with moderately impaired gastrointestinal function: A double-blind controlled crossover study. J Parent Ent Nutr [Suppl] 12:12S
107. Rubino A, Field M, Schwachman H (1971) Intestinal transport of amino acid residues of dipeptides. I. Influx of the glycine residue of glycyl-L-proline across mucosal border. J Biol Chem 246:3542–3548
108. Sampson B, Barlow GB (1980) Separation of peptides and amino acids by ion-exchange chromatography of their copper complexes. J Chrom 183:9–15
109. Schmitz J, Triadou N (1982) Digestion et absorption intestinales des peptides. Gastroenterol Clin Biol 6:651–661
110. Schultz SG, Curran PF (1970) Coupled transport of sodium and organic solutes. Physiol Rev 50:637–672
111. Semeriva M, Varesi L, Gratecos D (1982) Studies on transport of amino acids from peptides by rat small intestine in vitro: Synthesis, properties and uptake of a photosensitive tetrapeptide. Eur J Biochem 122:619–626
112. Shanbhogue LKR, Bistrian BR, Blackburn GL (1986) Trends in enteral nutrition in the surgical patient. J R Coll Surg Edinb 31:267–273
113. Silk DBA (1974) Progress Report: Peptide absorption in man. Gut 15:494–501
114. Silk DBA (1981) Peptide Transport. Clin Sci 60:607–615
115. Silk DBA (1986) Future of enteral nutrition. Gut [Special Suppl] 27:116–122
116. Silk DBA, Dawson AM (1979) Intestinal absorption of carbohydrate and protein in man. In: v Crane RK (ed) International Review of Physiology. Gastrointestinal Physiology III Vol 19. University Park Press, Baltimore, pp 151–204
117. Silk DBA, Marrs TC, Addison JM, Burston D, Clark ML, Matthews DM (1973) Absorption of amino acids from an amino acid mixture simulating casein and a tryptic hydrolysate of casein in man. Clin Sci Mol Med 45:715–719
118. Silk DBA, Perrett D, Stephens AD, Clark ML, Scowen EF (1974) Intestinal absorption of cystine and cysteine in normal human subjects and patients with cystinuria. Clin Sci Mol Med 47:393–397
119. Silk DBA, Perrett D, Webb JPW, Clark ML (1974) Absorption of two tripeptides by the human small intestine. A study using a perfusion technique. Clin Sci Mol Med 46:393–402
120. Silk DBA, Perrett D, Clark ML (1975) Jejunal and ileal absorption of dibasic amino acids and an arginine containing dipeptide in cystinuria. Gastroenterology 68:1426–1432
121. Silk DBA, Clark ML, Marrs TC, Addison JM, Burston D, Matthews DM, Clegg KM (1975) Jejunal absorption of an amino acid mixture simulating casein and an enzymic hydrolysate of casein prepared for oral administration to normal adults. Br J Nutr 33:95–100
122. Silk DBA, Nicholson JA, Kim YS (1976) Relationships between mucosal hydrolysis and transport of two phenylalanine dipeptides. Gut 17:870–876
123. Silk DBA, Chung YC, Berger KL, Conley K, Sleisinger MH, Spiller GA, Kim YS (1979) Comparison of oral feeding of peptide and amino acid meals to normal human subjects. Gut 20:291–299
124. Silk DBA, Fairclough PD, Clark ML, Hegarty JE, Marrs TC, Addison JM, Burston D, Clegg KM, Matthews DM (1980) Uses of a peptide rather than a free amino acid nitrogen source in chemically defined elemental diets. J Parent Ent Nutr 4:548–553

125. Sleisinger GE, Burston D, Dalrymple JA, Wilkinson S, Matthews DM (1976) Evidence for a single common carrier for uptake of a dipeptide and a tripeptide by hamster jejunum in vitro. Gastroenterology 71:76–81

126. Smithson KW, Gray GM (1977) Intestinal assimilation of a tetrapeptide in the rat. Obligate function of brush-border membrane aminopeptidases. J Clin Invest 60:665–674

127. Song I-S, Yoshioko M, Erickson RH, Miura S, Guan D, Kim YS (1986) Identification and characterisation of brush-border membrane-bound neutral metalloendopeptidases from rat small intestine. Gastroenterology 91:1234–1242

128. Souba WW, Scott TE, Wilmore DW (1985) Intestinal consumption of intravenously administered fuels. J Parent Ent Nutr 9:18–22

129. Stegink LD, Baker GL (1971) Infusion of protein hydrolysates in the new-born infant. Plasma amino acid concentrations. J Pediatr 78:595–602

130. Stehle P (1988) Bedarfsgerechte Bereitstellung von kurzkettigen Peptiden: Eine Voraussetzung für deren Einsatz in der künstlichen Ernährung. Infusionstherapie 15:27–32

131. Steinhardt HJ, Paleos GA, Brandl M, Fekl WE, Adibi SA (1984) Efficacy of a synthetic dipeptide mixture as the source of amino acids for total parenteral nutrition in a subhuman primate (baboon). Gastroenterology 86:1562–1569

132. Steinhardt HJ, Wolf A, Jakober B, Brandl M, Langer K, Fekl W, Adibi SA (1986) Efficiency of nitrogen absorption from whole vs. hydrolysed protein after total pancreatectomy in man. Clin Nutr [Suppl] 5:108A

133. Stengel A Jr, Ravdin IS (1939) The maintenance of nutrition in surgical patients with a description of the orojejunal method of feeding. Surgery 6:511–519

134. Tobey N, Heizer W, Yeh R, Huang T-I, Hoffner C (1985) Human intestinal brush border peptidases. Gastroenterology 88:913–926

135. Tweedle DEF, Spivey J, Johnston IDA (1973) Choice of intravenous amino acid solutions for use after surgical operation. Metabolism 22:173–178

136. Uram JA, Friedman L, Kline OL (1960) Relation of pancreatic exocrine function to nutrition of the rat. Am J Physiol 199:387–394

137. Vasile'v PS, Suzdaleva VV (1978) Preparations for parenteral protein nutrition and their significance in medicine. Sov Med 2:91–93

138. Whitecross DP, Armstrong C, Clark AD, Piper DW (1973) The pepsinogens of human gastric mucosa. Gut 14:850–855

139. Wiggans DS, Johnston JM (1959) The absorption of peptides. Biochim Biophys Acta 32:69–73

140. Windmueller HG (1982) Glutamine metabolism by the small intestine. Adv Enzymol 53:202–238

141. Wretlind KAJ (1974) Amino acid solutions. In: v Lee HA (ed) Parenteral nutrition in acute metabolic illnesses. Academic Press, London New York, pp 333–352

Is Milk Intake Still Adequate in Present-day Practice of Child Nutrition?

M. Kersting, W. Rühe and G. Schöch

Forschungsinstitut für Kinderernährung, Dortmund, FRG

Introduction

Recommendations for child nutrition have to allow for an adequate intake of high-quality proteins and of minerals, particularly calcium, to meet specific growth requirements. Among the convenient foodstuffs in our environment, cow's milk is an excellent source of both calcium and protein. Therefore, milk intake of children should be assessed from time to time.

Nutrition surveys

Two nutrition surveys, done about 15 years apart, were carried out on 1–14 year-old healthy children in Dortmund. Food intake was assessed with a weighed dietary record method; energy and nutrient intake was calculated from food composition tables. Study I took place from 1965–1979 comprising 310 children with usually 21–42 dietary record days for each child (3); study II from 1985–1987 comprising 275 children with three dietary record days for each child [2].

Results

In study I, average daily consumption of milk (e.g., whole, partly skimmed, pasteurized, and UHT) and milk products (e.g., yoghurt, sour milk, cheese) increased during childhood from 300 to 530 g (Fig. 1). More recently, only 300 to 370 g have been observed (study II). In study II, sweetened commercial milk products amounted to 15% (1–6 yrs) and 25% (7–14 yrs) of the total consumption of milk and milk products [2].

In study I, energy intake corresponded closely to the DGE recommendations [1] in nearly all age groups (Fig. 2). Recently (study II) energy intake has decreased and is now about 15% below the DGE recommendations for children older than 10 years. With increasing age, the contribution of milk and milk products decreased from 25 to 15% of energy intake in both studies.

In both studies (Fig. 3), protein intake was markedly above the already generous DGE recommendations [1]. Preschool children consumed about 50% and school-age children about 20% more protein than recommended. Protein from animal food amounted constantly to two-thirds of overall protein intake. From study I to study

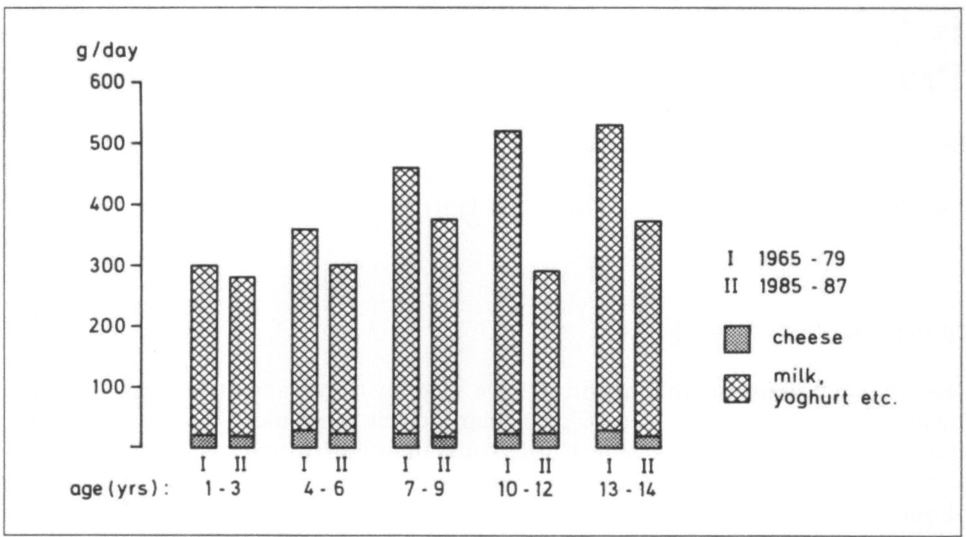

Fig. 1. Average daily consumption of milk and milk products of 1–14 year-old children observed in two nutrition surveys in Dortmund, done about 15 years apart

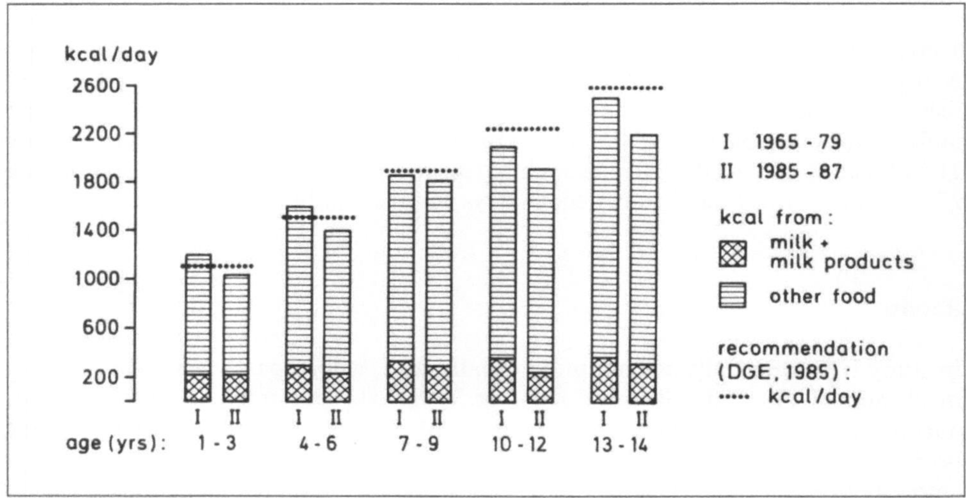

Fig. 2. Contribution of milk and milk products to average daily energy intake of 1–14 year-old children observed in two nutrition surveys in Dortmund, done about 15 years apart

II, the contribution of milk and milk products to protein intake decreased from 40 to 30% at 1–6 years, and from 30 to 20% at 7–14 years.

In study I, calcium intake corresponded to the DGE recommendations [1], with a contribution of about 75% from milk and milk products (Fig. 4). In study II, the decreased milk consumption from school-age onwards leads to decreased calcium intakes which may be 25% below the DGE recommendations. Only 50–60% of the present calcium intake stems from milk and milk products.

284

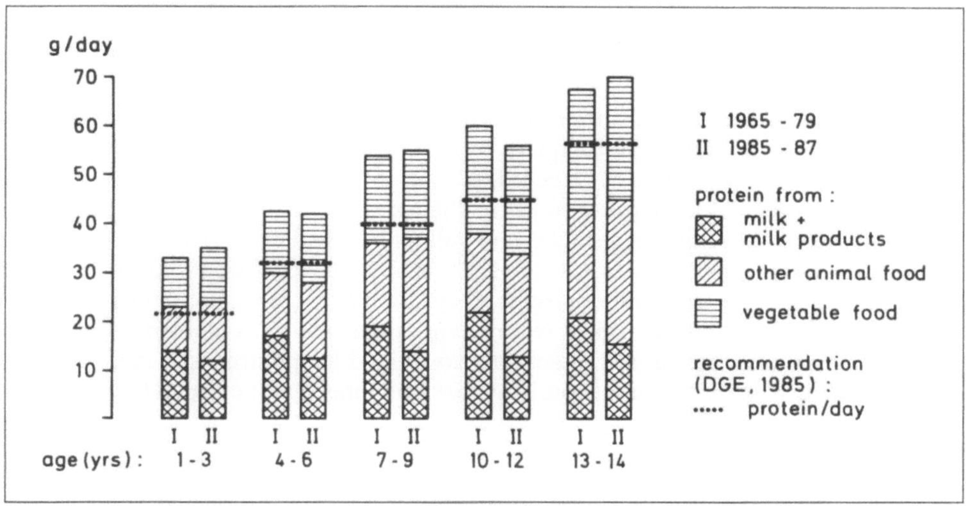

Fig. 3. Contribution of milk and milk products to average daily protein intake of 1–14 year-old children observed in two nutrition surveys in Dortmund, done about 15 years apart

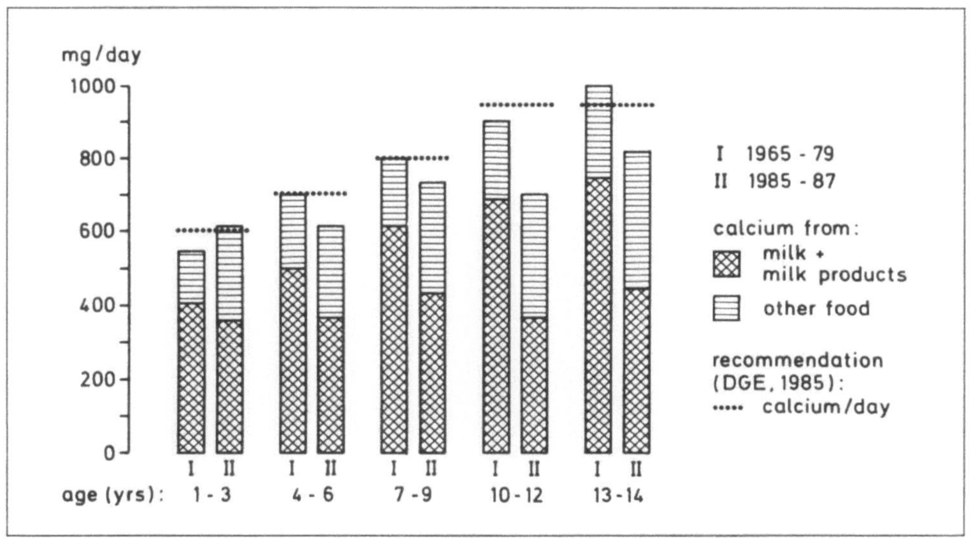

Fig. 4. Contribution of milk and milk products to average daily calcium intake of 1–14 year-old children observed in two nutrition surveys in Dortmund, done about 15 years apart

Discussion and conclusion

Energy intakes below the recommendations of nutrition boards in normal growing children of about the age of puberty, as in the recent study II, have been reported in British children, too [4]. The lower energy requirements are thought to be due chiefly to lower physical activity [4]. A low energy intake necessitates a sufficiently high nutrient density of the consumed food. It is therefore unfortunate that an

285

increasing proportion of commercial milk products consumed by children in all age groups contains substantial amounts of sugar which is virtually free from essential nutrients.

Present-day low milk consumption from school-age onwards does not lead to a reduction in the already superabundant overall protein intake because it is compensated for by other animal food. Regarding calcium however, the lower milk intake is not compensated for by other foods in the recent study. This observation needs additional consideration because the bioavailability of calcium from milk is higher than from most plant foods. Although the DGE recommendations for calcium intake [1] have generous safety margins, they should be approximated as closely as possible. Therefore, present-day recommendations for milk intake of children should be based on the quantities actually consumed by children 15 years ago – but with the precaution of recommending, at the same time, a reduction in the consumption of other animal food.

References

1. Deutsche Gesellschaft für Ernährung (1985) Empfehlungen für die Nährstoffzufuhr. Umschau, Frankfurt/Main
2. Kersting M, Schöch G (unpublished)
3. Stolley H, Kersting M, Droese W (1982) Energie- und Nährstoffbedarf von Kindern im Alter von 1–14 Jahren. Ergebn Inn Med Kinderheilk 48:1–75
4. Whitehead RG (1987) Nutritional requirements of healthy adolescents and their significance during the management of PKU. Eur J Pediatr 146:A25–A31

Enteral vs Parenteral Nutrition
in the Early Postoperative Phase
with Special Regard to Protein Metabolism

B. Lünstedt, J. Seifert

Department of General Surgery and Experimental Surgery, University of Kiel, Kiel, FRG

Introduction

Artificial nutrition in the postoperative phase is the method of choice for providing adequate nutrients to patients in the so-called post-aggressive metabolism. The method of artificial nutrition is not set and the doctor must decide between parenteral application of nutrients using a central venous catheter with its attendant problems, or enteral nutrition through a fine-bore feeding tube placed in the proximal jejunum. In a prospective randomized study of 60 patients after elective colonic surgery we assessed the value of early postoperative enteral feeding vs parenteral feeding.

Patients and methods

Sixty patients (aged 35–75 years) were randomized into two groups of different nutritional regimens after elective colonic surgery. Group I (EN) received from the first postoperative day until the 10th postoperative morning 2 000 ml/day of a diet containing carbohydrates, fat, and short-chain peptides (Nutricomb F Peptid) via a fine-bore feeding tube placed preoperatively in the proximal jejunum. Group II (PN) was treated with standard intravenous fluids consisting of carbohydrates and amino acids via infusion into a central vein. Calorie intake and the composition of artificial nutrition was similar in both groups (EN: 110 g peptide + AA, 370 g dextrose + polysacch., 60 g MCT + ess. FA., $\sim$2 000 kcal/day; PN: 70 g AA, 440 g glucose, 50 g LCT-fat emulsion, $\sim$2 200 kcal/day).

Samples were taken preoperatively, as well as on day 5 and day 10 postoperatively. Beside others, we investigated the albumin and short-living proteins (prealbumin and transferrin), the immunoglobulins, and the plasma fibronectin. The 24-h N-excretion was measured in the urine by an Antec® nitrogen analyzer.

Results

Both regimens were well tolerated by the patients. The short-life proteins and the albumin decreased up to day 5 in the postoperative phase in both groups (Fig. 1). Prealbumin as well as transferrin increased faster in the EN-group postoperatively then in the PN-group – at postoperative day 10 we measured 24.5 and 250, respectively, 18.0 and 210 mg/dl (Fig. 1).

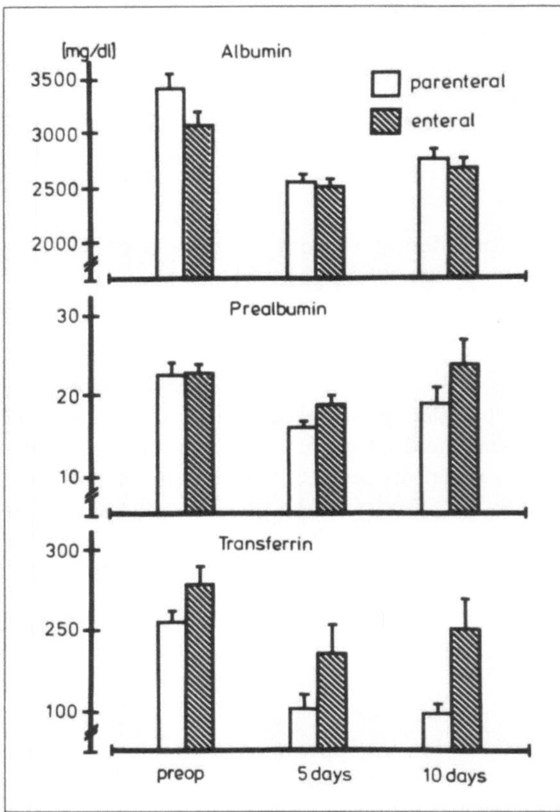

Fig. 1. Albumin and short-living proteins in the postoperative period

The plasma fibronectin and the immunoglobulin IgM showed significant higher levels at day 10 in the EN-group (39.8 and 340, respectively, 32.4 and 280 mg/dl) (Fig. 2). The other immunoglobulins show no difference in both groups. The five-day cumulative N-balance is similar in both regimens (Fig. 3).

Conclusion

In a prospective randomized study with 60 patients differences between enteral and parenteral nutrition were investigated especially with regard to short living proteins, immunoglobulins and nitrogen balance. Prealbumin as well as fibronectin and transferrin increased faster in enteral nourished patients postoperatively than in the parenterally nourished group. From the immunoglobulins only IgM was significantly increased after enteral nutrition. Both regimens were well tolerated in the early postoperative periode.

288

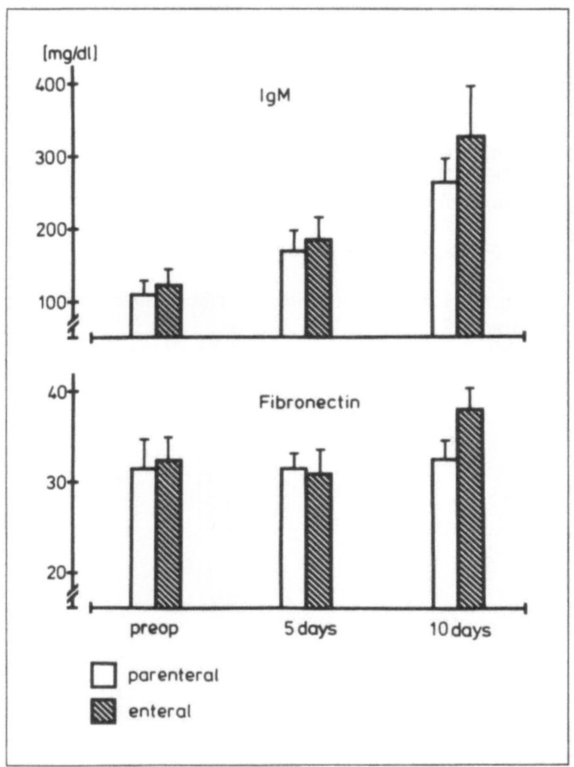

Fig. 2. Immunoglobulin-M and the plasma fibronectin, postoperative

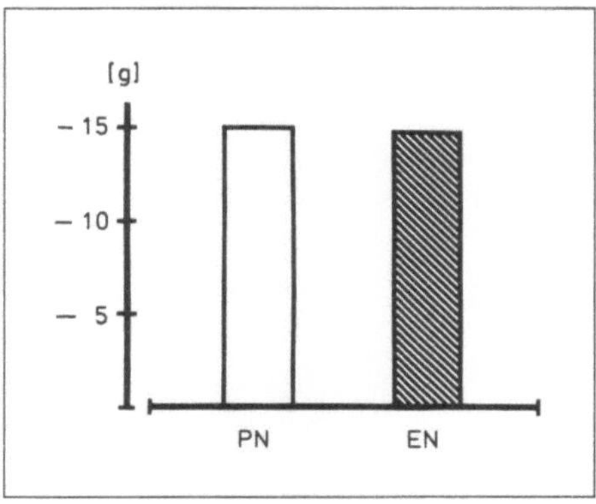

Fig. 3. The five-day cumulative N-balance, postoperative

References

Dietze G (1983) Fehler und Gefahren der enteralen und parenteralen Ernährung. Chirurg 54:18

Dölp R, Grünert A, Böhm A (1983) Enterale oder parenterale Ernährungstherapie – Eine vergleichende Studie. Infusionstherapie 10:286

Günter B (1983) Parenterale Ernährung als Langzeittherapie. Chirurg 54:12

Hausmann D, Mosebach KO, Caspari R, Feller D, Lippold R, Stoeckel H (1983) Enteral-parenterale Ernährung mit hohem Eiweißangebot bei schweren Schädel-Hirn-Traumen. Infusionstherapie 10:306

Heberer G, Schultis K, Hoffmann K (1976) Postaggressionsstoffwechsel: Grundlagen, Klinik, Therapie. Schattauer, Stuttgart New York

McArdle AH, Palmason C, Morency J, Brown RA (1981) A rationale for enteral feedings as the preferable route for hyperalimentation. Surgery 77:616

Merkle NM, Wiedeck H (1983) Die unmittelbar postoperative enterale Ernährung nach Colonresektion. – Eine Alternative zur parenteralen Ernährungstherapie? In: Schreiber (ed) Langenbecks Arch Chir Forum 1983. Springer, Berlin Heidelberg New York, p 273

Influence of Casomorphin on Plasma Lipid Levels and Lipid Secretion Rates

M. Pfeuffer and C. A. Barth

Institut für Physiologie und Biochemie der Ernährung der Bundesanstalt
für Milchforschung, Kiel, FRG

Introduction

Casomorphins are sequences of casein with opioid, morphine-like activity [1]. Casomorphins have been identified in the duodenal chyme of pigs following a casein meal [2]. Orally applied casomorphins have been shown to stimulate postprandial insulin release in the dog [4]. This effect was inhibited by naloxone.

There is a hypercholesterolemic and atherogenic potential of casein (as compared to vegetable protein, e.g., soy protein), which is most prominent in rabbits, less prominent in many other animal species and not present at all in man. In rats the rate of hepatic lipoprotein secretion has been shown [3] to contribute to changes of plasma cholesterol following different dietary proteins.

This paper addresses the question of whether casomorphins or any other sequences of casein with opioid activity are responsible for the increase of lipoprotein secretion rates and plasma cholesterol levels following casein as compared to soy protein.

Methods

We performed two experiments. In the first we fed rats sucrose-rich diets containing either casein (C), soy protein (S), soy protein plus β-casomorphin $(1-5)$ (0.7 mg/ 100 g diet) (SCM) or soy protein, casomorphin and naloxone (0.6 mg/100 g diet) (SCMN). In the second experiment we fed casein (C), soy protein (S), casein plus naloxone (CN) and soy protein plus naloxone (SN).

The SCM and CN groups in experiments 1 and 2 were the test groups, whereas SCMN and SN served as controls.

Male Wistar rats aged 4 weeks (40 g) were housed at 52% humidity and 21 °C. All animals consumed first the soy protein diet for $10-20$ days and were then kept on the experimental diets for exactly 10 days.

The experiment began on the morning of the 11th day at about 10.00 hrs after a 16 h fast. After taking a 0.4 ml heparinized blood sample from a catheter in the vena jugularis, 1 ml of 10% (v/v) Triton WR-1339 in saline was injected and four further samples were taken at 30 min intervals [3]. Cholesterol and triglycerides were determined enzymatically. Statistical calculations were done using Student's t-test.

Results

Plasma cholesterol levels were significantly higher with casein feeding as compared to soy (Table 1). Casomorphin or naloxone did not modulate plasma cholesterol.

Table 1. Plasma cholesterol levels and lipid secretion rates of rats given casein or soy protein in combination with β-Casomorphin $(1-5)$ or naloxone

	Dietary regimen	n	Cholesterol level mmol/l	Lipid secretion $mmol \times l^{-1} \times h^{-1}$	
				Cholesterol	Triglycerides
Exp. 1	C	9	1.72 ± 0.12 [b, c]	$0.72 \pm 0,06$ [a, e, k]	2.82 ± 0.14 [f]
	S	9	1.39 ± 0.07 [b]	0.43 ± 0.05 [e]	2.44 ± 0.21
	SCM	10	1.35 ± 0.08 [c]	0.48 ± 0.02 [a]	2.47 ± 0.11
	SCMN	10	1.50 ± 0.05	0.39 ± 0.03 [k]	2.12 ± 0.32 [f]
Exp. 2	C	9	1.87 ± 0.12 [d]	0.73 ± 0.03 [i]	1.84 ± 0.08 [g]
	S	9	1.46 ± 0.08 [d]	0.47 ± 0.05 [i]	1.42 ± 0.06 [g]
	CN	10	1.92 ± 0.09 [h]	0.74 ± 0.14	1.77 ± 0.17
	SN	10	1.43 ± 0.08 [h]	0.50 ± 0.04	1.53 ± 0.13

Values are mean $\pm$ SEM. Values sharing common superscripts are significantly different from each other at $p < 0.05$ (a–d), $p < 0.01$ (e–g) or $p < 0.001$ (h–k); n = number of animals

The injection of Triton WR-1339 was followed by a linear rise of lipid levels within the 2-h observation period in all groups of animals (Table 1). In both experiments cholesterol secretion rates of casein-fed animals were significantly higher than those of soy protein-fed animals. The same was true for triglyceride secretion rates in experiment 2. Neither cholesterol nor triglyceride secretion rates were influenced by the casomorphin or naloxone supplements.

Discussion

These experiments confirm former results [3] in that plasma cholesterol levels and both cholesterol and triglyceride secretion rates are significantly higher with casein as compared to soy protein.

These data show that β-casomorphin $(1-5)$ is not responsible for casein-induced hypercholesterolemia. The fact that naloxone did not modulate hepatic secretion rates suggests that other peptide sequences of casein possibly having opioid activity have no such effect, either. Furthermore, it becomes clear that casomorphin-induced insulin release is not responsible for casein-induced hypercholesterolemia.

References

1. Brantl V, Teschemacher H (1982) Opiatartig wirkende Stoffe in Milch und Milchprodukten. 2. Isolierung, Strukturaufklärung und biologische Wirkung der β-Casomorphine. Milchwissenschaft 37:641–644
2. Meisel H (1986) Chemical characterization and opioid activity of an exorphin isolated from in vivo digests of casein. FEBS 196:223–227
3. Pfeuffer M, Barth CA (1986) Modulation of Very Low Density Lipoprotein secretion is age-dependent in rats. Ann Nutr Metab 30:281–288
4. Schusdziarra V, Holland A, Schick R, De La Fuente A, Klier M, Maier V, Brantl V, Pfeiffer EF (1983) Modulation of post-prandial insulin release by ingested opiate-like substances in dogs. Diabetologia 24:113–116

The Absorption of Proteins From the Gut in Cyclosporin-A Treated Animals

J. Seifert, G. Axt, and P. Bonacker

Experimental Surgery of the Department of General Surgery, University of Kiel, Kiel, FRG

Introduction

Since proteins are absorbed from the gut, not only under physiological conditions, but also from patients treated with certain drugs, the influence of parenterally applied Cyclosporin-A, a potent immunosuppressive drug, was evaluated upon the absorption of human serum albumin (HSA).

Methods

Over a period of five days rats were treated i.p. with 15 mg/kg Cyclosporin-A daily. Thereafter absorption studies were performed and compared with similar studies in non-immunosuppressed animals. For this purpose animals were fed 1 g HSA dissolved in 0.9% NaCl solution. The uptake of HSA from the gut in blood and lymph was measured by immunological methods. It is possible to determine a concentration of a protein by an antiserum using the agar diffusion technique. If a sample containing HSA diffuses in agar with anti-HSA, a ring of precipitation will result. In this system the diameter of the precipitation ring is proportional to the concentration of the protein. With this immunological method we to determined intact human serum albumin in serum and in thoracic duct lymph of animals after enteral application of the protein. After five hours of absorption all animals were sacrified. The whole gastrointestinal tract was taken out and rinsed with saline; the content of HSA was determined in the rinsed fluid. From the originally applied dosage and the remaining content in the rinsed fluid the absorption rate could be calculated. The effect of immunosuppression was controlled by measuring the peripheral lymphocyte count. This parameter was markedly decreased. In immunosuppressed animals only 50% of the initial peripheral lymphocytes could be detected, which indicated an immuno-deficiency.

After treatment with Cyclosporin-A and enteral application of 1 g HSA the concentration of HSA in blood and lymph samples, which were taken in intervals of one hour over an observation time of five hours was controlled. Blood samples were obtained from the tail veins of the animals and lymph samples were collected from a drainage of the thoracic duct.

Results

Whereas the concentration of HSA in serum of untreated normal rats remains unchanged over the observation time at a level between 1 and 2 mg/dl, Cyclosporin-

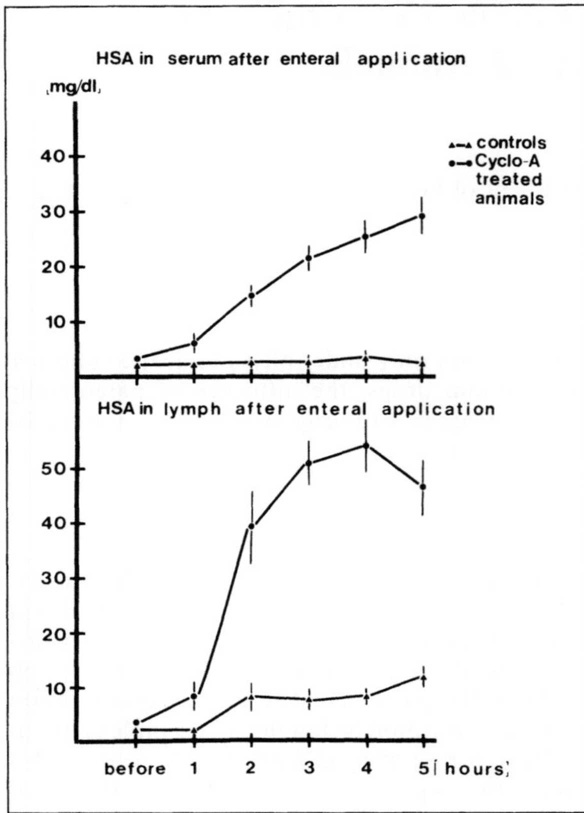

Fig. 1. Concentration of human serum albumin in blood and lymph of rats after enteral application of 1 g Cyclosporin-A; treatment was over 1 week 15 mg/kg daily. Control animals remained untreated

A treated animals show a linear and highly significant increase up to 29.2 ± 2.5 mg/dl (Fig. 1). The rate of increase per hour was 4 mg/dl in mean. At the end of the observation the albumin concentration in serum of Cyclosporin-A treated animals is 10 times higher than in control animals.

In principle, similar results could be observed with regard to the HSA-concentration in the lymph samples. Cyclosporin-A treated animals have significant higher HSA concentration in the lymph than control animals. Furthermore, it is obvious that the lymph concentration exceeds the concentration found in serum. This can be explained by the low flow rates in the lymph vessels.

The high concentration of HSA in serum and lymph of Cyclosporin-A treated animals correlates very well with the absorption rate of HSA from the gut. The remaining amount of HSA in the gastrointestinal tract of Cyclosporin-A treated animals is 20% lower than in control animals. With that the absorption rate is 20% higher in immunosuppressed animals (Table 1). The discrepancy between the 10-fold increase in serum and lymph and the 20% difference in the absorption rate can be explained by a slow metabolism or degradation of human serum albumin in rats. If absorption is faster than degradation serum levels must increase continuously.

294

Table 1. Remainder of albumin in the gut after enteral application and five hours absorption. The absorption rate is calculated from the applied dosage and the remaining content

	Cyclo A	Controls
	% of applied dose per ml	
Stomach	0.136 ± 0.064	0.172 ± 0.074
Duodenum	0.013 ± 0.005	0.120 ± 0.063
Jejunum/ileum	0.125 ± 0.058	0.190 ± 0.047
Colon	0.035 ± 0.002	0.03 ± 0.014
Total	31%	51%
Absorption rate	69%	49%

Conclusions

If the increased absorption of high molecular proteins in an immunosuppressed organism is a general observation and can be extended to all nutritional proteins it could be dangerous. Since nearly all proteins that come into the gastrointestinal tract are immunologically foreign proteins, the organism must react against these proteins with the induction of antibodies. An incompatibility to nutritional proteins would be the consequence.

Influence of Diet and Stage of Lactation on Taurine Contents in Milk

B. v. Blomberg, E. A. Trautwein, H. F. Erbersdobler

Institute of Human Nutrition and Food Science, University of Kiel, FRG

Introduction

The present study was conducted to determine if there are direct influences of conventional and lacto-ovo-vegetarian diets on breast-milk concentrations of taurine. Taurine can be obtained from the consumption of animal foods such as meat, meat products, fish, sea-food, and eggs. Plant foods on the other hand are devoid of taurine. Since milk and milk products are also low in taurine, a negative influence of a lacto-ovo-vegetarian diet on taurine levels in breast milk was expected. Additionally, changes in the taurine concentration during the lactation period were studied and taurine levels in human milk were compared to taurine concentrations in cow milk.

Methods

Breast-milk samples ($n = 224$) were obtained from 20 healthy mothers of term infants who gave their consent to the investigation. The mothers gave a written qualitative protocol about the food consumed on the sampling days. According to these written protocols and additionally by interviewing the women to confirm their usual nutrition behavior, the mothers were classified into three groups:
1) *Lacto-ovo-vegetarians* ($n = 6$), consuming a diet rich in vegetables, fruits, cereals, and also milk and dairy products, but no meat and meat products;
2) *Omnivores* ($n = 7$), consuming a high quality diet rich in meat, fish, milk, and dairy products;
3) *Omnivores* ($n = 7$), consuming a relative poor quality diet rich in fast-food of low quality, and high carbohydrates foods such as cakes and sweets.

Milk sampling started later than 30 days post partum and then breast-milk was collected at different stages of lactation (30–499 days post partum). The mothers were asked to express about 20 ml of milk by using a hand operated breast pump during the initial stage of nursing while they nursed the baby on the other breast. On test days a milk sample was expressed before (fasting level) and another one 3–4 h after the main meal (post prandial). More details about the experimental procedures are given elsewhere [2]. The milk samples were stored at –20 °C immediately after sampling and until analysis.

Taurine analyses were carried out on an automatic amino acid analyzer (LC 5000, Biotronik, Maintal, FRG) using a shortened analytical program with lithium citrate buffers and ninhydrin reaction for detection.

Results

Breast-milk protein content (Kjeldahl-N $\times$ 6,25) varied from 0.74 to 1.66 g/100 ml milk, whereas no striking differences existed between the fasting levels and the post prandial values. Moreover, the total protein concentrations in the milk samples of the mothers who consumed different diets showed no differences. The mean values of total protein in breastmilk were 1.07 g/100 ml in the lacto-ovo-vegetarians, 1.11 g/100 ml in the omnivores fed a poor quality diet, and 1.10 g/100 ml in the omnivores who consumed a high quality diet. Breast-milk taurine concentrations exhibited no differences between fasting and post prandial levels in the average of all lacto-ovo-vegetarian and omnivore mothers. Some single mothers, especially in the group of omnivores fed a high quality diet, showed higher fasting levels compared to post prandial values. The mean values of taurine concentrations were partially significantly different among the groups (Table 1).

The taurine values of the present study are in good agreement with the results reported by Rana and Sanders (1986), who also found significantly lower breast-milk taurine concentrations in vegans than in omnivores. Rana and Sanders did not differentiate between omnivores on a good or poor diet, although their range of variation (2.4–8.5 mg/100 ml) was in the same area as the data of the present study (3.8–9.4 mg/100 ml) for all omnivores. The taurine concentrations in the mothers who consumed a strict vegetarian diet (vegans) showed slightly lower values in the study of Rana and Sanders, which is understandable because their diet was completely devoid of taurine compared to the lacto-ovo-vegetarians of the present study.

Another investigation with the intention to determine an influence of different diets on the carnitine concentrations in human milk, performed at the same milk samples reported, also showed significant differences of L-carnitine contents in milk from lacto-ovo-vegetarians and omnivores [1].

As shown in figure 1 the taurine levels in breast-milk remained almost constant in the average of all mothers throughout the period of lactation. The breast-milk taurine contents in the lacto-ovo-vegetarians and the omnivores consuming a poor quality diet increased slightly during the lactation period, whereas the taurine levels

Table 1. Taurine content in breast-milk (in mg/100 ml); a comparison of the results of this study with the results of Rana & Sanders (1986)

	n	Present study		Rana & Sanders	
		Mean $\pm$ SE	Range	Mean $\pm$ SE	Range
Vegans		—	—	3.5 ± 0.4[a]	1.5–6.6
Lacto-ovo-vegetarians	39	5.0 ± 0.7	3.6–7.1	—	—
Omnivores, poor quality diet	34	4.9 ± 0.5	3.8–6.6		
Omnivores, high quality diet	44	6.5 ± 0.8[b]	4.8–9.4	5.4 ± 0.5	2.4–8.5

[a] significantly different from omnivores
[b] significantly different from the mean values for the lacto-ovo-vegetarians and omnivores fed a poor quality diet

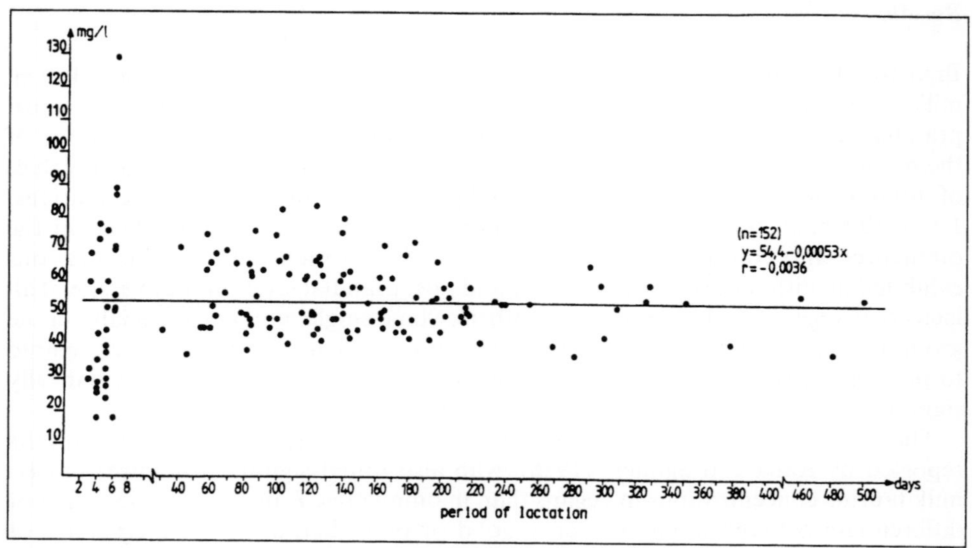

Fig. 1. Taurine content in human milk throughout the period of lactation

in the milk of the omnivores on a high quality diet showed a somewhat more distinct but not significant decrease.

In addition to the milk samples of the 20 mothers which were collected after the 30th day post partum, samples of the first days of lactation (mainly from the 3rd through 6th days post partum) from another investigation [3] were considered. The mean value of breast-milk from the first days of lactation (5.0 mg/100 ml) was in the same range as the taurine contents in the present study. The data of taurine content in the milk of the first days of lactation are also incorporated in Fig. 1. In contrast to the results found in human milk, the taurine concentrations in cow milk decreased considerably within the first weeks of lactation. The taurine concentrations in the colostrum of the cows were 410 mg/kg dry matter resp. 7 mg/100 ml milk, and were in the same range as the taurine concentrations in human milk [3]. In the first weeks of lactation the taurine content in cow milk decreased greatly down to 100 mg/kg dry matter. In cow milk of about 100 days post partum taurine concentrations of 30 mg/kg dry matter resp. 0.4 mg/100 ml were found (unpublished data).

References

1. Barth CA, Roos N, Nottbohm B, Erbersdobler HF (1985) L-carnitine concentrations in milk from mothers on different diets. In: Schaub J (ed) Composition and physiological properties of human milk. Elsevier Science Publishers, Amsterdam New York Oxford, p 229–239
2. von Blomberg B (1988) Der Tauringehalt der Muttermilch in Abhängigkeit von der Ernährung und dem Laktationsstadium. Agr Diss Kiel University
3. Erbersdobler HF, Trautwein E (1984) Determination of taurine in human milk, cow milk and some milk products. Milchwissenschaft 39:722–724
4. Erbersdobler HF, Trautwein E, Greulich H-G (1984) Determinations of taurine in milk and infant formula diets. Eur J Pediatr 142:133–134
5. Rana SK, Sanders TAB (1986) Taurine concentrations in the diet, plasma, urine and breast milk of vegans compared with omnivores. Br J Nutr 56:17–27

298

General Discussion:
Milk Protein and Clinical Nutrition

G. Schöch

Forschungsinstitut für Kinderernährung Dortmund, FRG

Milk is the most complete and, therefore, a very important food, but it is also a major cause of food allergies and of non-immunological food intolerances.

In this context, Dr. Grimble justly recalls the fundamental importance of the physiology of protein digestion and absorption pertinent for any discussion of milk in the context of clinical nutrition.

Even at the optimum pH of 1.5–2, which is not reached in the infant, gastric digestion of proteins is very limited. In the small intestine, the pancreatic endopeptidases trypsin, chymotrypsin, and elastase, produce oligopeptides which, from hexapeptides downward, are further cleaved by peptidases of the intestinal mucosa. Not only free amino acids are absorbed, but also and indeed preferentially, di- and tripeptides. These latter are mostly cleaved within the enterocytes, but they can also reach the portal vein and are then processed mainly by the kidneys, the liver and the muscle.

Given orally, oligopeptide solutions ideally containing fragments of about 2–6 amino acids in length have been shown to combine several advantages. They are no more allergenic. They are readily cleaved and/or absorbed even by a damaged mucosa, and they do not have the hypertonicity of an equivalent amino acid mixture.

As discussed by Dr. Grimble, parenterally given dipeptide, and possibly even oligopeptide solutions, offer the chance to increase the intravenous amino acid supply more than twofold without raising osmolality. Moreover, otherwise hardly soluble (Cys) or unstable amino acids (Gln, Trp, Tyr) can be included as needed.

On the other hand there remain potential immunological risks with the intravenous use of protein hydrolysates which can be circumvented by the restriction to pure amino acid mixtures.

The possibility to abolish or reduce allergenicity of milk proteins by partial hydrolysis has aroused widespread interest among clinicians during the last few years. In this context a rigorous distinction must be made between the treatment of manifest cow's milk allergy and the attempt to protect genetically predisposed infants from developing cow's milk allergy, that is, to take preventive measures. This distinction has not always been made with sufficient clarity by both users and producers of protein hydrolysates.

It is important to recall that all commercially available protein hydrolysates for oral use do still contain substantial amounts of oligopeptides with a molecular weight exceeding 1 200 daltons, corresponding to a chain length of about 10 amino acids, a value which is generally accepted as the threshold of antigenicity. The percentage of longer peptides is higher in products designed for preventive use than in those for therapeutic use. On the other hand the taste of a hydrolysate is worse the shorter the mean chain length.

Dr. Ferguson has explained in detail the different, fundamentally distinct mechanisms of food intolerance known today. Thus, food allergy has to be distinguished from food intolerance caused by non-immunological factors like enzyme deficiencies, as well as from psychogenic food aversion.

True food allergy is on the whole rather rare and can be proved beyond doubt only by elimination plus challenge. This is not always without risk. Therefore many cases of supposed "food allergy" are purely conjectural.

Cow's milk allergy is nearly always self-limited and is thereby, in general, restricted to the first two years of life. The clinical symptoms of cow's milk allergy are multifaceted and go from malabsorption syndrome to atopic eczema.

Small quantities of probably intact cow's milk proteins are absorbed by the mother and transmitted to the infant, who sometimes reacts allergically.

As a rule of thumb, however, breast-fed infants are generally well-protected against food allergies, whereas feeding heterologous protein, especially during the first days of life, implies an increased risk of sensitizing the infant. It is precisely during this first critical phase that many mothers do not yet produce enough milk to feed the infant to satisfaction.

In these cases a hypoallergenic hydrolysate is thought by many to enable a transient preventive feeding regimen which might significantly reduce the risk of sensitization before full lactation is achieved. The idea is that after an initial hypoallergenic feeding regimen followed by full breast feeding a subsequent feeding of cow's milk should be safe. By contrast, initial short-term feeding with cow's milk could initiate at first a latent and later, upon reintroduction of cow's milk, a manifest cow's milk allergy.

Dr. Ferguson has impressively demonstrated that antigen exposure, depending on circumstances, can elicit quite varied immunological reactions ranging from immunotolerance to anaphylaxis. Most antigens reach the organism via the gastrointestinal tract. The regular reaction is the local formation of secretory IgA by the mucosa whereas systemic immune reactions are generally suppressed. Oral antigen administration thus generally engenders oral tolerance, whereas parenteral application of the same antigen induces systemic sensitization.

If oral tolerance fails to develop for whatever reasons, a systemic production of serum IgM and IgG antibodies is generally harmless as long as the antigen is presented orally.

In atopic patients, whose mucosa is possibly characterized by enhanced leakiness throughout life, IgE antibodies are produced in addition. However, only in a fraction of cases are there concomitant clinical symptoms of a specific food allergy against the respective antigen, e.g., cow's milk protein.

Although food allergy does, in addition, induce cellular immune responses, delayed skin reactions have not proved useful in man for the diagnosis of food allergies.

The critical point of every immune reaction against oral antigens which are administered daily in large doses is the *regulation of the induction of the immune response*. In atopic disease there are disturbances of the IgE production by B cells, of T-cell immunity, and of antigen processing by the gut. Thus, oral tolerance depends on the fine tuning of a multistep process.

300

From this, Dr. Ferguson concludes that the problem in milk protein allergy is more likely to be in the individual infant's response than in the food itself.

The most important factor favoring the development of food allergy in infants is the immaturity of the infant's immunoregulatory mechanisms. Whereas large amounts of oral antigen suppress the production of IgE antibodies, multiple, extremely small doses of oral antigen as seen from animal experiments tend to induce IgE antibodies. These facts render the use of so-called hypoallergenic food for the prophylaxis of atopy theoretically problematic, since hypoallergenic food always contains traces of antigens. According to Dr. Ferguson one could even argue that the regular administration of large amounts of antigen might be more suitable for preventing allergies than the regular administration of minimal amounts.

However, field trials carried out to date with hypoallergenic milk hydrolysates in therapy and prophylaxis of food allergy do look promising.

The case of hyposensitizing food is far from closed. The refined immunological methods developed during the last few years have permitted, in animal experiments, a stepwise partial unravelling of the mechanisms described. It has been shown by these methods that the maturation of the digestive and immune system as well as the timing and doses of antigen administration are of decisive importance for the development of either immune tolerance or of one of the various immune reactions.

Thus there is an as yet unresolved inconsistency between anxieties based on theory and encouraging clinical observations concerning the oral application of protein hydrolysates in prevention and therapy of food allergy.

In this situation I consider it eminently important not to "jump to conclusions", but to progress by small and thoroughly controlled steps in the clinical evaluation of the promising new possibilities of milk and other heterologous proteins in hydrolyzed form.

List of Participants

Antila, Pirkko, T., Dr.
State Institute for Dairy Research
Osoite
SF-31600 Jokioinen
Finland

Barth, C. A., Prof. Dr.
Institut für Physiologie und
Biochemie der Ernährung
Bundesanstalt für Milchforschung
Hermann Weigmann-Str. 1
D-2300 Kiel 1
FRG

Beyer, H. J.
Institut für Milchwirtschaftliches
Maschinenwesen
TU München/Weihenstephan
Vöttinger Str. 45
D-8050 Freising-Weihenstephan
FRG

Bindels, J., Dr.
Nutritional Research
P.O. Box 1
NL-27100 Zoetermeer
The Netherlands

Bockelmann, B., Dr.
Institut für Mikrobiologie
Bundesanstalt für Milchforschung
Hermann Weigmann-Str. 1
D-2300 Kiel 1
FRG

Buchheim, W., Prof. Dr.
Institut für Chemie und Physik
Bundesanstalt für Milchforschung
Hermann Weigmann-Str. 1
D-2300 Kiel 1
FRG

Caroli, Anna
Istituto di Zootecnia Veterinaria
Via Celoria 10
I-20122 Milano
Italy

Crignis, G. L. de, Dr.
Institut für Chemie und Physik
Südd. Versuchs- und Forschungs-
anstalt für Milchwirtschaft
Technische Universität München
Vöttinger Str. 45
D-8050 Freising-Weihenstephan
FRG

Davidsson, Lena, Dr.
Department of Clinical Nutrition
Annedalsklinikerna
S-41345 Göteborg
Sweden

Dehn-Müller, Barutha
Institut für Humanernährung und
Lebensmittelkunde der Christian-
Albrechts-Universität Kiel
Düsternbrooker Weg 17
D-2300 Kiel 1
FRG

Dovc, P.
Biotechnical Faculty,
Zootechnical Department, University
Edvard Kardelj Ljublijana
Groblje 3
YU-61230 Domzale
Yugoslavia

Dübbert, D., Dr.
Milchwerke Westfalen eG
Bielefelder Str. 66
D-4900 Herford
FRG

Erbersdobler, H., Prof. Dr.
Institut für Humanernährung und
Lebensmittelkunde der Christian-
Albrechts-Universität
Düsternbrooker Weg 17–19
D-2300 Kiel
FRG

Evers, J.
Bundesministerium für Ernährung,
Landwirtschaft und Forsten,
Ref. 415
Postfach 14 02 70
D-5300 Bonn
FRG

Ferguson, Anne, Prof. Dr.
Western General Hospital
University of Edinburgh
Crewe Road
Edinburgh EH4 2XU
Scotland

Foissy, H., Prof. Dr.
Institut für Milchforschung und
Bakteriologie
Universität für Bodenkultur
Gregor Mendelstr. 33
A-1180 Wien
Austria

Frister, H., Dr.
Institut für Chemie und Physik
Bundesanstalt für Milchforschung
Hermann Weigmann-Str. 1
D-2300 Kiel 1
FRG

Frokjer, S., Dr.
Novo Industri A/S
Food Ingredients Team
Novo Alle
DK-2880 Bagsvaerd
Denmark

Fürst, P., Prof. Dr.
Institut für Biologische Chemie
und Ernährungswissenschaft
Universität Hohenheim
Garbenstraße 30
D-7000 Stuttgart
FRG

Gay, J., Dr.
Commission of European
Community
Rue de la Loi 200
B-1049 Brussels
Belgium

Georgi, Gilda, Dr.
Milupa AG
Bahnstraße 14–30
D-6382 Friedrichsdorf
FRG

Gislason, J.
Department of Nutrition
University of Uppsala
Box 5 51
S-75122 Uppsala
Sweden

Goldblum, R. M., Prof. Dr.
Pediatric Immunology/Allergy
University of Texas
Galveston, TX 77550
USA

Goujard, J.-P.
Trajectus Consulting Group
82 Boulevard des Batignolles
F-75017 Paris
France

Grimble, G., Prof. Dr.
Department of Gastroenterology and
Nutrition
Central Middlesex Hospital
Acton Lane
London NW10 7NS
UK

Gurr, M. I., Prof. Dr.
Nutritional Consultant
Milk Marketing Board
Thames Ditton, Surrey KT7 OEL
UK

Hagemeister, H., Prof. Dr.
Institut für Physiologie und
Biochemie der Ernährung
Bundesanstalt für Milchforschung
Hermann Weigmann-Str. 1
D-2300 Kiel 1
FRG

Hambraeus, L., Prof. Dr.
Institute of Nutrition
University of Uppsala
P.O. Box 5 51
S-75122 Uppsala
Sweden

Hatzold, T., Dr.
Milupa
Bahnstr. 14–30
D-6382 Friedrichsdorf
FRG

Jelen, P., Prof. Dr.
Department of Food Science
University of Alberta
Edmonton, T6G 2P5
Canada

Jonsson, B. N., Dr.
Sodinol Proteins Trading ApS
Bredgade 21
DK-4000 Roskilde
Denmark

Kay, H. W., Prof. Dr.
Daten- und Informationszentrum
Bundesanstalt für Milchforschung
Hermann Weigmann-Str. 1
D-2300 Kiel 1
FRG

Kersting, Mathilde, Dr.
Forschungsinstitut für Kinder-
ernährung
Heinstück 11
D-4600 Dortmund 50
FRG

Kinsella, J. E., Prof. Dr.
Department of Food Science
Cornell University
Ithaca, NY 14853
USA

Klostermeyer, H., Prof. Dr.
Institut für Chemie und Physik
Südd. Versuchs- und Forschungs-
anstalt für Milchwirtschaft
Technische Universität München
Vöttinger Str. 45
D-8050 Freising-Weihenstephan
FRG

Krause, I., Dr.
Institut für Chemie und Physik
Südd. Versuchs- und Forschungs-
anstalt für Milchwirtschaft
Technische Universität München
Vöttinger Str. 45
D-8050 Freising-Weihenstephan
FRG

Krawinkel, M., Dr.
Universitäts-Kinderklinik der
Christian-Albrechts-Universität
Schwanenweg 20
D-2300 Kiel 1
FRG

Lembke, A., Prof. Dr. Dr.
Institut für Virusforschung und
experimentelle Medizin
D-2420 Eutin-Sielbeck
FRG

Lönnerdal, B., Prof. Dr.
Department of Nutrition
University of California
Davis, CA 95616
USA

Lorenzen, P. Ch., Dr.
Milei GmbH
Rosensteinstr. 20
D-7000 Stuttgart 1
FRG

Lünstedt, B., Dr.
Allgemeine Chirurgie der
Christian-Albrechts-Universität
Arnold-Heller-Straße 7
D-2300 Kiel 1
FRG

Meisel, H., Dr.
Institut für Chemie und Physik
Bundesanstalt für Milchforschung
Hermann Weigmann-Str. 1
D-2300 Kiel 1
FRG

Millward, D. J., Prof. Dr.
London School of Hygiene and
Tropical Medicine
University of London
1 St. Pancras Way
London NW1 2PE
UK

Moerenhout, H.
Joyvalle C.V.
s' Gravenmolenstraat 10
B-1850 Grimbergen
Belgium

Van Moorsel, A., Dr.
DMV-Cambina B.V.
P.O. Box 13
NL-5460 BA Veghel
The Netherlands

Morunga, R., Dr.
Fresenius AG
Postfach 18 09
D-6370 Oberursel/Taunus
FRG

Nienhaus, A., Dr.
Verband der Deutschen
Milchwirtschaft
Schedestraße 11
D-5300 Bonn
FRG

Niepold, F., Dr.
Biologische Bundesanstalt für Land-
und Forstwirtschaft
Institut für Biochemie
Messeweg 11/12
D-3300 Braunschweig
FRG

Paulsson, Marie
Department of Food Technology
University of Lund Chemical Center
P.O. Box 7 40
S-22007 Lund
Sweden

Pélissier, J.-P., Dr.
Institut Nationale de la Recherche
Agronomique (INRA) Laboratoire de
Biochimie et Technologie Laitieres
F-78350 Jouy-en-Josas
France

Perraudin, J. P., Dr.
S.A. Synfinia-Oleifina N.V.
Rue de la Loi 15
B-1040 Brussels
Belgium

Pfeuffer, Maria, Dr.
Institut für Physiologie und
Biochemie der Ernährung
Bundesanstalt für Milchforschung
Hermann Weigmann-Str. 1
D-2300 Kiel 1
FRG

Plock, J.
Institut für Milchwirtschaftliches
Maschinenwesen
TU München/Weihenstephan
Vöttinger Str. 45
D-8050 Freising-Weihenstephan
FRG

Price, Gill M.
Nutrition Research Unit
St. Pancras Hospital
4 St. Pancras Way
London NWI 2PE
UK

Puhan, Z., Prof. Dr.
Institut für Lebensmittelwissen-
schaft Labor für Milchwissenschaft
Eidgenössische Technische Hoch-
schule, ETH-Zentrum
CH-8092 Zürich
Switzerland

Rampilli, M., Dr.
Phys. Chemical Sect.
Istituto Sperimentale Lattiero-
Caseario
Via Lombardo 11
I-20075 Lodi
Italy

Reeds, P. J., Dr.
CNRC, Dept. of Pediatrics
Baylor College of Medicine
Medical Towers Building, Suite 603
6608 Fannin Street
Houston, TX 77030
USA

Reimerdes, E. H., Prof. Dr.
Meggle-Milchindustrie GmbH & Co. KG
Megglestr. 6–12
D-8090 Wasserburg
FRG

Reiter, B., Dr.
2 Fosseway Crowthorne
Berkeshire RG11 7
UK

Renner, E., Prof. Dr.
Professur Milchwissenschaft der
Justus-Liebig-Universität
Bismarckstr. 16
D-6300 Gießen
FRG

Ribadeau-Dumas, B., Prof. Dr.
Laboratoire de Biochimie
et Technologie Laitieres
INRA
F-78350 Jouy-en-Josas
France

Rochette de Lempdes, J. B.
Directeur de Recherche
Gallia HPH
52/58 Avenue du Marechal Joffre
F-92000 Nanterre
France

Rüterjans, H., Prof. Dr.
Institut für Biophysikalische
Chemie der Johann Wolfgang Goethe-
Universität
Theodor Stern-Kai 7, Haus 75 A
D-6000 Frankfurt/Main
FRG

Schlimme, E., Prof. Dr.
Institut für Chemie und Physik
Bundesanstalt für Milchforschung
Hermann Weigmann-Str. 1
D-2300 Kiel 1
FRG

Schmitz, M.
Forschungsgesellschaft für
experimentelle Tierphysiologie und
Tierernährung mbH & Co. KG
Wiesenweg 10
D-2362 Wahlstedt
FRG

Schöch, G., Prof. Dr.
Forschungsinstitut für Kinder-
ernährung
Heinsrück 11
D-4600 Dortmund
FRG

Schoppe, Ina
Institut für Physiologie und
Biochemie der Ernährung
Bundesanstalt für Milchforschung
Hermann Weigmann-Str. 1
D-2300 Kiel 1
FRG

Schulz-Lell, Gisela, Dr.
Universitäts-Kinderklinik
Schwanenweg 20
D-2300 Kiel 1
FRG

Seifert, J., Prof. Dr.
Experimentelle Chirurgie der
Christian-Albrechts-Universität
Michaelisstr. 5
D-2300 Kiel 1
FRG

Sick, H., Prof. Dr.
Institut für Physiologie und
Biochemie der Ernährung
Bundesanstalt für Milchforschung
Hermann Weigmann-Str. 1
D-2300 Kiel 1
FRG

Sieber, R., Dr.
Eidgenössische Forschungsanstalt
für Milchwirtschaft
CH-3097 Liebefeld-Bern
Switzerland

Sievers, E., Dr.
Universitäts-Kinderklinik der
Christian-Albrechts-Universität
Schwanenweg 20
D-2300 Kiel 1
FRG

Simons, J. P., Prof. Dr.
Institute of Animal Physiology and
Genetics Research Edinburgh
Research Station
West Mains Road
Edinburgh, EH9 3JQ
UK

Sörensen, K. W.
The Danish Government Research
Institute for Dairy Industry
Roskildevej 56
DK-3400 Hillerod
Denmark

Spik, Genevieve, Prof. Dr.
Université des Sciences et
Techniques de Lille
Flandres-Artois Laboratoire de
Chimie Biologique
F-59655 Villeneuve d' Asco Cedex
France

Swaisgood, H. E., Prof. Dr.
Department of Food Science
North Carolina State University
Raleigh, NC 27695
USA

Syvaeoja, Eeva-Liisa, Dr.
Valio Finnish Co-operative
Dairies Association
P.O. Box 176
SF-00181 Helsinki
Finland

Trautwein, Elke, Dr.
Institut für Humanernährung und
Lebensmittelkunde der Christian-
Albrechts-Universität Kiel
Düsternbrooker Weg 17
D-2300 Kiel 1
FRG

Van Eckert, Renate
Forschungsinstitut der
Ernährungswirtschaft
Blaastraße 29
A-1190 Wien
Austria

Waard, H., de, Dr.
Department of Nutrition
Netherlands Institute for Dairy
Research
P.O. Box 20
NL-6710 BA Ede GLD
The Netherlands

Wit, J. N. de, Dr.
Nederlands Instituut voor
Zuivelonderzoek
P.O. Box 20
NL-6710 BA Ede GLD
The Netherlands

Young, V. R., Prof. Dr.
Department of Nutrition and Food
Science Massachusetts Institute
of Technology Laboratory of Human
Nutrition
Cambridge, MA 02139
USA